G

Parallelism in Hardware and Software

REAL AND APPARENT CONCURRENCY

Prentice-Hall Series in Automatic Computation

George Forsythe, *editor*

Prentice-Hall International, Inc., *London*
Prentice-Hall of Australia, Pty. Ltd., *Sydney*
Prentice-Hall of Canada, Ltd., *Toronto*
Prentice-Hall of India Private Limited, *New Delhi*
Prentice-Hall of Japan, Inc., *Tokyo*

Parallelism
in
Hardware
and
Software

REAL AND APPARENT CONCURRENCY

HAROLD LORIN

Systems Research Institute
International Business Machines Corporation

PRENTICE-HALL INC.

Englewood Cliffs
New Jersey

© 1972 by Prentice-Hall, Inc.
Englewood Cliffs, New Jersey

10 9 8 7 6 5 4 3 2 1

ISBN: 0-13-648634-7

Library of Congress Catalog Card Number: 74-172889

Printed in the United States of America

CONTENTS

1627456

PART 3 Multiple Machines

11 MULTICOMPUTER CONFIGURATIONS, 153

PART 4 Input/Output, Multiprogramming, and Operating Systems

12 CONTEMPORARY I/O SYSTEMS, 187

13 PERFORMANCE OF PROGRAMS IN A UNIPROGRAM ENVIRONMENT, 200

PART 5 Multiprocessing

PROLOGUE

The book is based upon a series of sixteen hour-and-one-half lectures that I deliver over an eight-week period. The first eight lectures are organized into classes called "Parallel Operations In Computing Systems." Parts 1, 2, 3, and 5 are discussed during this period. The second eight lectures are organized into a class called "Multiprogramming Scheduling and Resource Allocation," covering Parts 4, 6, and 7.

Students may take either or both parts. When a heavy number of students in the second class are carryovers from the first (and I know this before the beginning of the first) then some rearrangement of material is undertaken so that Parts 4 and 5 (almost) exchange places. The first course I think of as a survey of the ways that parallel or apparently parallel functions can be built into hardware. There is also a demonstration of a fundamental software mechanism for inducing parallelism when a potential for parallelism truly exists in the system. This course has been called "Hardware for Programmers." The second course I think of as being a discussion of ways of simulating parallel operations with software techniques; I sometimes privately call it "Simulating Multiprocessors and Usable Machines."

I tell my students that my goal is to give them a background sufficient to enable them to avoid panic when they are suddenly called to a meeting at which it is going to be announced to them that they are about to participate in the design of a multiprocessor, a multiprocessor operating system, or a multiprocessor compiler, or to undertake an application on a highly parallel machine. In addition to the lecture notes, each student receives approximately fifteen articles (in each course) selected from the Sources and

Reading list. At the end of the course they have in their possession what I consider the basic literature of the field and experiences in speculating about what might be done as well as in hearing about what has been done.

A number of published papers have developed from the course. Students are encouraged to undertake papers in order to avoid responding to a list of specific questions based upon the text. The penalty for avoiding creative thinking is rather high.

I get very heterogeneous groups of students with varying professional interests and backgrounds. The reason for the broad range of material is to present to each student a large number of topics in which he may develop a special interest. The pace of the class is easily adjusted to the development of special interests. Successive classes are never the same. Some classes are preoccupied with "practical" problems, and we spend much time in developing opinions about the various scheduling strategies in OS/360 versus other systems. Other classes are less pragmatic and invent machine designs of their own.

One reason why there is relatively little OS/360 material or a detailed discussion of IBM's present product line is the need for our IBM-employed students to see that issues that they think are closed and rightly resolved by all of us are indeed still open issues for workers in other parts of IBM, other manufacturers, and universities.

From time to time one sees an insight emerging from a class session. For example, in discussing the close-follow of multiprocessors and speculating on techniques for reducing contention, one student made the point that a mechanism presented in the text could be used to transform any generalized symmetric multiprocessor into an ILIAC IV class machine. He suggested providing two sources of instruction stream: (1) Memory for multiprocessor operation of "private" code; (2) The processor interface mechanisms for instruction broadcast where all but one processor would be put into an instruction fetch suppress mode.

The spirit of the text is speculative, experimental, and informal, just as the courses are. There are several reasons for this. Primarily I feel that the experience of the field is such that workers who are not in any sense mathematicians and who are uncomfortable with more than a minimum amount of formalism have made valuable contributions in the past and should continue to do so. The ideas of this business are simple and should acquire formidable notation only when necessary. I have been thanked by more than one holder of a master's degree in mathematics or statistics for the narrative style of the text.

It has been my observation that my students like to hear general concepts followed by very detailed discussions of alternatives and solutions. It amuses them to walk through an emerging system, speculating on alternatives and possibilities as they go. This is the way they learn because this is the way they make their living—this is the way they create.

The students are members of the industry with three or more years of experience. They are, as it were, about to emerge or already very professional. At the edge of going into a specialization very deeply for the first time or looking for another, they can be remarkably naive about areas with which they have not been involved despite their backgrounds in other areas. Some are very competent application programmers who really do not know how an interrupt system works or why one uses buffers. Some are very competent engineers who really do not know anything about compilation or the problems of operating systems. They can also be remarkably sophisticated very quickly when introduced to a new area.

One of the reviewers of the draft of the text commented that there are times that the discussions almost seem like "IBM coffee break" talk. This is exactly the spirit of the book. "Coffee break talk" provides an area for tremendously effective exchange of information and insight among professionals. More important for the reader of this book who is about to enter the industry or undertake to advance in it, a sniff of the coffee in the cubicles of those long corridors in manufacturer laboratories and centers serves the invaluable purpose of showing a little of "what it is like."

ACKNOWLEDGMENTS

To the authors of the 449 articles in the literature of hardware and operating system design which I have collected and (in some sense) perused during the two years of developing the courses that are the bases of this book I owe a serious debt. To those not represented in the Sources and Readings section I apologize with the explanation that I have listed only those articles that I think to be especially easily accessible to a reader with a reasonably adequate library.

I owe gratitude to my colleagues on the Systems Research Institute faculty for many hours of exchanging ideas, particularly Messrs. M. A. Seelye, P. Davies, and P. Sterbenz (who first laughed openly at the concept of writing an algorithm for "N" processors). I am in the debt of Mr. J. T. Martin for his very helpful suggestions and his careful review of the first drafts.

To the Misses Katherine LaRusso and Katherine Chandri for their patient translation of my improbable neo-Cyrillic handwriting into English type face I owe a great deal.

I extend my appreciation to J. B. Dennis and E. C. Van Horn for permission to reproduce a section of their article, "Programming Semantics for Multiprogramming", published in the *Communications of the ACM*, Volume

9, No. 3, March 1966, pp. 143–155 (Copyright © 1966, Association for Computing Machinery, Inc.). Appreciation for this same permission is also due J. P. Anderson, author of "Program Structures for Parallel Processing", *Communications of the ACM*, Volume 8, No. 12, December 1965, pp. 786–788 (Copyright © 1965, Association for Computing Machinery, Inc.).

Of course to the atmosphere at SRI I owe the most. The encouragement of the management, the library support provided by Miss Doris Schild, the general excitement generated by classes of highly motivated students, themselves serious professionals in the field, and the influence of a faculty directly involved in influencing the direction of this industry as well as in developing their own understanding must not be undervalued.

The students in my courses during Classes 27 and 28 of SRI struggling with me and offering invaluable commentary and constant challenge are the true authors of this text. To them I offer my most heartfelt appreciation.

HAROLD LORIN

1

Fundamental Concepts
of Coexistence
in Time

Chapter 1

A BASIC ANALOGY

1.1. CONCEPTS OF COEXISTENCE IN TIME

When we think of the things going on together in "parallel," we intuitively include a large variety of working situations. Our general concept is rather broad and includes notions of many people working together in some way on the same task or on different but related tasks. It also includes the notion of an image of one person doing more than "one thing at a time."

The diversity of specific situations to which the concept of parallel activity applies is reflected in the diversity of design of computing systems. The richness of the human imagination in organizing activities for speed and reliability is reflected in the rich variations in the architecture of systems that attempt to reflect concepts of the separability, commutativity, and parallelizability of work.

The first discussion that we shall undertake here will attempt to reflect in a simple human work situation some of the multivarious forms of parallel activity and roughly to indicate computer system analogies. We wish to introduce in this way the flavor of those considerations that are relevant to the design of systems which are, in some sense, parallel and to provide an overview of various organizations.

1.2. ESSENTIAL ELEMENTS OF A BASIC PARALLEL SITUATION

Let us visualize a television repair shop that employs two repairmen. Each man is identically trained with exactly the same competence, and either can repair any set with exactly the same efficiency. Each man is provided with a workbench and a complete and duplicate supply of all necessary tools and all replacement parts. There is no communication between the men as they work. Each works in exactly the same manner as he would alone and is independent of the activity or even the presence of the other.

Let us initially assume that at 9:00 A.M. one morning both arrive at work to find a television set in need of repair on the benches. Each begins to work and each works without stopping until his set is repaired. By some coincidence, each finishes repairs at exactly 5:00 P.M. when it is time to leave.

We feel the essential concurrency of activity very directly. We think of the parallel activity of the two repairmen and concurrent repair of the two sets in the time period 9:00 A.M. to 5:00 P.M. This is a very basic notion of parallelism. To define it in terms of this example we would say that in a given time interval two independent tasks were being performed by two entirely independent agents and that the activity is continuous in the interval. Figure 1.1 is a graph of the work of repairmen 1 and 2 in the

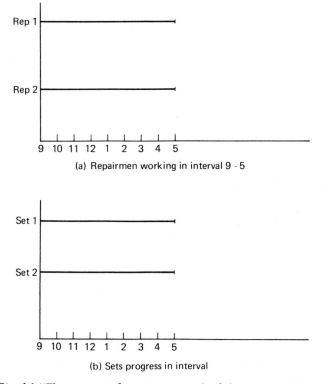

(a) Repairmen working in interval 9 - 5

(b) Sets progress in interval

Fig. 1.1. "The concept of concurrent work of the processor is separable from the concept of concurrent progress of the job."

interval and a graph of the progress on the repair of sets 1 and 2 in the interval 9:00 to 5:00. In this model the graphs are identical, but the fact that separate graphs are possible is an indication that the activity of the repairman is a concept separable from the progress of the task.

There are a number of significant elements in this model as described so far. We have four primary notions with which to deal: (1) the task(s) to be done, (2) the repairman, (3) the tools and parts (resources) used in performing the task, (4) the time for the task.

1.3 IMPACT OF DEFINITION OF TASKS

We notice that the definition of a task here is "repair a set." This is a gross task definition which hides a number of definable subtasks such as "remove chassis," "test tubes," "inspect connections," "replace tube," "test result," etc. The definition of the task at its current level has a critical influence on our appreciation of concurrent activity and upon the apparent homogenity of the model. In the definition of work for both human and computer, the level at which a cluster of elemental activities is defined as a task has definitive impact on our appreciation of the performance of a system. It determines how we will measure the usefulness of the system by providing the units whose completion we will count and by revealing the interfaces between subtasks.

Fundamental in a definition of tasks is a definition of dependence between them. Clearly, in this situation the repair of set 1 is independent of the repair of set 2. There is no precedence or ordering relationship such that there is a logical dependency of one task to the other. Such dependencies do, of course, normally exist both in human and computer work. Certainly among the subtasks of the repair of a set there is an imposed logical ordering relating some, but not all, as there is in the classical COMPILE-LOAD-GO sequence of the computer environment. The more logical order that exists between tasks, the less opportunity for concurrent performance.

1.4. DEFINITION OF "REPAIRMAN": THE "PROCESSOR"

The definition of a repairman is intuitive in a human work model. The repairman is our processing agent. Here he is a totally capable processor capable of executing any subtask that is necessary for the repair of a set. We view this "processor" as a very complex aggregation of capabilities, some sequential, some themselves parallel (he can pull out two tubes at a time, he can reach for a tube while reading a manual, etc.), without more detailed appreciation of his functional components.

In the discussion of systems we may characterize them as symmetric or asymmetric. Since we have indicated that each repairman is capable of doing exactly what the other can do, we have here a symmetric system. In a symmetric system no particular skill or capability is associated with any processor(s) in the system. No job or parts of a job can be performed faster, more efficiently, or at less cost by one processor in a system than by any other.

An implication is that the processors of a symmetric system are highly intelligent. By intelligence we mean the range of system activities that a processor can undertake, the independence with which it can perform them, and the amount of local decision making a processor can undertake with regard to any job it takes on.

As with living creatures, the concept of intelligence in a machine is a relative one. We talk about highly intelligent units like channels or devices when they are more capable than most current members of their class. Similarly, we talk about highly intelligent children and beagles when they are more capable and independent than other children or beagles. Channels, children, and beagles are stupid, of course, compared to CPU's, adults and poodles.

1.5. NATURE OF RESOURCES

The third element of the model—the resources, or the collection of parts, tools, and materials used in accomplishing the task—concerns us for two reasons: first, because their availability and accessibility limit the performance of the repairman, and if we were taking a closer look at the system performance and associated costs, we would be concerned with efficiency here. Second, it is not always obvious in all situations when a given thing is a resource or part of a processor. The separation we make here is supported by our notion of an anatomically and biologically human repairman whom we easily see acquiring and releasing resources (tools and materials).

In a computing system it is not always obvious where a "processor" ends and where its "resources" begin. In fact, a shift of perception in this case is a fundamental of new design concepts. For example, does one consider the primary core storage of a System 360 Model 65 a part of the processor or a resource available to it? Currently, we think of memory as a resource and have opened up many possibilities in design because of it, but at one stage in the development of computing systems a processor was conceptually inseparable from its memory.

1.6. TIME—THE FOURTH ELEMENT

Finally, we must think about time. This is the element of concurrency that most directly affects our understanding. We can never, by its very nature, invoke a time-independent notion of parallelism. The definition of a time-frame is a fundamental in the comprehension, analysis, or evaluation of any work system.

Let us notice that our concept of concurrency need not be as rigid as that which we have so far defined. The repair of both sets need not have started at precisely the same time, nor ended at the same time. Nor is it necessary that both activities be continuous. Notice Fig. 1.2. We see here that our definition of interval is critical. The time period from 9:00 to 5:00 includes subintervals during which we find sometimes both men working, sometimes one working, sometimes neither working. If our perception of time is limited to eight-hour intervals, we apprehend concurrency exactly as we did in Fig. 1.1. We would consider a perfect parallel case to exist here.

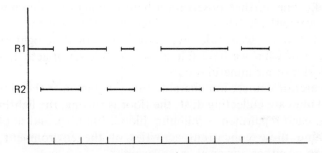

Fig. 1.2. Apprehension of parallelism sensitive to time scale. How fine is the measurement/perception unit? Do we see the gaps?

If our time reference is finer, however, we become uneasy with the intuitive conception and begin to search for a way to describe more precisely the degree of "effective true parallel operation" in view of the fact that we sometimes find concurrency and sometimes not.

By effective true parallel operation we mean the amount of time that processors are truly working simultaneously. Our appreciation in this regard is a function of our understanding of the relative speeds of work agents and time for tasks and at what time level it is appropriate to view them, and this understanding, of course, relates to the fineness of our definitions of processors and tasks.

"Same timeness" is naturally a function of our standard of time. A most rigorous definition of "same time" would be that, given any measurement, no matter how fine (e.g., the time it takes light to cross a chronom), a competent observer equidistant from two processes would discover multiple things happening at any sampling. But we rarely insist on this precision and settle for apparently appropriate, loosely appreciated intervals. In human historical terms we are quite content to say that while Athens built her naval empire, Sparta consolidated her position on the Peloponnesus, yet if we looked at these activities starting at one instant in 480 B.C. and determined every 150 milliseconds whether these processes of expansion and consolidation were indeed going forward, we would find many samplings where relative specific activity was indeed quiescent. It is, of course, absurd to measure historical processes in milliseconds, and, therefore, we do not do so, but we still speak of "parallel" developments in historical frameworks. We are content with a relaxed definition of what constitutes "going on at the same time."

1.7. BREADTH OF SYSTEM DEFINITION

Let us make one further observation before going on to computer analogy. If we were asked, "How many things are going on at the same time in the interval 9 to 5?" we would say, "Two." We would be right only in the limited sense of what we have felt to be a usable set of activities relevant to some cluster of meaningful events.

There are, of course, innumerable things going on at the same time in the shop. Tubes are collecting dust, the floor is rotting, the lighting system is working, each repairman is pumping blood, blinking, and smoking. But we determine these concurrent activities of the environment and the repairmen to be either universal or irrelevant.

We might very well expand our image to include as a member agent the power system without whose performance no work could be done. Our conceptualization of the intelligence of our repairmen roughly ascribed to them total capability, yet it excluded the ability to generate electric power. Clearly, in any situation there are supportive services being provided by some agencies, and these may or may not be included in our understanding of the system.

This is particularly true in computer installations where we may think in terms of the simultaneous operation of all elements of a computer system, or of all elements of a computer room, or of all elements of all machines and programmers, systems analysts, key punchers, etc. which operate simultaneously as part of the computing environment. The system analyst defines the breadth of his universe hoping to include as much as he can understand, and to exclude as little of what he should understand as possible.

1.8. THE COMPUTER ANALOGY

We have been clearly speaking of an installation where two computing systems exist. They are functionally identical; let us say that both are System 360 Model 50's with identical core size and I/O configurations. They are entirely "stand-alone" with no communication between them whatsoever. Each system has a complete library of procedures duplicated on private packs or tapes, and each system has an independent supply of packs and tapes to be used as output or scratch vehicles.

We define the analogy to the repairman to be the Model 50 processing unit, the resources to be all core, drums, disc drives, channels, control units, procedures. The time interval may be a single shift, and the task is to sort files of large size. Either system is capable of performing the sort on any data set in an identical fashion; no resources are shared in any way. We view parallel operation as the sorting of two files by two independent systems. The supporting services include, as well as power supply, such things as off-line conversions, perhaps.

1.9. THE NATURE OF THIS SYSTEM

Let us look at Fig. 1.3 to discover the significant elements of what we have described so far. We are interested in the number of processors. Since we have restricted our overview to two repairmen and two Mod 50's, we have two processors in each case. The two processors are *identical* and *independent.* They are *anonymous* in the sense that it makes no difference to the user requiring a sort which processor is going to run it. In the literature of parallel processing our model might be called a *symmetric independent multicomputer system,* where the word "computer" is used to suggest that the duplication of facilities includes a total computing system. The term "processor" is becoming restricted to the arithmetic-logical-control sections (the CPU without memory) or even smaller units.

It is the fact that we have two active agents that makes the parallelism so obvious to us. In our discussion so far we have only two tasks. We say, therefore, that we have as many processors as tasks to be performed. This is a very crude statement unless one visualizes that there are only two sets to be repaired in the shop, that they arrive two at a time, and that they arrive only when both repairmen are free. The concept introduced here by the enumeration of the number of tasks that can be performed together and the number of performers becomes quite useful later on. At this point we have restricted the population of tasks to be performed so that we may delay consideration of the mechanism of task selection and initiation, since variations in those mechanisms are profound in their impact on organization.

	Repairmen			Computer
Number processors	2			2
Number tasks	2			2
Task name	Repair set			Sort
Time frame	8 hours			8 hours
Utilization				
Repairman 1		System 1		
Repair set 1	8	Sort file 1		8
Other	0	Other		0
Repairman 2		System 2		
Repair set 2	8	Sort file 2		8
Other	0	Other		0
Total		Total		
Repair	16		Sort	16
Other	0		Other	0
Gross utilization		100%	100%	
Effective utilization		100%	100%	
Task completions		2	2	
Average time		8	8	
Cost				
Repairmen, each		$30	1,000	
Resources		$20	200	
Total		$100	2,400	
Cost per repair		$50	per sort $1,200	
Thruput increase by addition of processor 2				
Effective degree of parallelism 100%				
Cost increase 100%				

Fig. 1.3. Performance of two symmetric independent parallel systems. Independent tasks with different data.

Fundamental to the operation of the repair shop and the center we have described is the fact (mentioned above) that not only are the processing agents independent, but the tasks they perform are equally so. There are two basic reasons why two things cannot be done together: either there is only one processor and he can do only one of these things (as defined) at a time, or the things have a relationship such that one cannot begin without the completion of the other. In our case, tasks are defined as identical but with a different input, and with no interaction whatever.

1.10. PERFORMANCE OF THIS SYSTEM

The remainder of Fig. 1.3 has to do with the performance of the system we have described. We are interested in being able to determine whether or not the presence of an additional repairman or computing system is economically worthwhile or if indeed we are receiving proportional increase in work relative to our increase in expense.

Initially we wish to know whether two processors have given us an

honest day's work or whether through their own misdoing or the misdoing of others their total productive hours have not been what they might be. We must first define a reference in time over which we will measure their performance. We content ourselves with one eight-hour day, but determine that we will measure activity in terms of total hours or parts thereof in which there is activity. The total available system time is 16 hours (2 × 8). We are interested in how many hours are actually worked and of those hours how many are actually involved in work on a given set (or sort). In both the repair shop and computer room we charge by units of time, and consequently we need to know how many hours are chargeable for accounting as well as for performance analysis reasons.

We find from Fig. 1.3 that 8 hours of each repairman's time was actually spent working. This comes from Fig. 1.1, where activity is represented as continuous. The utilization of each repairman was therefore 100 percent. The utilization of both repairmen totals 16 hours or also 100 percent of available time.

We must further distinguish, however, what time was actually spent repairing. There might have been activities legitimately work related that did not contribute to the repair of assigned sets. The necessity for repairing a tool or for getting additional materials or parts would be included in utilization of the repairman but not chargeable or contributive directly to the repair of sets 1 and 2. These might be thought of as overhead activities. Counterparts in the computer system might be disc or tape mounting time, operator or operating system time. It is because of these general overhead activities that the concurrency of activity of a processor may not coincide with concurrent progress on given jobs. Hence we have a separation of graphs in Figs. 1.1 and 1.2.

We might call the amount of time spent in directly repairing the "effective utilization." In an independent concurrent model the difference between utilization and effective utilization is due to local problems with each processor. No delays or overheads are introduced by one processor which affect the other. The "effective degree of parallelism" is the amount of time that both processors are working on the assigned task and in our model is equivalent to the limit of effective utilization.

A relevant concept here is the time cost of parallelism. In this example this cost is nothing. No delay or overhead is experienced, because the two processors are working together in the shop or in the computer center. The facility's cost, of course, is enormous—double the cost of the repair capacity.

We can go farther to make throughput and turnaround statements about the model, observing that turnaround time is 8 hours and daily throughput is two sets. From this we can compute the cost per repair. When we have computed the cost of both repairmen and the number of sets repaired each day, we can determine if doubling of capacity has resulted in a doubling or less or more than a doubling in result. Certainly we would be very critical of an organization if two times the capability produced only 1.3 times the result.

1.11. MOTIVES FOR THE SYSTEM

Ultimately we face the question of why we hired another man, why we installed a second duplicate system. We may find diverse answers of the following kind:

1. The amount of sets coming into the shop is so great that I must increase my capacity to repair them. Actually, I do not need to double it, but it is impossible to acquire less than one full man and I am satisfied that the first man cannot work faster.

2. The first man is not loaded but I must have a man to fill in, and I find the economics of emergency help untenable and availability unsure.

3. The economic utility of reducing the amount of time it takes a set to be repaired by having two men available and reducing waiting time for a set more than makes up for the additional cost, since I can charge premium prices for fast service.

4. There are certain days of the week when my load is much heavier than on other days, and I must still expend an average of 2 hours on a set.

These reasons are all directly translatable into computer terms— loaded system, backup, priority express capability, peak loading capacity. We notice that the fundamental reason for acquiring an independent processing agent was the inability to make the single processor go faster.

In conclusion, we might note that we have so far no appreciation for the utilization of resources. The amount of time that a soldering iron was used or the amount of time a tube tester was used during the day could be measured, and the utilization might very well lead us to some considerations about the necessity for full duplication of all resources and the conjecture that perhaps some resource sharing might be undertaken to reduce the cost of the concurrent capability without reducing the throughput.

Chapter 2

SOME INTERACTION
INTRODUCED

2.1. WORK ASSIGNMENT

So far our two repairmen have been sufficiently independent that there is really no difference between them and two men working in different shops. The consideration which links them is that they draw from a common pool of nonoperative television sets delivered for service to their organization. These sets arrive at some rate, and we must extend our model to include the mechanisms by which a set is received, enqueued for service, and finally "bound" to a repairman for service. This is a first representation of the scheduling allocation problem, the solution of which has a profound impact upon the design of a parallel system.

Let us first postulate that we now expect a repairman to finish more than one set per day, and consequently he must, during the day, undertake new tasks. The essential question is to what extent the repairman schedules his own work and to what extent it is scheduled for him. If there is a separate scheduling agent, to what extent should he be considered a part of the concurrent situation? This is currently a significant area for computer systems, and it is not clear to what extent computers are self-scheduling or that we fully appreciate the limits and responsibilities of scheduling done by center administrators, operators, and operating systems.

2.2. SHARING AND ACCESSING THE INPUT TABLE

One approach is to provide a large table on which customers leave their sets. If the customer is not familiar with the repair facility, we might say that the presence of two processors is unknown to him and the names of the processors are not known; that is, the processors are anonymous to the user. The collection of television sets in this case forms a truly common pool, any one of which may be worked on by any repairman. It is possible, of course, that certain customers are able to express a preference for a given repairman and so to indicate on the form they fill out when they leave the set. The set remains on the common table but is marked for a specific processor.

Alternatively, a separate table might be provided for each repairman; customers would leave their sets by choice on one or the other. If this is so, then there is not a true shared pool of work to be done. In queuing theory terms our common table to which both have access is a multiserver situation with its attendant service and waiting time characteristics, and the separate table represents single servers.

We are interested in the common table approach, but we do not exclude the possibility that for some reasons some sets on the table are earmarked for a given repairman. If there is no intervening activity between the customer and the repairman, we consider that the processors are *self-scheduling* and that the act of scheduling is to walk over to the table and select a new set for repair. The basis for selection might be simple (the next in line on the table) or rather complex, involving customer priorities, rush jobs, etc. The critical point is that the repairman schedules his own work and removes a set from the pool, making it unavailable for the other man to select.

The scheduling is dynamic in the sense that each visit to the table is a complete scheduling event involving a review of all sets and the selection of one of them by the selection criteria. (This review may consist simply of recognizing the first on the table closest to the repairman.) No selection activity that preplans the day's work is undertaken.

We have a *dynamic self-scheduling* system. We must recognize that some of the working time spent every day by the repairman is devoted to selecting a set, and the activities may now be shown as in Fig. 2.1. We notice that at points A, B, and C both repairmen are selecting new sets. This possibility introduces the concept of potential "contention" for a resource.

2.3. GROUND RULES FOR CONTROLLING SHARING

We have shown that the two men can select simultaneously and have implied that no delay in the work of one man is caused by his having to wait for the other to select a set. In the human situation this is a quite natural occurrence, but it is because of the nature of the devices less usual

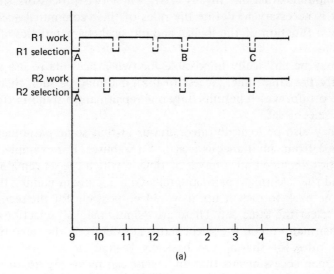

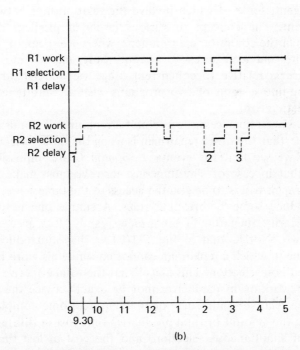

Fig. 2.1. (a) Asynchronous work and selection-independent agents. Three instances of potential contention. (b) Delays due to contention for single service facility.

in a computer situation. In any case, when two processors share something, it is necessary to define the rules of their communal access. These rules are a function of the logical and physical characteristics of the thing they share.

It may be physically impossible for two processors to use a resource at exactly the same time. Access to the table may be such that there is a narrow corridor which permits only one repairman to stand in front of the table to select a set.

It may also be logically incongruous in that some pathological result may come from simultaneous usage of a resource. For example, if the act of selection included an interval of time in which the set remained on the table and the selecting repairman prepared a document naming the set and then came back to pick it up, it would be possible for the second repairman to select the same set. These are "simultaneous" selections, because both repairmen were in the select status at effectively the same time.

The rules for sharing are basically "open" access or "constricted" access. Open access means that any agent can make any use of a resource regardless of the use of it by any other. Constricted access implies a set of constraints, some of which involve an appreciation of the need and privilege of the agent, some of which involve the relationships between more than one agent. The concept of various levels of privilege in usage is familiar to us in the computer environment, where we speak of "read only access," "write and read access," "execute only access," etc. The constraints on access relative to concurrent usage we understand from the problem in on-line systems of two programs wishing simultaneous access to the same data record.

When we constrain access, we prohibit a repairman from usage for the period of time that the other repairman is using the table (selecting a set). If we do so, we see that the events A, B, and C are impossible in the system and that in case of simultaneous need we must make a decision about which repairman is to be granted access to the table.

We introduce some tie-breaking rules. A simple one is to define a continuing priority such that if repairmen 1 and 2 tie, access is always granted to two. Notice that in Fig. 2.1(b) we have introduced delays, periods of time in which a repairman cannot continue his work because he is "interlocked" on selection. This means that the total effective output of the two men working in parallel cannot be exactly twice the output of one man. The relative significance of this requires some complex evaluation involving the amount of total productive time lost in this fashion (the probability of simultaneous selection) and the cost of lost time vis-à-vis the cost of providing two tables or having a wider corridor.

We must also note that we must be careful with our notion of simultaneity. The requirement is merely that at some point in the interval during which repairman 1 is selecting, repairman 2 wishes to select. It is not necessary that they, at precisely the same instant in time, develop the need to select a new set. One way of reducing the loss of productive time is to define the selection procedure in such a way that it is shorter or to define selection subtasks in such a way that some of them may indeed be done in parallel. This reduces interlock time to those subtasks which indeed cannot be done together.

2.4. ASSIGNING A SPECIALIZED FUNCTION AMONG PEERS

Before going on to our computer analogy, we shall add one more element. Let us conjecture that the process of receiving a set into the shop requires the participation of the repairman. He must approach a counter at the other end of the table and formally accept the set from the customer, preparing a receipt, listening to the complaint, etc. This is another activity that interferes with the work of repair. It is necessary to determine exactly which of the men will undertake this activity. Since these men are absolute peers, sometimes one will receive the set and sometimes the other.

We have so far no communication between the men to allow for any dynamic consultation or negotiation between them, and we have no general basis upon which to establish a rule. Let us simply establish that on Monday, Wednesday, and Friday, repairman 1 performs this function and on the other days, repairman 2. We have many alternatives to this rule, but they all either violate the perfect peer relationship or require active communication between them, which, for the moment, we exclude.

2.5. THE COMPUTER ANALOGY

We postulate two complete and separate systems, completely independent with the exception that they share a disc drive on which are recorded task descriptions. Whenever a new task comes into the center, it is thrown immediately into a card reader (a System 1 reader on Monday, etc.), and the associated processor is interrupted to read the task description into the common disc. We may recognize the fact that the act of placing a task in the task pool may interlock the act of selecting a new task for the other system.

Whenever a system wishes a new task, it attempts to access the disc with task descriptions and is either successful or unsuccessful at gaining this access. If successful, it selects a new task and while doing so establishes some condition that will preclude the other processor. The exact mechanism is a detail of the interface. Since there is no program communication between the processors, it would probably be a hardware interlock of some sort. Its essential feature is its "passivity." By this is meant that no action on the part of the other system is required, and indeed the interlocked condition will go unnoticed by the nonaccessing system unless it attempts access. The interlock condition is not in and of itself a significant event in system performance.

Such systems, dynamically self-scheduling common task pools, have been defined in large installations. The fundamental advantage is the ability dynamically to balance loadings of two computer systems by allowing them to acquire tasks as they need them. In some measure the burden of prediction is removed from human schedulers. In a cost sense the sharing of the common disc realizes some economy.

2.6. NATURE OF THIS SYSTEM

We have been describing the loosest possible coupling of two independent systems. The symmetric nature of the system has been preserved; the general time frames, task definitions, and resources of our first situation have gone unchanged.

We have made some fundamental decisions about the system in our reaction to task acceptance and selection. We have determined that both these activities are within the intelligence of our two-agent system, and that they will take time off to do it. Further, we have arbitrarily allocated time intervals in which one processor will perform a service function for them both.

The particular service function performed (accepting sets) is not truly a cost of the duplication of repairmen, since this function would be performed in any case even if only one man were involved. Task selection, however, potentially delays one repairman and is, therefore, a true penalty of the relationship between them. We have shown a general principle related to our earlier statement that two things cannot proceed concurrently if there is no agent to perform them or if there is a logical dependence between them. We extend this to include a case where a processing agent is free to perform but performance involves his access to a given resource that is unavailable. In this case the table of broken sets is not simultaneously available or the disc drive is not.

The sparseness of resources potentially introduces delays and restricts the performance of two sharing elements. The degree of delay is a function of how sharply they contend and what penalty in time is involved in loosing the contention.

Sometimes there are overriding operational advantages to pooling a resource which more than compensate for contention delays, not only in cost, but also in time and performance. The ability for each repairman to access any set rather than only sets on his table is a true advantage. If this were not true, one repairman might experience periods of idle time because his table was empty although there were sets to be repaired in the shop.

In fact, one possible way of looking at sharing and the contention thereby implied is that a basically nonpooled resource exists which is private to some agent. The interval of privacy, however, is reduced to a smaller time period so that there is greater flexibility in the system because of a greater discretion in resource usage. The resource is private, not for the life of the system, but for discrete intervals of time. It is possible to simulate the availability of a duplicate and either increase performance or reduce cost. The sharing of a resource, therefore, has two effects on a system. On the one hand, it introduces contention, which is a potential cause of delay; on the other hand, it introduces the possibility of flexible reallocation of resources over short periods of time, which may increase throughput over a given resource basis.

2.7. PERFORMANCE OF THIS SYSTEM

If we look at Fig. 2.2, we see that resource A is shared and its cost distributed. Because of contention on resource A (the table) a system hour is lost, and the utilization of the repairmen falls off from 100 percent. We assume that there is no effective alternative activity, and they are idle during selection interlock. The cost of the two men and resources divided by the number of sets they repair gives a cost/performance measurement that can be compared against (b), where there is no contention. (We are assuming that there is no problem of unavailability of broken sets on the private table).

We notice that the increase in cost of the resources due to the new table does indeed eliminate idle time and increase the number of sets repaired. However, the cost of repairing each set goes up. Finally we see in (c) that our sharing is worthwhile in this case because by sharing the table we invest only 1.6 times the cost of the original man in acquiring a new one but increase output by 1.8. Unless there is some reason for insisting on 26 sets a day, we do well to share.

Closely connected with our decision to pool is our fundamental deci-

(a) With selection contention

    System cost   Available hours 8 + 8 = 16

Processor 1	$ 30
Processor 2	$ 30
Resource A	$ 20 (shared)
Resource B-1	$ 5
Resource C-1	$ 5
Resource B-2	$ 5
Resource C-2	$ 5
	$100

Processor 1			Processor 2	
Repair sets	6:30 (13 sets)		Repair	5:30 (11 sets)
Select sets	1:00		Select	:55
Wait to select	0:30		Wait	:30
Accept	0		Accept	1:05
Total repaired	24			
Total repair time	12:00		Cost per set repaired = 4.17 (100/24)	
Total select	1:55		Loss due to contention = 1 hour	
Total wait	1:00			2 set repairs
Accept	1:05			
	16:00			

(b) No system contention

    System cost

Processors	$ 60
Resources A	$ 40
Resources B	$ 10
Resources C	$ 10
	$120

Processor 1			Processor 2	
Repair sets	6:55 (14)		Repair	5:55 (12)
Select	1:05		Select	1:00
			Accept	1:05
Total repaired	26		Cost per set repaired (120/26) = 4.66	
Total time repaired	12:50			
Select	2:05			
Accept	1:05			
	16:00			

(c) Stand alone

Processor and resources	$60	
Repair sets	6:30	(13) sets
Select	1:00	
Accept	0:30	

(d) Analysis

1. Increment in cost for no contention	2x
2. Increment in cost for contention	1.6x
3. Increment in work for no contention	2x
4. Increment in work for contention	1.8x

Fig. 2.2. Cost performance.

sion that no specialization is needed for the acts of acceptance and selection and that no specialized capability needs to be introduced into the situation in the form of a scheduler. In the next chapter we introduce some alternatives to this approach.

Chapter 3

INTERACTION
EXTENDED

3.1. ASYMMETRY INTRODUCED

We have now two trains of thought active for us. One is the relationship between the repairmen with regard to selecting work, and the other is the concept of pooled resources. They are closely connected concepts, since we see that the functional capability of selecting work is associated with access to the table and that the performance of the repair shop is affected.

Let us change our situation with regard to scheduling and postulate that both the acceptance of jobs and the assignment of work is always performed by one of the men, whom we now consider to be senior. Consistent with this "seniority," it is the senior man who is always responsible for accepting sets. Accepted sets are put on a table at acceptance time by the senior man, and it is always and only he who removes a set from the table when either man requires a new set.

3.2. NEED TO COMMUNICATE

To support this organization of work we require that there be a new capability between the two men. They must be able actively to communicate with each other. The junior man must be able to register a request directly with the senior man, and the senior man must interrupt his work on some basis to service the request of his colleague.

We need some rules to determine on what basis the junior man may interrupt the senior man and how quickly the senior man will respond.

The degree of effective simultaneity of work in the shop will be determined by these specific rules. See Fig. 3.1. Figure 3.1(a) shows the graph of parallel activity if the junior repairman can interrupt the senior repairman only when the senior repairman is himself ready for selection, or if the junior man can interrupt at any time but is not serviced until the senior man finishes a task. This causes a considerable idle time in the junior path while he waits for a new set to be brought to him; however, it tends to make the selection process more efficient, since two sets can be selected together at less than twice the effort to select them independently.

Figure 3.1(b) shows the effect of allowing the junior man to interrupt at any time and to gain immediate service. His idle times are reduced, but an element of inefficiency is introduced into the work of the senior man. He selects one set at a time, and he must experience some delay in recovering his work when he returns from selection. Even in this situation, however, we should probably define some activities by the senior man that cannot be interrupted, for example, when he is soldering a connection.

Asymmetric junior/senior (à la B5500)

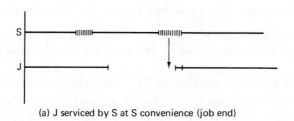

(a) J serviced by S at S convenience (job end)

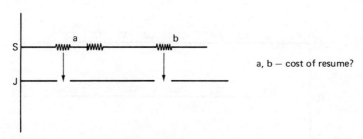

a, b — cost of resume?

(b) J serviced by S at J convenience

Fig. 3.1. Who is the master?

3.3. IMPACT OF SPECIALIZATION

We have taken a step here toward abolishing the symmetry of the system and introducing an asymmetric organization where particular activities are permanently assigned to one repairman because of some particular characteristic of his own—in this case, seniority. This may be done because it is felt that the senior man is inherently faster and can sustain his rate of work on sets while performing the secondary activities. We might adopt this functionally specialized system because we feel that the acceptance and scheduling of sets requires some capacity for judgment that only the senior man has.

We may extend this functional specialization considerably farther in many directions. For example, we may define the secondary and support activities of the senior man extensively so as not to allow him any time to repair sets at all, but to spend his days entirely in the activity of accepting sets, scheduling, ordering parts, performing postrepair check-out, etc. This implies a larger view of the activity of a repair shop to include functions beyond the repair of sets that must be performed on each set. This extended view carries with it two considerations:

1. The resources that each man uses to perform his work are no longer identical.
2. Our image of parallel activity must be extended to include activities which are related to a goal but which are not identical.

The effect on us is to collapse our resources of tools and to reorganize our resources of parts and materials so that they are wholly available to the junior man.

Further, from Fig. 3.2 we see that we have expanded our concept of a task into further functional activities—receive and schedule, repair, and check-out—and we have defined work for the two men on this functional basis. They no longer work independently and in parallel on different sets but cooperatively on the same set, performing different functions. This

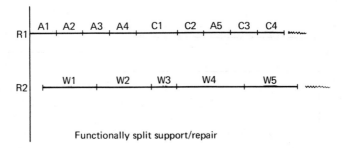

Functionally split support/repair

Fig. 3.2. Cooperation on a single job. A=accept; W=repair; C=checkout.

further implies that the activity is no longer entirely asynchronous but that they are dependent upon each other in order to proceed with work. Conceptually one might reverse the roles of the two men and nominate the junior man to perform the special secondary functions on the assumption that the single most difficult and time-consuming functional task is the repair of sets and that what one wishes to do is to unburden the most capable repairman.

3.4. COMPUTER ANALOGY

We have described what is essentially an asymmetric multicomputer system. In our first remarks we introduced an "active" connection between two processors; this may be a channel, or direct interrupt, capability over which it is possible for one processor to "shoulder-tap" the other. We have two stand-alone systems that are no longer entirely identical. We might envision a 360 model 44 and a model 65 connected by a channel-to-channel interface. They may or may not share a mass storage device.

The term *channel* in this text is used to describe an electronic path between a processor unit and an I/O subsystem. Control signals and data travel across this path between units at either end. IBM-oriented readers will recognize that the devices that IBM calls channels are more than passive paths. They have certain local control and decision capability built into them. This is a result of the way IBM has chosen to distribute functions in its I/O subsystem and is not of specific interest at this point.

Let us postulate a configuration like Fig. 3.3(a). We have added multiple 44's "behind" the 65, as we could add multiple junior repairmen. If the 65 does all job reception and assigns work to 44's and writes outputs, we have truly added a functionally specialized processor in a master relationship to all 44's. The 44's operate as computational slaves. The resources that the 65 and 44's have associated with them are certainly not identical. How many and what type of devices and storage associate with each 44 is a function of what it needs to perform computations, those of the 65 to support system I/O activity.

It is possible, if the 65 has access to all necessary resources, for it (in addition to performing service functions) to undertake computational activity of certain types, just as it was possible for the senior repairman to continue to repair sets. Perhaps he would now repair only certain kinds of sets, with certain kinds of problems, either the simplest or the most complex, depending upon what elements of competence had caused him to be regarded as the senior man.

An actual configuration is that of Fig. 3.3(b), where a 50 serves as a

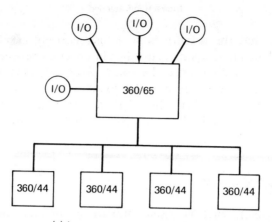

(a) Large system controller backed by
computing slaves

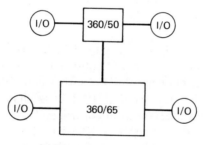

(b) Support processor serves
computational processor

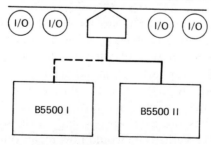

(c) Identical processors with one
nominated as master

Fig. 3.3. Asymmetric computers.

"support processor" to a model 65. The 65 is a slave to the 50 in the sense that the 50 receives job requests and schedules work on the 65, receiving from the 65 files of output for printing. Loosely coupled asymmetric multicomputer systems 7044/94 DCS and 360 ASP will be described in more detail in a later chapter.

What we have observed of particular note in this chapter is the introduction of functional specialization and asymmetry in a concurrent system. We notice that the master/slave concept which we initially introduced in the scheduling mechanism can be extended to include a number of secondary utility functions in the work environment.

3.5. NATURE OF THIS SYSTEM

The concept of asymmetry and functional specialization is one that occurs repeatedly throughout the design of systems organized for concurrent operations and does so at all levels. The gross asymmetry of a master/slave computer "hook-up" is merely one representation of the concept that finds at the other extreme of fineness the development of functionally specialized units within "single" processor systems. These are semiautonomous subprocessors that are traditionally floating-point arithmetic units designed to perform these operations at high speed and essentially in parallel with other operations.

The concept of semiautonomous is derived from political science, where a semiautonomous political unit is one in which certain decisions can be made by local government within a loosely defined national policy. A state of the United States is a semiautonomous unit which, within the framework of the Constitution, has broad local self-determination but which cannot undertake an independent foreign policy.

In a computer system a semiautonomous unit can be conceived to be a floating-point unit which, under the general policy set out by the central processor to "perform a float operation," determines for itself which one it is to be and how it is going to go about it.

Semiautonomy in a system provides for a specialized subunit to take work out of the main path in small functional chunks and perform it in parallel with the main path. The effectiveness with which it does this is a function of its intelligence, which determines how much it can take over and how independently it can operate.

In machine systems one infers intelligence and degrees of autonomy in a unit from the number of registers it may have and the amount of storage space, control circuits, etc. In general, we find that almost all "symmetric" systems contain some asymmetry in the system if we descend low enough in our definition of a function or wide enough in our image of the system.

Chapter 4

MORE ON
RESOURCE SHARING

4.1. RESOURCE SHARING

Up to this point we have observed concurrency across two repairmen, each working at either the same function on different sets or working in parallel at different stages of the repair of sets. We have alluded to resource sharing apropos of the facility that holds sets to be repaired. It is possible for these repairmen to share other resources as well, and the degree to which resources are shared and which specific resources are shared and by what rules affect the appearance and performance characteristics of a repair shop as well as a computer center in very profound ways.

4.2. MOTIVES FOR SHARING

Essentially, the decision to reduce resource duplication is based upon an appreciation that the amount of work going through the shop and the pattern of arrivals and repairs is such that as many sets can be repaired by the two repairmen if they share tools, parts, and materials as can be repaired if each is privately equipped. This is a concept based upon resource utilization and availability that we have not looked at deeply.

The concept of utilization is familiar to us from our appreciation of the amount of time in one day that each repairman did useful work and the amount of time that he was idle. This "idle" status we saw developing because of delays forced upon each other by the two men.

We could go farther, however, and investigate the amount of time each man used his soldering iron, oscilloscope, tube rack, etc. during the day. We might conclude that we could reduce the amount of resources, reducing cost per repair without reducing the number of sets we repaired.

For example (Fig. 4.1), we notice that the utilization of a soldering iron for each man over the course of a day is quite low, 72 minutes out of an available 960 minutes of soldering capacity. We easily observe that if we had one soldering iron easily available to each man we could fulfill the gross soldering requirement for the shop. However, we notice that there are instances of overlap in time when the repairmen both desire to solder.

We have an instance now where the unavailability of the resource over a specific period of time is going to delay the work of the repairmen despite the very low utilization of the soldering iron. As utilization increases for a resource and as more resources are shared, the delays caused by contention of this type tend to be more serious. The contention may be so serious as to cause performance to decay to a level not only under the potential productivity of two repairmen, but to a level below the productivity of one.

Analyses of contention are necessary for any sharing decision about a resource. A full understanding of contention and its effects may be gained through simulation or through analytic probabilistic models developed for

Rep 1	Time	Rep 2	Time
9:30 – 9:37	7	10:14 – 10:17	3
10:23 – 10:31	8	11:04 – 11:15	11
1:05 – 1:12	7	1:06 – 1:09	3
1:30 – 1:33	3	2:15 – 2:23	8
4:10 – 4:14	4	3:25 – 3:31	6
4:47 – 4:49	2	4:38 – 4:48	10
	31		41

Available 480

31/480   41/480

Instances of use	6	Instances of use	6
Average use	5.2	Average use	6.5

Parallel Usage

1:05 – 1:12, 1:06 – 1:09
4:47 – 4:49, 4:38 – 4:48

Time lost to 1 if shared   to 2 if shared
4:47 – 4:48   1 minute   1:06 – 1:12   6 minutes

Fig. 4.1. Usage of the soldering iron.

the understanding of the system. What is necessary to determine is the probability that contention will occur, the expected length of delays, and the impact on productivity of delays. Whether or not a given level of contention is tolerable depends not only upon the level of productivity in terms of the number of sets repaired during a day but upon the economic utility of a given rate of set repair as opposed to the cost of the duplication and private availability of the resource.

There is a further aspect of resource sharing. This is the concept of a general "pool" of resources available to each repairman as he needs it. Consider that with private soldering irons and no communication between repairmen, it is possible for a malfunctioning soldering gun to make it impossible for one man to complete the repair of some number of sets during the day. He would be forced to be idle (if no sets could be repaired without soldering) or to reduce the number of sets that are selectable for repair by him. Since he may often not know in advance which sets will require soldering, he may build up a collection of incompleted sets or he may try to apply a substitute technique. If he does this, he may be said to be operating in a "degraded" mode, using a technique that enables him to complete repair but more slowly. However, the soldering iron of the other man is not in constant use, as we have seen, and could be made available to the repairman whose iron is not functioning. This might introduce contention delays, and in this sense we would think of the repair shop sharing the iron as also in a "degraded" mode. The sharing may be done in the sense of "Yes, you may use my iron," with possible preemption by the owner of the iron. The concept of particular identification of an iron with a man may be abandoned entirely, and the collection of irons thought of as a common pool to which each man has equal inherent right to any member of the pool. This right may be modified by any priority associated with the task being performed.

With materials other than soldering irons it is possible to think further in terms of the relative usage. If we think of electric outlets as a resource and envision 10 of them forming a common pool and each man dynamically acquiring and releasing outlets for his set, his testing equipment, his soldering iron, the coffee pot, etc., we have a usable image of a facility that is very significant in resource pooling. This is the capacity of each active agent to expand and contract his utilization of the resource as his requirements change.

The acquisition and release of resources from a pool has a number of implications for performance, as we have seen, but there are also implications for the reliability of the working unit. We have a motive for making resources commonly accessible beyond resource cost minimization and beyond performance optimization. This motive is the assurance that the "system" will be able to operate, even if on a degraded basis, and to fulfill some defined minimum requirements for productivity.

4.3. COMPUTER ANALOGY

We have in one sense been discussing the configuration problem—the determination of population of resources which is to be associated with a processor in order to achieve necessary performance and reliability. The resources of a computing system are memory, channels, control units, and devices, and we are concerned with providing a sufficient number of various resources to meet the expected needs of the system in the performance of its work load.

We are also concerned with the distribution of the resources in terms of accessibility. In computing systems a processor may contend with itself for access to a resource. For example, if we assign two tape devices to a single control unit and this control unit to a single channel, it is impossible for the two drives to be accessed simultaneously. This could only be achieved by increasing the resources by providing another channel and control unit or by developing a capability of defining alternate paths dynamically.

In the two-processor situation analogous to our repairmen, the fundamental resource problem is to determine what paucity of resources one can maintain while approximating the performance of two independent systems fully duplicated (or essentially duplicated).

What is commonly understood to be a "generalized multiprocessing system" includes a high degree of sharing, including the sharing of all primary directly addressable (corelike) memory. Each processor has full addressability to all locations and commonly identical addressability. (However, with a rather common technique for interfacing memory and processors, it is possible to define "private" memory addressable by only one processor. Other privately addressable stores may exist in the processor as registers or larger private working areas. The reason for this and its impact on performance are described in a later section.)

In the sharing of memory the concept of specific ownership is entirely abandoned, and the assignment of space to a processor is dynamic from a pool of available locations. This is not always the case in the sharing of channels. The concept of ownership of a channel is often preserved in a multiprocessing system, and it is impossible for a processor to address the channels of its colleague. Often the link between the processors may be a channel; this, of course, may be commonly referenced.

Where the link between processors is more direct, as is typical in multiprocessors, the shoulder-tapping capability may be used for one processor to request another to undertake a read or a write on its behalf. Let us imagine that two processors are sharing a large disc file and each has a channel to it. Processor one finds its channel inoperative or busy on another device. It may interrupt processor two to request it to read the device on its channel, transferring data to common core. This "request" may be forced interruption to I/O or a less imperative arrangement that allows the second processor some options.

In a previous section we discussed the performance of auxiliary functions on one processor or another, limiting our interest to task selection. Considering I/O as an auxiliary function, we can add a dimension to our discussion. If we consider that each processor has private I/O, then there is no question that each processor must submit its own I/O commands and field its own interrupts. If, however, I/O is shared on a pooled basis and each processor can acquire devices dynamically on any channel in the system and can address any channel, it becomes possible to raise a question as to whether it is absolutely necessary that the processor that has "enqueued" an I/O request on a list of yet to be processed requests actually submits the request to the channel and, further, whether it is necessary for the submitting processor to field the interrupts (assuming an interrupt system) that will occur.

The concept of "most interruptible" becomes more specific here. It should be possible to assign the fielding of interrupts to the "most interruptible processor," i.e., the processor that will cause the least decay to productivity if it is interrupted. This may be based upon the priority of the task currently assigned.

Another technique is to assign all interrupt handling to the processor that is executing control coding, for example, task selection. While executing selection the processor becomes the "control processor" and dynamically acquires the function of interrupt handling until it leaves control status.

4.4. SPACE- AND TIME-SHARING

The concept of sharing and the techniques of sharing are fundamental to all computer system design. There are two forms of sharing in a computer system—time-sharing and space-sharing. Space-sharing occurs on storage devices that have collections of addressable units that can be allocated in partitions of various sizes to different logical structures ("files," "data sets"). Two processors space-share a device when data relevant to the current tasks of each processor are recorded on the device.

The processors, while space-sharing memorylike devices, may time-share the access to them. The concept of time-sharing most fundamentally challenges our ideas of simultaneity and true concurrency and displays most graphically the relativity of our intuitive notions.

Perhaps the best illustration of time-sharing is a multiplexing channel servicing three card readers. We observe easily that the three devices appear to be operating in perfect simultaneity as if each had a separate channel to a memory from which characters were being taken and in which (reader) characters were being deposited.

We are talking about three objects with characteristically different cycle rates: a memory unit with a cycle time of two microseconds character fetch, capable of delivering 500,000 character times per second, the channel, which has a delivery rate of 50,000 characters per second, and a 1000 cpm (cards per minute) reader, which passes a card in 60 milliseconds approximately.

Let us assume a column reader that reads a column at a time and presents the character collected through a control unit to the channel. We may include a two-character buffer area to hold characters temporarily. Let us further assume that the total number of column times passed through the read station is 80 plus five preceding columns representing space between cards. Therefore, the total number of columns passed in 60 ms is 85, of which data transfer occurs on 80. This means that time to read a column is .70 ms. Every .70 ms for 56 ms, the reader will present an image to the channel; it will then suspend for around 4 ms space between cards. The effective character delivery rate of the card reader is 1400 characters per second. The total transfer rate of all readers is 4200 characters per second.

They will all operate in parallel by "time-sharing" the multiplexer channel. The possibility of "time-sharing" is based upon the fact that our effective unit of time, "visible time," as it were, or "virtual time," is second- and minute-oriented. We are content to observe perfect simultaneity if we read 3000 cards per minute on three readers. We achieve some true simultaneity in that each reader is physically moving and sensing cards, but we do not have simultaneity with regard to the transfer of data over the channel. At this point we have imposed a sequence of single character transmissions, but on a time basis such that the illusion of concurrency is preserved.

Time-sharing basically occurs when asynchronous units have disparate rates of service such that the basic cycle of the slowest unit (to whose performance we are oriented) is maintained as if it had free and uncontended use of the service of the faster unit. The faster unit is used in a sequential way so as to create the illusion of there being multiple units. This is possible because of the relative speeds of the slow and fast units.

4.5. MULTIPLEXING—A SIMULATION OF SIMULTANEITY

There are a number of different ways to multiplex. What we shall describe here is called fixed time division multiplexing. The multiplexor can receive 50,000 characters per second but is called upon here to process only 4200, so we have an underutilized channel capable of running many, many readers.

We shall divide the second into a number of fixed time slices. We know that the fundamental character delivery rate of a reader is .70 ms and that the character processing time of the 50,000 cps channel is .02 ms (and let us say that that includes memory access). In order to service the three card readers it is necessary only for the multiplexor to be available to each reader every .70 ms for .02 ms.

In fixed time division multiplexing, we shall define a time slice to be .02 ms and we will define a slot for each reader, such that we have a three-stage cycle that occurs in fixed sequence.

We say that there are a T_1, a T_2, and a T_3 slot corresponding to each reader. The multiplexor visits a reader and waits .02 ms at each one. If there is a character to be transmitted, it goes down the line to core; otherwise, the channel is idle for that time slice. At the end of the slice the channel visits the next reader for .02 ms, etc., completing a scan every .06 ms. Reader 1 is visited at time .00, .06, .12; reader 2 at .02, .08, .14; reader 3 at .04, .10, .16, etc. Notice that there is no contention with this form of time-sharing, since priority is binary and time-oriented. When a character is present it must be unloaded within .70 ms, or it will be overlayed by a successor. We are well ahead of the game even without a buffer. Figure 4.2 illustrates the multiplex function.

The hardware time-sharing is replicable in essence in software, where we talk about "multiplexing" the attention of the CPU. The fundamental scheduling mechanisms of time-shared systems like TSS/360 or MULTICs are software approximations of multiplexing and afford the illusion of simultaneous operations in exactly the same way as our multiplexor sustained the illusion for the card readers.

Apparent parallelism based on disparate cycles in complex systems is the basis for multiprogramming in any form. If we consider CPU execution circuitry as a resource, then we see that multiprogramming systems implement the concept directly through the definition of a "virtual machine" that time-shares the execution circuitry of the hardware basis.

In any given system, depending upon the level of detail at which one looks, parallelism will be either real or apparent. There will be true concurrency based upon separate resources and the simultaneous use of them.

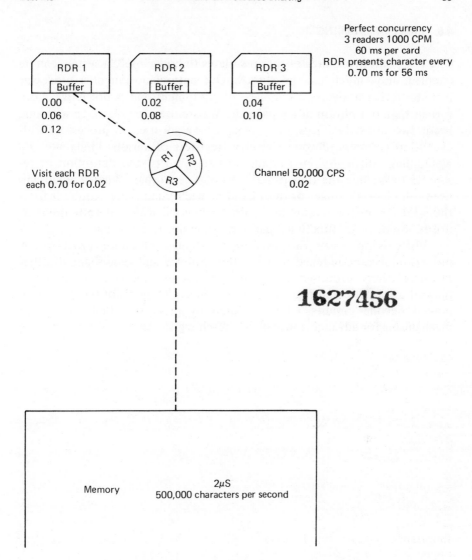

Fig. 4.2. Time-sharing. Time division multiplexing.

There will be apparent concurrency based upon time-sharing, the sequential use of resources in such a manner as to provide the illusion of separate resources. Mechanisms of time-sharing will be distributed across hardware and software boundaries through hardware and software scheduling and priority mechanisms.

4.6. LIMIT OF SHARING

One last point in the area of sharing relates to how closely a system can be coupled, how much can be shared. Our natural notion of a processor includes rather a diverse assemblage of equipment and is not much more formal than our notion of a repairman. We are used to repairmen with one head, two arms, two legs, etc. apiece as we are used to processors with execution counters, decode circuitry, registers of various kinds, etc. We would have difficulty envisioning an extension of the definition of resources to include the fingers and toes of a repairman. When a repairman needs an eleventh finger or third hand he will dynamically acquire it from the pool. There is a minimum repairman beyond which, despite the poor utilization of many functional parts, we shall not fractionalize.

There is no reason, however, to insist on a minimum processor. A processor may be reduced to a location counter and time-share all other facilities. Many proposed system designs for multiprocessing envision a pool of arithmetic and logical units, as well as I/O, pools of registers, even pooled decode circuits, all time-shared by processors that are merely mechanisms for advancing instruction fetch registers.

Chapter 5

PARALLELISM
IN A
SINGLE PROCESSOR

5.1. DOING TWO THINGS AT A TIME

To this point we have postulated the general theme and some variations of two men working together. We shall now investigate some of the concurrent elements of a single repairman performing his job. This topic is suggested to us by the fact that we often use phrases that suggest that an individual is indeed doing "two things at the same time."

When we speak of "doing two things" apropos of the repairman, we might be making a time-sharing reference to the fact that he is interleaving activities on named tasks during the time interval we are consciously (or unconsciously) considering. He might have undertaken the repair of two sets. To understand on what basis he might choose to do this, it is necessary for us to refine further our image of the task of set repair. It is reasonable for us to do this, since we are talking about a finer level of activity on the part of the repairman in terms of smaller functional elements.

Let us postulate that there are two sets available for repair; from the complaints associated with each, it appears that one will require a good deal of prerepair observation because it begins to display its symptoms only after it is on for 15 minutes. The other displays its symptom immediately (it does not go on). The repairman takes both sets and plugs in the delayed malfunctioner (set 1) while he begins to disassemble the other (set 2). Every two minutes approximately he looks at set 1 to see if it has misbehaved.

From Fig. 5.1 we can observe a sequence of events which qualifies as doing two things together. What is being time-shared is the attention of the repairman who interleaves attention between one television set and another. However, he can be looking at set 1 while pulling out a tube from set 2, so he is not totally devoted to any set at any time. What is time-shared is his attention.

While this interleave of attention is going on, there is true simultaneous use of the resources: electrical outlet, tube tester, bench space (space-sharing). We observe that a single processor can grant different tasks simultaneous use of different resources for subtasks. At a later point, he may alternate use of the tube tester, time-sharing this resource between the two sets. Whether we perceive something as occurring in parallel depends basically on the fineness of the interleave relations to our time frame, as we saw from the discussion of the multiplexor.

In addition to this activity we observe other forms of parallel activity: He can remove two tubes at a time, one from each set. He can decide which tube to pull next while he is observing the reading of the test of a tube.

The human is a highly parallel machine. His essential parallelism comes from the fact that he is a collection of asynchronous subprocessors with highly distributed intelligence throughout the subsystems. This means that once a general action is initiated it can proceed under largely local direction, leaving the "central" machanism free to go on to the next action. It is not clear whether simultaneous seeing, hearing, arm movement, decision making, etc. is to some extent time-shared or whether we have truly concurrent asynchronous independent mechanisms to support these parallel activities, but it is clear we can organize ourselves to enhance or reduce parallel capability.

Single worker, time-shared activities

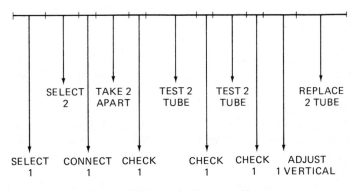

Fig. 5.1. Doing two things at a time.

   Descending another level, we perceive parallelism involving the repair of a single set, where the cluster of subtasks relevant to set repair can be mapped upon the parallel capability of the repairman. He can perform together those subtasks that are not logically dependent or physically dependent. He essentially "parallelizes" a sequential operation by finding opportunities for doing things together.

5.2. COMPUTER ANALOGY

The collection of functional capabilities that we call a central processing unit (CPU) can be identified and elaborated in a similar way to our perception of a single repairman.

   In a general way we talk of multiprogramming as the interleaved attention of the system to tasks during a given time frame. We have seen that this is a time-sharing concept. We have introduced the notions of resource sharing and contention and propose that the selection of jobs for a multiprogramming mix is similar in spirit to the selection of two sets for "simultaneous" repair by the repairman.

   The repairman was interested in minimizing the resource and attention requirement contention of the two sets. He was roughly aware of his own parallel capacities, as one is aware of the parallel capacity of a system, and picked sets that would map neatly onto his abilities. The "attention" of the repairman is analogous to the control circuits of a CPU, where in a multiprogramming environment different programs are "active" (stabilizing instruction sequences) at different times.

   The repairman time-sliced his activity in two-minute intervals between the sets; a multiprogramming system may time-slice in this way. In addition, however, he was sensitive to the occurrence of an event—the beginning of malfunction in set 2—which would have caused him to change his attention pattern. Multiprogramming systems are also "event"-sensitive; in fact, current systems are predominantly I/O event-sensitive.

   The true simultaneity in the repairman's activities can be likened to the ability of a computing system to read and write while it is computing internally. The ability of a repairman to pull a tube for set 2 while looking at set 1 is analogous to a CPU granting a memory access to one program while another is in control. The repairman's ability to remove two tubes at a time is equivalent to the parallelism implicit in the width of channels and memory interfaces, where we "pull" 2, 4, 18, or 36 bytes "at a time."

   Much of our perception of a computing system as an aggregate of subprocessors with distributed intelligence comes from our appreciation of I/O subsystems, where we observe in many systems the ability for a subprocessor to undertake complex I/O operations, including control, branching, and error checking, without the participation of the CPU.

Our perception need not be limited to this area. We can localize many functions in a computer system, and by this localization increase the parallel capacity of the "CPU." The localization involves increasing the intelligence of units in the CPU by providing localized instruction interpretation and local storages.

In large measure in current systems, the I/O concurrency is able to support either the "parallelizing" of a single program or multiprogramming. The addition of concurrent capability within a CPU is generally intended to support the "parallelizing" of a single task.

5.3. SYSTEMS AS A COLLECTION OF FUNCTIONS

We had previously observed parallelism by two processors over two or more tasks either by concurrent performance on different data or by specialized performance of different functions related to the same task. We find now that time-sharing of resources and true subsystem simultaneity can result in parallelism in the performance of a single processor over more than one task or within a task. We further find the parallelism to be of the same two fundamental types—doing different things together or doing different elements of the same thing together. Our appreciation of which is occurring is founded in our concept of the elements of performance—the basic functions that are to be performed.

By viewing a computer system as a collection of functions we can observe opportunities for parallel performance. Our appreciation of the load of each function determines how much power or how many functional units we shall place in the system.

Once we establish the concept of a system as a collection of functions, we see how critical the definition of a function is. This is again to say that the scope of task definition determines our image of a proper size of a system building block. The size of the building block determines the nature and number of interfaces and consequently the flexibility and complexity of the system.

5.4. MAPPING FUNCTIONS ONTO SYSTEMS

See Fig. 5.2(a). We have here a chart that represents the ordering relationship between ten subtasks of a task. This ordering is the logical ordering such that it would be possible, given processors and resources, to process these subtasks in this order. Time is loosely represented horizontally, but we can specify time for each subtask, and we do so in Fig.

5.2(b). This leads to Fig. 5.2(c), which shows the potential time intervals of each subtask as well as its earliest possible starting time based upon its precedence relationship with its predecessors. Figure 5.2(b) also shows us, however, what the resource requirements for each subtask are to be. By consequence of these requirements and the contention arising therefrom, we shall not be able to execute in the 19 time units that would have been required if the system had two B resources. Notice that the schedule in Fig. 5.2(d) reflects the operation of A and B over time with the contention between 2 and 4. Figure 5.2(e) shows the schedule if two B's had been available.

Note in passing here that the charts of Fig. 5.1(a) need not have been interpreted to mean 2-3 AND 4-5, 8 AND 9, but might have had the traditional flow chart meaning OR with 1 and 7 decision boxes. Figures 5.2(f) and (g) show the operation of the system following two of the possible paths. One notices that (f) is as long as (e) and that (d), although two units longer, accomplishes 25 time units of work in 21 units of time. Figure 5.2(i) shows that 25 units of work could have been accomplished in 14 units elapsed time if the resources were held the same but prece-

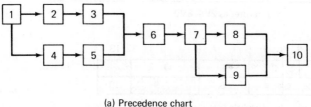

(a) Precedence chart

Activity	Resource	Time
1	A	2
2	B	3
3	A	1
4	B	2
5	A	1
6	B	3
7	A	4
8	A	3
9	B	4
10	B	2
		25

(b) Time/resource map

	Start	Latency to
1	0	2
2	2	5
3	5	6
4	2	4
5	4	5
6	6	9
7	9	13
8	13	16
9	13	17
10	17	19

(c) Earliest starting times

Fig. 5.2.

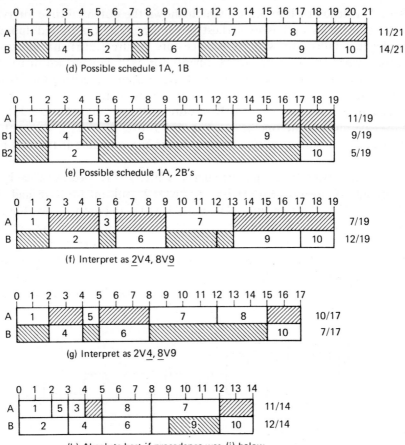

(d) Possible schedule 1A, 1B

(e) Possible schedule 1A, 2B's

(f) Interpret as 2V4, 8V9

(g) Interpret as 2V4, 8V9

(h) Absolute best if precedence was (i) below

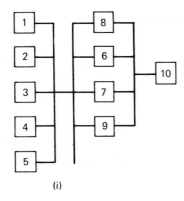

(i)

Fig. 5.2. (cont.)

dence relationships between the tasks were relaxed. As precedence relationships relax, we have considerably more flexibility in scheduling the system. In general, in a computing environment we say that two tasks have no ordering relationship if the output of one interacts in no way with the input of the other.

Aside from showing the impact of precedence and resource on the performance of a system, the truly interesting thing about the collection called Fig. 5.2 is that we have said nothing about the nature of A, B, or the "subtasks." The charts are applicable if A and B are computing systems in an asymmetric organization, in which case we might think of the numbered tasks as "runs" in an application. They are equally applicable if numbered subtasks are subroutines of the same run, in which case we could conceive of A as being a CPU and B as being a special-purpose matrix processor attached to the system. Further, they are applicable if we think of numbered boxes as being individual instructions, of A and B as being elements of a CPU (adders and indexers) involved in execution, and finally of the boxes as being stages of execution of the same instruction and A and B as being instruction decode and store/fetch units.

5.5. LEVELS OF PARALLELISM

What we have said is that the general concepts of resources, sharing, and precedence that we have been discussing are relevant and appropriate at a number of different levels of parallelism:

1. Parallelism between jobs
2. Parallelism between runs in a job
3. Parallelism between subroutines in a run
4. Parallelism between instructions in a subroutine
5. Parallelism between stages in an instruction

We generally associate different mechanisms for achieving concurrency at these levels. Parallelism between jobs and runs we think of as the domain of a given run through subtasking mechanisms. Parallelism between instructions we consider to be the domain of compilers, which optimize instruction sequences to minimize contention (register optimization, for example). Parallelism between stages is considered a hardware feature.

It is, however, true that the scheduling of work at all levels is subject to the same sort of analysis and involves the same techniques. A compiler optimizer is a scheduler as much as is the scheduling element of an operating system.

5.6. EXPLICIT AND IMPLICIT PARALLELISM

There are two further introductory considerations. First is the point at which parallel possibility is perceived. If a piece of code is written with expressions being used to denote parallel activity, we think of that code as being explicitly parallel. If such expressions are not in the code but opportunities for parallel execution are discovered by the compiler (unwinding loops, etc.) or actually by hardware (instruction look-ahead and "pipelining"), we think of "implicit" parallelism. Above the coding level, above the level of separating out elements that can be performed together, at the job level, we find all parallelism explicit in the sense that the very definition of a job defines its independence in precedence from other jobs.

5.7. COST OF PARALLELISM

A second consideration is the cost of achieving parallelism. We introduced the concept of cost initially by describing interference or delays due to resource contention. There are other costs, however. Specifically, the details of each system may involve a certain amount of work in order that concurrent operation may be achieved. For example, in a multiprocessor the initiation of a task on a processor involves all of the work of task-switching. This preparatory work is a cost of parallelism, since it occupies the attention of the processor for a period of time in system overhead introduced solely to initiate a parallel path. There are lower bound constraints, therefore, on the size of a parallel unit that it is profitable to run on a given system.

This is why certain levels of parallelism are associated with specific system types. Manifestly it is not profitable to initiate 50 microseconds of task establishment code to enable a processor in a multiprocessing system to execute a 3 μs sequence in parallel with a 5 μs sequence. Therefore, the size of the unit of work (the subtask) that is profitably done in parallel is rather large in a multiprocessor.

In highly parallel single processors with the asynchronous distributed intelligence subsystems to which we have been alluding, the unit of work is considerably smaller (an instruction stage) because there is essentially no cost of parallelism, no overhead involved in starting up a successor instruction. The cost here is the expense of designing and implementing the capability.

As we describe the architecture of parallel systems, the scheduling, compilation, and language approaches to their use, we shall have constant reference to these ideas of the point of perception of parallel potential and the cost of initiating parallel operations.

2

High-Speed
Single-Stream
Systems

Chapter 6

DEVELOPMENT OF
DESIGN CONCEPTS

6.1. SPEEDING THE SYSTEM

In the early days of computing, there was in the "commercial" world much concern that a processor keep up with its input-output devices, that is, that peripheral units could be kept active and perform at "rated" speeds. Even "large-scale commercial" machines like the UNIVAC I and UNIVAC II were characteristically "processor bound" in that the rate at which data could be processed caused the interval of time between I/O requests to be such that devices quiesced and, therefore, did not maintain their data delivery rate. On the other hand, there was a large enough population of "I/O bound" jobs that a pressure was felt to increase the speed of the I/O devices and refine the techniques of I/O-CPU interfacing.

In "scientific" work a severe pressure for faster CPU's was felt, since many jobs involved complex calculations or many iterations over data, so that the total elapsed time of a job was intolerably long. Since the jobs were inherently processor bound, faster CPU speeds were seen as a solution to a problem where the CPU was an observed bottleneck.

The pressure for more "raw" speed has brought tremendous fruits in the form of the current and expected technologies. It has also brought about a situation where for machines of the now-considered medium- to large-scale (but not in any sense "super") class (e.g., IBM system 360 model 65 or 75) we have almost universal I/O limitation and a great concern for increasing the rate of still lagging, although much improved, I/O subsystems.

6.2. RELIEVING I/O LOAD-HARDWARE BUFFERING

Very early in the history of our art we observed that in addition to raw speed the performance of a system could be improved by unburdening the processor by reducing the amount of attention that it must give to certain functions.

In a 1955 machine the processor encountering an I/O command was completely involved with the I/O operation until its completion. This included start time, transfer time, any malfunction detection, and attempts at retry. To read a 1000-character block from a 15,000 CPS tape truly took the CPU .81 ms. It excluded any other operation, including other I/O.

Early hardware buffering systems in UNIVAC I and II, IBM 650, DATATRON 205, and others reduced this burden somewhat, engaging the attention of the CPU for only the amount of time required to initiate the order and then to unload the buffer. Data streamed into a buffer, which was characteristically the size of a defined physical record (80 columns, 720 characters from tape, etc.), imposing fixed record sizes on devices. When the buffer was filled, the transfer to main store was initiated by a reference to the buffer perhaps associated with a new I/O command.

Some systems provided a test buffer instruction to determine if the buffer was filled; other systems interlocked the processor on an unload buffer issued before the data transfer was complete. A measure of simultaneous capability was achieved by allowing the I/O devices to operate and collect "blocks" of data while the processor could undertake other business. Simultaneous operation of more than one I/O device could be achieved by providing special buffers for each device (IBM 650) or special input and output buffers (UNIVAC I and II).

Since the rate of transfer of data from a card reader or punch tape unit into/from buffers is characteristically slower than the transfer of the buffer to main store, the time spent by the CPU was reduced to buffer-store transfer time plus initial command submission. The distribution of local stores to act as transient repositories for data in order to reduce CPU access and handling time is a technique much used in current CPU design, but ironically not at the CPU/I/O interface to any significant extent.

6.3. PROCESSOR/MEMORY SEPARATION-CYCLE STEALING

The advance to the 1965 system involved a number of observations and the development of some concepts involving the relationship between I/O, CPU, and main storage.

Fundamentally the goal is to allow I/O to operate with a minimum dependency upon the CPU. This allows the CPU potentially to undertake other operations. The potentiality for this is or is not realized, depending upon the availability of work for the CPU to undertake. Once we have provided paths (resources) on which some things can simultaneously occur, we shall remember from Chapter 1 that it is still necessary to have something to do on resource 2 that is logically independent of the completion of resource 1's activity.

Whether there is something to do depends upon either the layout of a particular program or, in a multiprogramming environment, the availability of another independent program. The fundamental observation is similar to that of early hardware buffered systems—the burden on the CPU can be reduced if external equipment is provided to pick up some of the function. In hardware buffering the function that was picked up "outside" was the collection of data from the device.

In the "memory-sharing" interrupt interface the external resource is a channel that is capable of interpreting its own private set of orders as well as completing the interpretation of the system instruction that initiates an I/O function. This serves to reduce further the amount of time a CPU must involve itself in I/O by relieving it of the function of selecting, addressing, and readying devices by assigning that function to the channel.

However, the most dramatic observation inherent in this interface is the basic conceptual separation of a processor from its memory, allowing it and other units to "time-share" access to this resource on some basis. This enables the channels to access memory independently of the CPU on a cycle-stealing basis. This introduction of contention for the resource of memory is a necessary precondition for tightly coupled multiprocessing. "Cycle-stealing" introduces the possibility for variable-sized data structures due to the removal of fixed buffer constraints. It introduces the ability to share memory cycles between a CPU and many I/O elements, and it introduces the possibility of maturation at the CPU memory interface.

With a 1000-character record on the identical tape drive of 1955, the interference with CPU activity caused by data transfer is only .5 ms at maximum. This will occur only if each time the CPU wishes to access memory it finds it active for the channel. The local store capacity exists in such systems traditionally to collect as many characters from a device as can be deposited in memory at one reference. The model 50 can store and fetch four characters (bytes) on one memory reference, and consequently a four-byte buffer is provided with the selector channel.

If indeed things are to happen asynchronously, it is important for there to be a mechanism for reporting events between paths. The interrupt mechanism serves two purposes. It turns the synchronization and interaction control between paths into an active rather than a passive mechanism, and it removes responsibility for coordinating action from the programs running on the system.

Individual systems differ widely on how much the processor must do to handle an interrupt once it occurs. They primarily differ in the amount of information that they give, determining how much work the CPU must do to determine exactly what has happened. Software handling of interrupts is a true burden and a legitimate "cost of parallelism" for the CPU.

The concept of asynchronous event reporting as embodied in the I/O-CPU interface is very much generalized in multiprocessing and multiprogramming systems, where events other than I/O completion trigger true or simulated interrupt conditions.

6.4. CENTRALIZED SYSTEMS

The separation of a CPU from its I/O and from its memory are first steps in the maturation of the computing system. It is a true evolutionary step, transforming the system from a form of "pincushion" to a kind of a "tree." Figure 6.1(a) shows a concept of a machine where the CPU is a "pincushion" into which all functions are deeply imbedded in the form of pins. Nothing happens at a pin head unless it is directly and totally controlled from the cushion. Only "one thing at a time" can occur in a given time frame.

Even with this system, however, if one casts the time frame properly, one can speak of parallelism. The mere presence of two tape units implies the phrase "Tapes 1 and 2 are created in parallel," even though the writing of one completely precludes the writing of the other.

6.5. HIERARCHIC SYSTEM

The organization of Fig. 6.1(b) is basically what current computer people are familiar with. It involves a hierarchy of specialized units, each with some local store and limited intelligence. Units down the tree tend to be highly specialized, capable of performing reasonably sophisticated functions in their area but useless outside of it. An IBM I/O channel of contemporary design can autonomously control I/O operations more complex than CPU's of the first generation, but it recognizes and performs no other functions. The intelligence of a unit is limited by the scope of the total functions of the system that it can perform. In an hierarchical system there is a pyramid-like distribution of general capability.

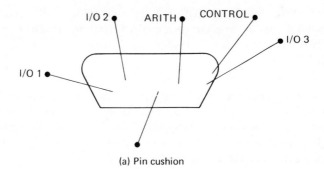

(a) Pin cushion

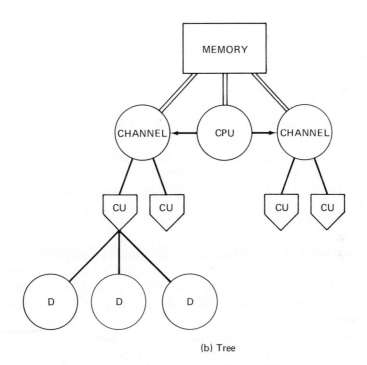

(b) Tree

Fig. 6.1. Concepts of two machines.

Each parent unit performs an interpretation of function to determine to which member of the tree it relates and passes it off to that member for continued interpretation or execution. For example, the CPU interprets an order to undertake I/O on a given channel and determines to what channel to send the information necessary to perform the operation. It is then released on this partial interpretation. The channel receives relevant control data, independently accesses any additionally required data, and sends whatever is needed to a control unit to which it initiates contact. The control unit addresses the device. There is potential time-sharing at all levels.

Different systems differ, of course, in the amount of intelligence and degree of asynchrony of each level. This difference is manifested by such details as where addresses and local buffers are kept and the degree of interpretive power at each level. The principles, however, are the same for systems like the UNIVAC 494, UNIVAC 1108, IBM 7094, and larger members of the IBM 360 line.

6.6. SUMMARY

We discuss in a later chapter some of the finer variations of capability in the I/O/CPU separation and their implications. What we have wished to observe here in terms of the most commonly understood area where they have been applied are the fundamental techniques of unburdening a facility by distributing local intelligence to functionally specialized units in a loosely arranged asymmetrically parallel hierarchy. We have particularly examined the reduction of the central processing unit to a function distributor rather than performer in the case of the I/O. The implications for reliability, configurability, and performance in the machine of Fig. 6.1(b) have been exploited further and further in the last decade.

Maturation is a process of growth and development, a coming into being of capabilities that lie latent at an early stage in the development of a child or a computer system. The development of high-performance machines is a result of the continuing elaboration of interfaces. Initially an insight is gained into the separability and independent development of elements previously thought of as inseparable. A processor and its memory is one example, a processor and its I/O another.

Once the separation is made, the new elements mature; they acquire extensions and functions to enhance their performance and their intelligence. The maturation at the CPU memory interface has given rise to interleaving techniques, to hierarchies of directly addressable storage, and in the future to functionally active memories, perhaps.

Chapter 7

THE FUNCTIONS
OF A
PROCESSOR

7.1. LOOKING INWARD

The early separation of a processor from direct I/O control was facilitated by the recognition of I/O as a significant and definable function of a
system. This recognition was enabled in part by the dramatically different
time spans which distinguished I/O from "internal" functions and in part
by its obvious reliance upon devices actually physically external to the
CPU. These devices had characteristically quotable inherent speeds of
their own, and the need for techniques to "balance" a system could be
easily seen.

This naturally led to the consideration that there might be in the mass
of functions left to the CPU, differing in complexity and character from
shift register to floating-point divide, other functions that might be critical
enough to justify the provision of some asynchronous limited intelligence
resource for their performance.

The development of highly parallel single-processor systems was motivated by the observation that the unburdening of the CPU with regard to
I/O was not relevant to the performance of computing systems on many
large and significant jobs involving massive manipulation of relatively little
data. What was needed here was a "super-machine" initially LARC and
STRETCH and currently the CDC 7600, Burroughs 8500, and IBM 360
model 195 and in the near future machines with computational capacities
in multiples of even the potentials of these massive systems.

Within any given technology certain things are logically and economically feasible to attempt; the technology limits the raw speed of a processor. Further, each technology feels that it has achieved a kind of limit in
raw speed beyond which it must yet be shown scientifically feasible to go.
We are at such a point now.

The speed at which electrical current will pass down a line meaningfully limits the size of systems when one is talking about nanosecond or less processors. The problems associated with proper miniaturization, in terms of heat generation, maintenance, and economics of production are not sufficiently solved to allow for a great leap forward with raw speed. The natural alternative is the introduction of more and more parallel capability into a processor.

This solution is primarily oriented toward the faster processing of a single large problem rather than the concurrent processing of many jobs. Parallelism is obtained at the instruction level or below, and the user of the system remains (or may remain) largely unaware of the asynchrony and overlap achieved by the components of what he views as "the processor."

7.2. FUNDAMENTAL ELEMENTS OF A PROCESSOR

Processors have traditionally been perceived to have two fundamental "cycles" of operation: an "I" cycle when an instruction is acquired and stabilized in the processor and an "E" cycle when the particular function is actually performed.

In Fig. 7.1 we represent the performance of an ADD instruction in a machine with one arithmetic register (RA) and no indexing cycle. The first five activities, from the development of the address of the next instruction to the decoding of the operation, form the I cycle. The last four activities form the E cycle, and the entire operation I and E can be viewed as a single line of two segments. The reference to storage is represented by the arrow to S.

We can perceive a system executing in this way as consisting of three functional elements: an I unit, an E unit, and a store S. This separation of the hardware into semiautonomous boxes is a conceptual representation of the acknowledgedly separate functions or cycles of instruction handling. On lower-performance and "traditional" machines the "I-time and E-time" concepts are not supported by hardware in this way. Much circuitry is shared, and the resolution from I to E is seen as a progression of status over time for the total processor.

The recognition that I and E could be organized by function into separate subprocessors the reader will, at this point, understand to be a precondition for the recognition of potential parallelism.

The I unit contains an instruction counter, an instruction register, an operation register, and a decode matrix, which is associated with the operation code register. This decode matrix is very simply presented in Fig. 7.2 as a series of signals excited by the specific bit pattern in the operation code register which provide signals on lines that direct appropri-

ADD R, MEM

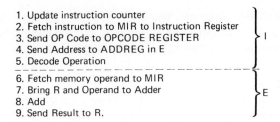

1. Update instruction counter
2. Fetch instruction to MIR to Instruction Register
3. Send OP Code to OPCODE REGISTER
4. Send Address to ADDREG in E
5. Decode Operation

6. Fetch memory operand to MIR
7. Bring R and Operand to Adder
8. Add
9. Send Result to R.

Fig. 7.1. Basic processor functions.

Decode/encode

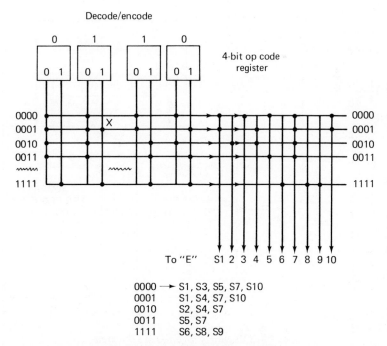

$0000 \longrightarrow$ S1, S3, S5, S7, S10
0001 S1, S4, S7, S10
0010 S2, S4, S7
0011 S5, S7
1111 S6, S8, S9

Fig. 7.2. Simple instruction matrix.

ate units to perform. For example, the code 110111 generates signals to prepare for a storage access, and to prepare for use of rA and the adder.

The E unit contains an arithmetic result register, an adder, a multiplier, and an address register for operand fetch—all elements involved in the actual performance of instructions.

The machine that we are beginning to define here is a single address machine with memory references possible from the E unit; therefore, I and E share some register interface to memory. This we will call a Memory Interface Register (MIR).

Figure 7.3 shows S, I, and E in a system. It is, of course, not unusual in smaller machines for the registers in I and E to be in storage rather than being implemented as "hard registers" in the units of the processor. This is a speed/cost decision. In high-performance machines these registers are invariably implemented in the I and E units.

In the sense that both I and E have local registers, they can be said to be functionally specialized units with local stores, the local store in this case being merely the small collections of registers not in S.

The critical characteristic of these registers is that they operate much more quickly than store. From the CDC 6600, IBM 360, RCA Spectra, and UNIVAC 1108, we are familiar with the fact that register to register operations require less time than memory reference operations. This is an

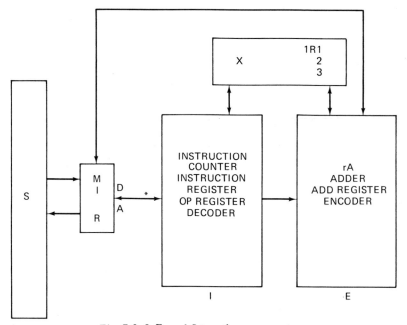

Fig. 7.3. I, E, and S together are a system.

obvious consideration in programming multiple register machines and is an obvious concern of compilers operating for these machines.

If we were to elaborate this machine, we would place index registers in it and provide additional registers for arithmetic in E. Notice that it would not be obvious where to put the index registers. I requires them for address calculation for instructions, E for operands. One solution would be to provide a separate indexing unit, X, available to both I and E. Indeed, in some processor organizations the basic split between I and E is not made, but I and E are one unit with an asynchronous X.

The split of work, the definition of units, and the size and nature of local store are truly as diverse as human imagination.

7.3. POSTULATED PERFORMANCE OF S, I, E

I, E, and S have characteristic rates of performance, which we shall postulate. Let us say that S requires .375 μs for the delivery of 36 bits to MIR from S and .375 μs residence in MIR for the reading of 36 bits into S. We further stipulate that on a read of S by I or E it is necessary for S to restore the location of data read, so that although data are available in MIR in .375 μs an additional .375 μs interlock is imposed. A read followed by another read requires 1.125 μs before both data references are available. This is a common feature of core memories and of current systems, although randomly addressable NDRO (nondestructive readout) memories operating in nanosecond ranges are certainly upon us.

Let us break up the functions of I and E into their component stages (see Fig. 7.4). We see that the cycle of I is fixed and independent of the

FLOAT ADD	I, MICROSECONDS (μS)	E	
Update IC	.050		
Fetch instruction	.375		
Send to op code	.050		
Decode and send address	.225	Receive signals	.225
		Fetch operand	.375
	.700	Bring to adder	.050
		Add	.200
		Return RA	.050
			.900
I only .475			
I-E .225			
E only .675			

Total add 1.375 microseconds

Fig. 7.4. Basic stages of an instruction.

instruction that it is executing. This is a simplification, since many machines do not have fixed decode cycles but are decode-sensitive to instruction. For example, some machines have "multistage" instructions, such as multiply, which are implemented by a retention of the basic instruction in the instruction register and the incrementing of that register for each stage (UNIVAC Solid-State). Other machines have a separate stage counter that is augmented and concatenated with the instruction register. On these machines for those instructions the decode time is not static.

However, a basic feature of a high-performance machine would be the avoidance of retention of op codes in I for the purpose of instruction staging. The movement of an instruction out of the I cycle, out of a physical position in an I box as quickly as possible, is a goal of these machines. We shall assume for this text that multistage instructions are implemented so that they impact only the execution unit.

In addition to the fixed nature of the I cycle, we see that it is dominated by the storage reference required to fetch the instruction. This is fundamentally true of processors of medium to large scale. They are capable of generating memory references much more quickly than memory can respond.

The E cycle is, of course, completely dependent upon the instruction to be performed. Whether the instruction requires a memory fetch and the complexity of the operation itself have definitive impact on how long the E unit is concerned with execution of an instruction. This variability of instruction times, either seen as in the "processor" surrounding the I and E units or in the E unit itself, is one of the reasons for the extreme difficulty involved in describing or measuring the capability of a processor.

7.4. MIPS AND I RETIREMENT RATE

We have lately begun to characterize processors in terms of a MIPS rate (millions of instructions per second). Since the time it takes to execute an instruction is variable, the MIPS rate cannot be a precise statement of capability. There are a number of significant rates in which we may be interested. One is the maximum rate at which instructions may be performed if each instruction is exactly the time of the minimum cycle of the machine. This is a directly interesting rate.

We would be interested in the MIPS rate of the I unit. Our interest in this derives from our observation that I defines the rate at which instructions can be gathered and presented to E for execution. The rate at which instructions pass through I is the maximum instruction processing rate of the machine. An optimum machine would relate I, E, and S to sustain that rate.

7.5. EVALUATING E: PROBLEMS IN PERFORMANCE DETERMINATION

For evaluations of E (or, when E and I are undifferentiated, of a "processor") we often use some basic coefficient of usage for instructions or instruction groups. For example, if we define basic machine functions to be ADD, SUBT, MULTIPLY, DIVIDE, and COMPARE it is possible to determine what amount of time one expects the machine to be involved in executing each function over a given time.

Such coefficients can be empirically developed by running "representative" code on the processor and counting instruction or function occurrence. The expected "instruction" time is equal to A_i * Add Time + A_j * Subtract + A_k * Divide + ..., where A is the summation of the occurrences of the function. This number gives a quite rough indication of the basic speed of a processor, and the components may suggest inadequacies about the content and form of the instruction set, the register population, and the general architecture of the machine. Its use for us in the text is as a statement of the "mix" of instructions that flows through the machine in its characteristic processing. For example, in Fig. 7.5 we have listed some specific instructions that are performed on yet another hypothetical machine. This machine is hypothetical in the sense that its associated timings are based upon conveniently rounded times for these instructions taken from the processor manual of one machine, plus a charge for

Instruction	Time Each, μs	Number of occurrences	Total (sec)	a
FLOAT ADD	1.6	2,250,000	36	.06
FLOAT MPY	2.4	1,000,000	24	.04
OTHER	1.6	109,200,000	180	.30
FLOAT DIV	8.5	705,882	6	.01
FIX ADD	1	12,000,000	12	.02
FIX MPY	2	3,000,000	6	.01
INDEX MODIFY	.8	112,500,000	90	.15
FIX DIV	10	600,000	6	.01
CMP	1.5/1.0	4,000,000/6,000,000	6/6	.01,.01
LOAD	1	78,000,000	78	.13
STORE	1	96,000,000	96	.16
SHIFT	1	54,000,000	54	.09

MIPS = 798,000
Average instruction time   1.25   /instruction
Instruction per μs   .706 instruction/μs

Fig. 7.5. Mix for hypothetical machine—10 minutes (600 seconds).

indexing not felt on that machine, but that the percentages associated are derived from studies associated with a direct competitor.

The true "hypothetical" nature of such a combination can be directly experienced by any attempt to evaluate the performance of machines of divergent architecture using some mix approach of this type. Variations between architecture may render the attempt useless. One standard mix, the Gibson mix, was developed from an observation of the amount of time the IBM 709 spent performing certain standard functions. Extensive samples of 709 code were run to determine the A's for the population of 709 coding across compares, divides, multiplies, etc., where each function included associate loads and stores.

In attempting to judge whether an IBM 360/65 or a UNIVAC 1107 outperformed a 7094 using Gibson mix figures, one discovered that any result could be obtained depending upon whether one assumed direct register operation in these multiregister machines or assumed 709-type coding patterns. Further, the assumption about concurrency of function and the conditions under which it might occur seriously changed the results. It is very difficult to compare the power of machines of different architecture with a standard oriented toward a specific design.

7.6.　BALANCING A MACHINE

What Fig. 7.5 suggests to us is that it is possible to determine internal functions in a machine and to measure the significance of these functions. The very term "instruction mix" calls to mind terms like "job mix" for multiprogramming, and implies that in a manner similar to the balancing of jobs by characteristic resource usage (I/O bound, low core vs. compute bound, high core, etc.) it is possible to balance a machine design to optimize for expected instruction mixes.

It is not clear that one can optimize a new architecture based on mixes observed for another architecture. However, it is certainly possible to speed up a machine without a general increase in raw speed by determining what impact local improvements would make on the MIPS rate.

A dramatic example of speed-up within an architecture is the CDC 6600 and the CDC 6400. One machine (the 6600) has been carefully designed to optimize certain functions and to achieve parallel operations in arithmetic; the other, with exactly the same appearance to a programmer, has not. These functional optimizations to achieve performance within an architecture imply a further division of the E unit into more semi-autonomous components; this is characteristic of many high-performance machines, including the ones we shall discuss.

7.7. I, E, S RELATED; INCREASING UTILIZATION OF ELEMENTS

Converting the figures we have developed into rates for E, I, and S, we establish that the maximum rate at which S can deliver 36-bit words is roughly two per microsecond, if an equal number of reads and writes is assumed. This is 2,000,000 per second. We also see that the maximum rate at which instructions can pass through I is 1,400,000 per second, and through E (if we assume that E average retirement time is .78 μs), roughly 1,280,000 per second.

If we were charged with making this a faster machine, how might we go about it without increasing its raw speed? To undertake a speed-up of this machine we might first study the relationship between I, E, and S in more detail. We see that of the 1.25 μs average instruction time, .475 μs are spent in I only, .225 μs in I-E, and .550 μs in E. This is shown on the graph of Fig. 7.6(a).

With I and E to this point viewed as sequential cycles, we see that I is idle during the time E executes its instructions. This idleness of I for .550 μs reduces its delivery rate of instructions to 798,000 from 1.4 million, which results (from Fig. 7.6) in the MIPS rate of this machine. In terms of utilization we see I used .70/1.25 or 56 percent of the time.

Since E receives only 798,000 instructions against its potential throughput of 1.28, it is going to be idle 38 percent of the time. This idle figure is computed by dividing either 1.28 into .798 or 1.25 (total instruction time) into the .78 μs actually used by E for an instruction. With regard to memory, I-E will, of course, never contend for S and will generate 1,600,000 maximum references, less than the memory can provide. We also notice that since we are sequential in I-E, there can never be contention delays on S.

What we would like to do is to increase the utilization of E so that we approach its maximum throughput potential. At this point we take the step we have been preparing for and anticipating throughout the chapter. We perceive I and E as asynchronous overlapping elements so that the I cycle of one instruction can be overlapped with the E cycle of another. We hope to achieve the idealization of Fig. 7.6(b).

If we could achieve this, then we would be using E 100 percent of the time, achieving 1.28 million instructions per second. I would be utilized much more heavily to deliver these, being idle only for the .075 μs between the time that it has readied itself to decode the next instruction and the time that E is free to receive it. Of each 1.25 instruction time I would be busy for .475 $\times$ 2 or .950 for instruction address generate, fetch, and stabilize, and .225 for the decode, for a total of 1.175. Unhappily, we shall not achieve the ideal of Fig. 7.6(b) for two reasons.

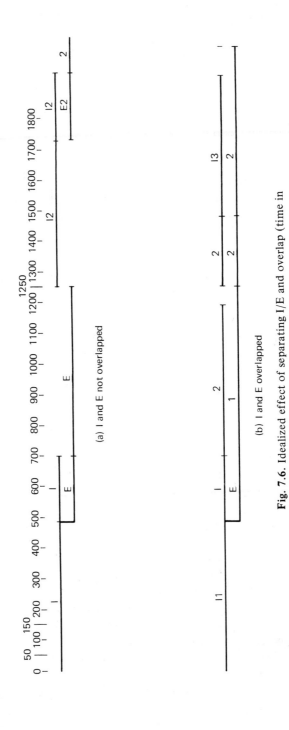

Fig. 7.6. Idealized effect of separating I/E and overlap (time in nanoseconds).

The first reason relates to the statistical distribution of instruction times that actually exists on the machine. Naturally we expect that if E encounters a nest of longer instructions its throughput will go down, just as in multiprogramming we expect a nest of longer jobs to reduce throughput for the system. The effect of a longer instruction in E is also to reduce the utilization of I, since I will be interlocked on E's availability for longer periods of time.

This is an interesting example of what we have seen before with our repairmen. I and E are independent, but they share resources—the decode, the op register, and the address register—and, therefore, they may delay each other. A further interpretation is that they are only partially asynchronous. Among the subtasks they perform (Fig. 7.5), some must be performed jointly, and there is a dependence of one on the other for this joint effort.

If we compress Fig. 7.5 into Fig. 7.7 we can get a measure of the potential variability of performance for this machine.

We see that for 56 percent of the instructions executed in the mix E time is less than I-alone time by .175 ms or .375 ms, so that we shall not be idle on I but on E. For all other instructions I will be idle from .325 up until 8.825 μs. If we encounter long strings of 1 μs instructions, loads, shifts, stores, fixed adds, or compares with no transfer, we shall for that interval perform as a 1.43 MIPS machine with 100 percent utilization of I and idle time on E. If we encounter long strings of floating-point instructions, performance will be considerably reduced from this level for those intervals.

Our appreciation of the rate of the machine is entirely dependent on our concept of the mix. This would be a slow machine for a heavy floating-point usage program in an installation. It is interesting to conjecture whether one expects admixture of "long" and "short" instructions in the same program or characteristic dominance of one type in different programs.

Instruction time, μs	%	I time	I-E	E	I idle
1	41	.475	.225	.300	−.175
1.5	1	.475	.225	.800	.323
1.6	36	.475	.225	.900	.425
2	1	.475	.225	1.300	.825
2.4	4	.475	.225	1.700	1.225
8.5	1	.475	.225	7.800	7.325
10	1	.475	.225	9.300	8.825
.8	15			.100	−.375
	100%				

Fig. 7.7. Percentage of time in instructions of various lengths; I/E interference.

A more critical reason, however, for our not achieving the performance one hopes for in Fig. 7.6(b) is that when we operate in this fashion it is possible for the combined operations of I and E to generate a much higher demand on S than S can possibly fulfill. Worse, I and E will tend to contend heavily, with I going after the instruction at roughly the same time that E goes after an operand. The effect of this can be seen in Fig. 7.8. In this simplified representation using average instruction time, we can see how E constantly interferes with I's attempt to access the next instruction and how this lock-out of I from S causes a delay in next instruction fetch so that E must idle.

	I	E	DELAYS I	DELAYS E
0-225	Start execution of instruction one			
225-275	Update IC	Fetch operand		
275-600	Idle on memory	Fetch operand	325	
600-775	Fetch instruction	Execute		
775-975	Fetch instruction	Idle on nest inst.		200
975-1025	To opreg	Idle on next inst.		50
1025-1250	Start execution of instruction two			

Fig. 7.8. Effect of memory contention. Lockout of I delays availability of next instruction for E.

The impact on the machine is to extend the I cycle and the E cycle to 1.025 μs, and thus to reduce the machine to below 1 MIPS (970,000) and to reduce utilization of I to 68 percent and of E to 74 percent. The overlap of I and E has achieved an increase in utilization and instruction throughput, but not at the level we would hope for. It would seem reasonable for us to investigate further ways of speeding the machine. Most profitably investigated might be a way of reducing memory interference delays.

7.8. MEMORY DELAY REDUCTION

We have to this point assumed one arithmetic register in E and following from this we have been led to assume that the instruction set of the processor is such that every instruction contains a reference to memory. However, we already have one function (SHIFT) that does not involve such a reference. Nine percent of the instructions executed by E will not cause a contention with I, and we have overstated our contention mildly.

A further immediate candidate for eliminating memory reference might be the index modify function. This can be done by either an immediate instruction that contains the value to be added to IX, eliminating the reference to store for the operand, or by loading an increment value into an extended index register. This would add another 15 percent of executed instructions that do not reference memory and consequently do not cause a contention between I and E.

We might further choose to execute all indexing in I, passing to E addresses that are already modified. The implication of this is that there are classes of instructions that need never be passed on to E but are locally executed in I. The branching instructions might well be a member of this class. Notice that by removing some executions from E and placing them in I we tend to balance I and E rates. If E were a real bottleneck, we might find other ways of placing additional functions in I.

A further solution to the memory problem is to provide additional registers in E in an attempt to reduce the number of loads and stores that E must execute. Traditionally these registers have been addressable, and their proper use has been left to the programmer or to the compiler.

Associated with additional registers is usually a set of registers to register instructions that permit, in addition to the collection of intermediate results in multiple registers, direct register arithmetic or logic. The result with proper coding or code generation is indeed to reduce memory references and, consequently, contention delays.

This approach is common, not only on high-performance machines, but in small and medium machines as well—for example, the IBM 360/Model 50, UNIVAC III, UNIVAC 1050, and UNIVAC 490. An interesting approach in the 360 line is to give the appearance of a family of registers in all machine models, but actually to separate registers from memory only in the middle of the line and upwards.

The Model 30, for example, contains the addressability to 16 general-purpose registers that are actually core locations. The intent here is to preserve architectural compatibility for the product line. A programmer who very carefully plans register usage for the 30 will receive small reward for his efforts, but massive payback for the 50 and above. The increase in registers is an increase in the private resources of E which, as we would expect, increases its capacity for independent operation.

A further solution for decreasing the contention on memory is to provide for multiple addressable units to be presented from store on a single access. There are a number of ways of doing this. One way, on our machine, is always to deliver a 72-bit two-word structure to the MIR and provide a place for it in I and E. I would always have a successor instruction available, and E would use both the words some percentage of the time.

An alternative is to provide the equivalent for I only by reducing the size of the instruction word so that two instructions of 18 bits are deliverable to I on each memory request. The reduced instruction might rely on external support from a register to complete addresses.

Not uncommon are variable-size instructions, as in the IBM 360 and the CDC 6600. The 6600 packs from four 15-bit to two 30-bit instructions in a single word. Figure 7.9 shows the impact of providing two instructions per fetch (a form of parallelism like picking up two tubes) from I.

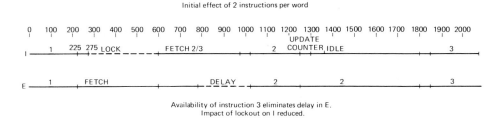

Fig. 7.9. Short instructions help a little.

We have reduced idle time in E by avoiding the instruction fetch delay for instruction 3, and we have reduced the impact of lock-out in I, so relative to our initial 36-bit instruction machine we are performing faster. We are not, however, performing as fast as we would perform on a machine with 18-bit instructions and no memory interference.

The two schemes we have presented, proliferation of local registers and multiple instructions per word or multiple addressable units per reference, are commonly found in medium to large machines; however, the fundamental and significant technique for reducing interlock is memory interleaving.

7.9. EFFECTING MEMORY SPEED-UP BY MEMORY INTERLEAVES

To this point we have thought of S as one homogeneous collection of addressable storage that can produce a single or multiple unit of information at a single interface or accept units at that interface. When S is busy serving a request, we have seen that all other requests are "locked" out.

Large systems have attempted to reduce memory delays by dividing total S into "banks" of storage, where each bank contains a subset of the memory addresses. This is simply shown in Fig. 7.10. The subdivision of memory allows the two banks to operate in parallel, delivering words to

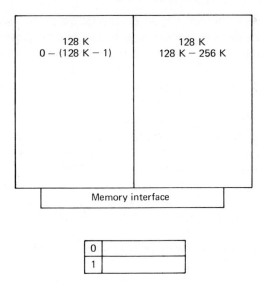

Fig. 7.10. High bit overlapping.

requestors asynchronously without lock-out and interference. Since each bank can still service only one request at a time, lock-out still occurs when they contend for access to the same bank.

The effectiveness of banking memory depends, therefore, in large measure on the distribution of addresses being generated by asynchronous units in the system. Notice that in our situation no increase in speed would be achieved if all instructions fetched by I and all operands fetched by E had addresses in the same unit. Considerable speed would be achieved if I and E never referenced the same unit.

Consider in Fig. 7.11 the effect of placing all instructions in 0-128K, the I bank, and all operands in 128K-256K, the E bank. We see that we achieve the ideal of Fig. 7.6(b), full utilization of E and an effective 1.28

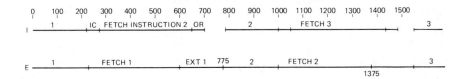

Independent bank references eliminate contention

Fig. 7.11. Achieving Fig. 7.6 with high bit interleave.

MIPS machine. Whether or not it is realistic to posit such a separation depends on many factors: the average size of the programs, the levels of multiprogramming, hardware and software memory management techniques, and the flexibility of addressability derived therefrom. We must also remember that throughout this discussion we have discounted memory access requests from channels.

The banking organization described so far is "high-bit" banking in the sense that the selection of a memory bank depends upon the high-order bits of an address. A 16-bit address structure providing for addressability to 65,536 words would address the low bank as 0 to 32,767 (0111111111111111) and the high bank as 1000000000000000 (32,768) to 1111111111111111 (65,535) where addresses in the high bank are interpreted as bank 1 relative 0-32,767. This high-bit scheme is implemented in the UNIVAC 1107.

An alternative to this approach is "low-bit" banking, or the technique from which the term "interleaving" derives. This scheme organizes memory into independent banks where (in simplest form) all even addresses are in one bank and all odd addresses are in the other. The effectiveness of this is again determined by the specific pattern of addresses generated during execution. The banks are referenced by the presence or absence of a 1 in the low-bit position of the address word. The schemes merge toward each other as more and more bits are taken from left or right to serve as bank indicators. Highly modular memories of 4096 word banks are provided on the 6600.

In some machines (UNIVAC 1108 and IBM 360/75 among them) the two approaches are provided. Memory is sold (on the 1108) in 65K banks, where each 65K bank is high-order bit overlapped, but within each 65K bank there is odd/even interleave. On a 128K system, therefore, a four-way interleave is possible—one odd from bank 1, one even from bank 1, one odd from bank 2, one even from bank 2.

The intent of all variations of the underlying principle is the same, to provide the greatest achievable relief from memory contention delays consistent with the general performance/cost goals of the system. Very high-performance systems do not find interleaving or banking memory sufficient to achieve performance requirements.

Chapter 8

MOVING A
SINGLE PROCESSOR SYSTEM
TO ITS LIMIT

8.1. LOOKING MORE DEEPLY AT I/S

We have achieved parallelism between E and I to a high degree, and
we see that with this parallelism we can increase utilization and instruction throughput. Since I and E are roughly balanced, we might not
be further concerned with the bottleneck possibilities between them
for the moment, but look even more closely at another relationship, that
between I and S.

Of the 700-nanosecond I cycle, 375 nanoseconds are involved in the
interface between I and S. If we look closely at the process of fetching, we
can see it as being the delivery of an address by I to the interface and the
subsequent delivery of data by S. During the interval between this S is
fully utilized, but by descending a level in our appreciation of the fetch
task, we can see that I can be said to be idle, waiting for the delivery of
data. Figure 8.1(a) shows this. I is truly busy only 100 out of 375 nanoseconds.

This 275-nanosecond delay is a nontrivial fraction of total I time
involved in instruction processing, and this leads to the question of
whether additional things could be done to speed up this machine without
increasing its raw power.

We have previously stated that the rate of processing instructions on
the machine was limited by I. Regardless of the speed that E can achieve
(encountering only short instructions) I can deliver only 1.43 MIPS. This
rate of I now seems to be limited, not by its own character, but by its
dependence upon S. If S responded instantaneously, then the process time

(a)　　I-S Interface

DELIVER ADDRESS TO MIR	50	Nanoseconds
WAIT	275	
TAKE DATA TO INSTRUCTION REGISTER	50	

(b)　　Reduced I cycle

DELIVER MIR ADDRESS	50
WAIT	275
TAKE DATA	50
UPDATE IC / SEND IR ⟶ OCR	50
DECODE AND SEND ADDRESS	225
	650

Fig. 8.1. A finer look at instruction fetch.

in I for each instruction would be reduced to 425 nanoseconds and its basic delivery rate would be increased to 2.35 MIPS. Further, I could be made inherently capable of generating requests to S at a faster rate than S can respond. This is typical of large-scale machines that are truly memory-bound. We suspect, therefore, that something can be done at the S/I interface.

8.2.　A CLOSER LOOK AT I; A NEW GOAL FOR THE MACHINE

We might look further into I itself and observe its basic operations. We have represented five basic functions. It would seem that with current structure we might be able synchronously to update the instruction counter either while sending a fetched instruction to operation code register or during decode time. This would reduce our I time to 650 nanoseconds and give us an I cycle like Fig. 8.1(b).

We could have chosen to update the counter in the S wait time, but let us assume that it is just as easy and economical to run the increment and distribution to OP register at the same time, since no common circuitry is involved.

Of the five functions of Fig. 8.1(b), we observe that the dominating function is the actual decoding of the instruction and that the others are organized to make this event possible. We should like to investigate the possibility of organizing I so that the 225-nanosecond decode time is the true limit of the machine. That is, we wish to organize the functions of instruction fetch, address generation, Op code register transfer, etc. so that the effective elapsed time of an instruction delivery is the time to decode the instruction.

The goal of the design of the machine is to deliver a new instruction for decoding so as to keep the decoder utilized 100 percent of the time.

With a decode time of 225 nanoseconds we could achieve a MIPS rate of 4.44 on this machine if E could keep up. Beyond that we might reorganize the decode function itself to reduce its time without increasing circuitry speeds.

We see that apart from wait-for-delivery time the decode cycle is indeed the largest single function that an I unit performs. It is possible to perform the total 150 nanoseconds of address delivery, data delivery, and op code delivery in less than 225 nanoseconds of decode time.

One transformation we could make on I is completely to rearrange its relationship with E and S, distributing functions in an entirely different manner. Perhaps we could devise new structures such as an asynchronous store and fetch unit or a decoder unit. Let us roughly keep the integrity of I in this discussion, choosing rather to elaborate it and then reduce it. What would an I require to deliver instructions to E at a rate near 4.44 MIPS?

8.3. SPEEDING THE S/I INTERFACE

We notice that I and S are elements with fundamentally different speed characteristics. The classical way of interfacing synchronous devices of differing characteristic speeds is to interpose a buffer between them. We saw this at the I/O interface and will see it again when we discuss software buffering. This buffer serves as a repository for data at the interface in such a way as to "smooth" the operation of the asynchronous elements.

What would be the effect of interposing an eight-instruction buffer between I and S so that whenever I referenced S, eight instructions would be pulled out of S and deposited in I? Initially we see that the total time to fetch eight instructions is reduced because of the saving of address deliveries on seven instruction acquisitions. Let us further postulate that the buffer serves as a stack of IR's and that op code, operand distribution proceeds from the buffer.

Allowing the variation that will naturally occur in E performance, we can see how the availability of instructions in I will enable it to keep up with E when E is encountering short instructions and to avoid contention with E when E is using longer instructions.

We should introduce one further point connected with looping and branching. The effectiveness of the buffer depends upon the usability of the instructions that are in it. The best possible case is where no branching occurs, and all instructions are executed in sequence. When the first instruction is an unconditional branch beyond the range of the eight present instructions, then the transfer times of all the instructions that have followed is absolute lost time. A further problem occurs when a conditional branch is encountered and it is necessary to inhibit the advance of instruction reference until it is resolved which direction will be taken.

A problem that develops in connection with branching is the advancing of instructions in the buffer in the period of time between the encountering of a conditional transfer and the point at which the decision is made. This develops because of the possibility that, after passing off an arithmetic instruction to an E, the I unit selects a branch on condition instruction as next in the buffer and cannot decide what to do until the E completes the arithmetic. One can inhibit instruction advance, or one can discover a technique for allowing "tentative" advance. The instructions past the transfer point are advanced with an interlock condition that is removed if the transfer is not made. When the transfer is made, a signal to delete these instructions is generated.

8.4. CONCEPT OF RECENT HISTORY

The presence of eight instructions in a local store provides a further possibility. This is the use of the store as a "look-aside" device. The fundamental concept of look-aside is that the recent experience of the system is meaningful in predicting its short-term future requirements. High-speed systems have multiple occurrences of the concept at different design levels. Its application here is that the collection of eight instructions not only provides a repository of instructions to be executed, but, as the system advances through the buffer, it represents a collection of instructions just recently executed. If it should occur that an instruction at the end of the buffer transferred control to an instruction at the top of the buffer, then use should be made of the fact that the next instruction is already available in the buffer. The perfect situation is the eight-instruction loop, where for the duration of the loop the machine is iterating on instructions in the buffer, bypassing memory access for new instructions until the loop is "fallen through."

8.5. VARIATIONS IN BUFFER FILL

A variation in buffer usage is automatically to bring in new instructions behind the current instruction. The buffer might be thought of really as two buffer spaces partitioned at some point. As the processor brings up instructions from the buffer, it passes that point and replacement of the executed instructions is automatically begun. This results in a constant stream of instructions filling the buffer, tending to distribute access times for instruction fetches. When this automatic replacement is used, a special "short loop" mode is used to inhibit the replacement of instructions when the machine is in a short loop.

The provision of an instruction buffer partitioned into two parts increases the probability that I will generally have instructions available to it and will not be dependent upon a cycle from S. By allowing S to stream instructions into the buffer at its rate and I to pick them up at its rate, we have provided asynchrony between I and S, adding an element of parallelism.

Figure 8.2 shows the buffer between I and S. I and S are roughly in balance here. S can deliver four instructions in 1150 nanoseconds; I can process four instructions in 1100 nanoseconds. I is "executing" instructions in buffer positions 1-4 while S fills 5-8, and then they switch. What is necessary is some circuitry to interpret the fact that the instruction counter is pointing to an address whose contents are truly in the I-S buffer. Further, we need a mechanism to convert addresses to relative buffer addresses. Such circuitry and capabilities are common to high-performance machines of the IBM 360/91 and CDC 6600-7600 class.

An even more flexible arrangement for buffer control is available in IBM's 360/91 and 360/195. In addition to the location counter pointing

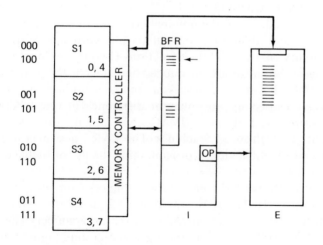

I cycle
    1. SEND TO OP, UPDATE IC 50
    2. DECODE   225   PROCESS 4 INST = 4 x 275 = 1100

S cycle
    1. START 50
    2. DELIVER 4 INST = 1100
       TOTAL = 1150

Fig. 8.2. I, E, and S with interleave, buffer, and multiple AR's.

to the current instruction word, there are two additional pointers to the buffer.

An upper bounds register indicates the instruction word that has most recently been brought to the buffer, and a lower bounds register the oldest word in the buffer. Buffer fill involves bringing in instruction words from memory under the control of the upper bounds register, which is independently advanced. The buffer is filled ahead of execution until the upper bound is seven buffer positions ahead of the instruction register. Instruction register progress is slowed by the existence of multiple instructions in each instruction word.

The fetch of the instruction word is made conditional upon the traffic to core and the relative position of the upper bound pointer to the location counter. The priority of an instruction fetch is reduced vis-à-vis other references to a bank when the upper bound is more than three positions away from the location counter. The fetch is inhibited, of course, for loop mode. The attempt is to maintain instructions ahead of the location counter by bringing in words when it is convenient for the system to do so.

We might further insure the availability of instructions for I by utilizing our ability to have multiple instructions per "word," per true unit of delivery from memory. The CDC 6600, the IBM 360/91, and the 195 do deliver such multiple instructions. In the CDC 6600, they do so by having instruction lengths smaller than the nominal 60-bit word size; in the 91 by delivering a larger unit than the nominal 32-bit word and further by having the possibility of two 16-bit instructions (two bytes) in a "word."

Two very significant ideas are introduced here. One is the provision of a storage capability between a processor and a main memory which reduces the delays caused by the inherent difference in speed between processors and stores. As early as UNIVAC LARC, such stores were considered important to high-speed machines. The LARC designers rejected the idea because of the difficulty of programming the use of such a device intelligently.

This brings us to the second consideration—the need for addressability of stores at various levels. The IBM 360 Model 85 provides an enormous intermediate store that is entirely invisible to the programmer. The Burroughs 5500, 6500, 7500, and 8500 are organized around populations of registers of which only one is visible and even inferentially addressable by the programmer or the compiler. Our buffer is transparent to the programmer; he can make no direct use of the registers except to attempt to organize short loops.

If the architecture of a system provides fast automatic working storage that is optimally used by hardware, a question arises as to the utility of multiple addressable registers under programmer control. The author predicts that future machines may reflect the multiple addressable register concept for reasons of compatability but that there will be a tendency in the future for the "registers" to be absorbed in high-speed local storage of significant size. A perfect candidate for this design is the UNIVAC 1100 series, which actually implements registers in a local control store (originally 600-nanosecond thin film on the 1107), which is addressable either from the register address portion of an instruction, or from the memory address portion. What is unique in the IBM 360/85 design is a machine algorithm for forwarding and replacing data in local store, entirely eliminating the LARC anxieties.

With our current machine organization we have increased the instruction delivery rate of I to 3.63 MIPS. We have now a serious imbalance between I and E, which forces us to reconsider our relationship here and to investigate what can be done to help our actual execution of instructions approach the rate of instruction delivery.

At this point we have a number of parallel operations in the system. S is delivering instructions to the instruction buffer, while I is setting up for decode, while E is either executing or receiving operands from another bank of S. We wish to adjust the rate at which things are done by further increasing the parallel potential of the machine.

8.6. EXECUTION SPEED-UP

There are two basic approaches to speeding up the operation of the E activity without increasing the raw power of E boxes. One is to increase parallel E capability so as to allow the acceptance of more instructions per second from I, and the other is to relieve E boxes of some of their functions. We shall investigate both, beginning with an increase in parallel capability within an E activity.

The E box that we have been discussing executes all instructions in the system with the exception of I/O. It has at this point multiple arithmetic registers that allow, if a programmer attends to it, a reduction in memory references. However, it can only execute one instruction at a time. Perhaps we might select a class of instructions that could be executed in parallel with other instructions. We would want this class to be significant enough in occurrence so that its diversion from the "main stream" would have a significant impact and also provide significant utilization for the hardware that we are to provide for its execution. Further, we would want each individual occurrence to have significant execution length.

The decisions we make here will have a profound impact on the register organization of our machine as well as on the instruction set, the addressing structure, and the compilers of the system. For the moment, let us investigate the I-E possibilities with the constraint that we do not yet wish a major reevaluation of the split of work between I, E, or other potential functional units.

If we look at the instruction list of Fig. 7.5 we find that LOAD, STORE, and INDEX-ing modification are the greatest users of E time, accounting for 44 percent of usage. These short individual execution times are compensated for by their enormous frequency. On the other hand, the FLOATING POINT arithmetic instructions, while accounting for only 13 percent of utilization, are extraordinarily long when they do occur. If we optimize for the first group we shall achieve a general increase in the throughput of the machine; if we optimize for the second we shall achieve an increase for those problems that have high occurrence and no increase for those problems that do not.

8.7. MULTIPLE E BOXES

We could increase the execution throughput of this machine by adding another E box entirely. Figure 8.3 shows this system with two E boxes. Two problems arise. The relationship between the single I box and E must be modified so that I has some way of selecting the E to execute the next

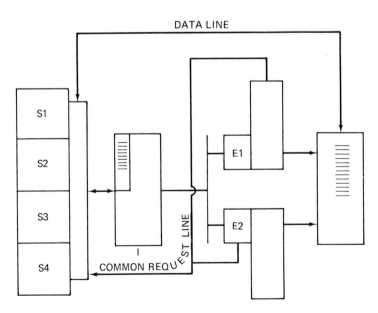

I probes for free E.
E'S share registers.
E'S share memory request line.

Fig. 8.3. Two E boxes.

instruction. This could be done by placing a signal at each E and allowing this signal from an E box to indicate when it was ready to execute another instruction. I would probe E_1 and then E_2 to determine which one was ready. If it found neither ready, it would hang on instruction distribution until a free signal was found.

The second problem concerns the addressable arithmetic registers in the system. Both E units must have access to them, so they may no longer be private to a unit; they must be collected into a common box equally accessible to E's. Since E's will be working asynchronously, it will be necessary to develop some synchronizing capability for the shared resource of registers so that logical dependency in instruction sequence is preserved.

As soon as we introduce the parallel execution capability of two E's, we must provide for rational sequence preservation. For example, the sequence in code shown in Fig. 8.4 might result in the storing of R_1 by E_1 before E_2 can finish the add as well as a fetch by the ADD before the end of LOAD, unless an interlock mechanism is provided.

One solution to this problem is to have a bit associated with each register. This bit is set to 0 whenever it is free for reference and to 1 whenever it is "interlocked." All instructions like add, subtract, multiply, and divide, which take the contents of R_1 somewhere but which also refer to R_1 as a destination, set the interlock bit at the time of fetching contents and reset it at the time of delivery results. In the interval any referencing instruction is forced to wait for the reset of the interlock bit. All instructions that refer to the register as a destination similarly set the interlock until completion. The arrival of data in the register is what actually causes the bit to be reset.

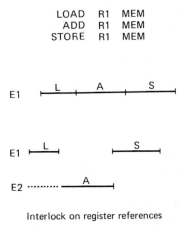

Interlock on register references

Fig. 8.4. E synchronization.

As with any shared resource, interlock control will cause potential delays in execution but preserve logical dependence in instruction sequences. Figure 8.4 shows that there is no potential for parallel execution in the three-instruction sequence because of the logical dependency between them. This is an instance of the problem of asynchronous data reference that one sees writ small and manipulatable by trivial hardware. When the problem is writ large, as in the reference of a data base by multiprogramming or multiprocessing, the solutions are neither obvious nor inexpensive.

A two-E-unit system would increase the instruction execution potential of the machine to 2×1.28 MIPS, or 2.56 MIPS. It would do this at a nontrivial cost. The true utility would depend upon the infrequent occurrence of long logically independent streams of code. Further memory conflict beween E's might preclude them from realizing potential utilization. These problems are common to any introduction of parallelism and are not unique in the two-E situation. Surely here we would tend to favor a low bank memory interleave, since the asynchronous E units would tend to be addressing in the same area of store at the same time. In any case if we are unhappy with the MIP rate we can add a third, fourth, or nth E box to absorb the instructions that I can distribute.

8.8. INTERINSTRUCTION INDEPENDENCE

One feature of the relationship between E's that is interesting to note is the fundamental anonymity of E to the instruction sequence. It is an anonymity like that of the two repairmen, whose names and presence are unknown to the customer delivering the set.

Further, each repairman's $(n + 1)$th specific set to repair is independent of his nth set's particular characteristics or malfunction. Similarly, the $(n + 1)$th instruction of an E is independent of what his nth instruction was. An E unit might be called an interinstruction independent processor in that no interaction exists in the unit itself between the output (result) of an instruction and the input of its predecessor. This is particularly not true when the registers are part of the E unit and is achieved particularly when they are removed. Other examples of interinstruction independent processors will be shown.

8.9. LOOKING AT THE E/S RELATIONSHIP

We had previously, in the context of I and S, felt that the delay caused to I by the relatively slow S was critical. We might suspect the same to be true in the relationship between E and S. Of the average 780 ns in which an instruction is processed in E, 375 ns are involved in fetching operand data from S.

We have earlier provided multiple registers to help reduce storage contention. We note here that they will speed instructions through E without reference to the contention problem, because a register reference instruction need not depend on the basic S cycle.

We consider ways of improving the efficiency of the E-S interface. One obvious path is to explore a solution like the one we found for the I-S case, a buffer unit that would help to balance the rate at which E could generate requests and the S could deliver data. Balancing is more complex in this case, however. First, the rate of request is much more variable for an E than for an I, since there is a distribution of instruction execution times where I process times were constant. Second, the distribution of addresses generated by E will not be as regular as those generated by I.

8.10. CACHE

One approach is to provide a massive intermediate storage between the processor and main store. This is an approach of the IBM 360/85. A 16K to 32K byte store with an 80 ns access time is provided locally between both I and E and main storage. The intent here is to provide sufficient storage that the processor might characteristically behave as if it had an 80 ns main store.

The fast store is invisible to the programmer with an automatic machine algorithm controlling the contents of the store dynamically. The local store, called the buffer or the "cache," is organized into sectors of 1024 bytes, where each sector is in turn divided into 16 "blocks" of 64 bytes each. Each sector is associated with a 1024-byte sector in main store through a sector address register.

Since there are fewer sectors in the cache than in the main store, cache sectors become dynamically associated with main store as main store references are made by the program. The problem of determining which sector to assign to a main storage sector which has been referenced but which does not have a current cache association is handled by a hardware algorithm.

Each sector has a dynamic "binding" priority based upon the recency of its last reference. A reference here is meant to be an execution or read, since a store reference has the property that main store is immediately updated. On a store an inspection is made to determine if the referenced address is in the cache, and if it is not, a write to main store is initiated. If the address is in the cache, the cache is updated and a write through to main store occurs without any modification of the recency ordering of the sector.

Only a read or execute reference, therefore, can cause a reordering of the recency list or the dynamic reassignment of a cache sector. When such a reference is made to an address not in cache, then the least recently used cache sector is reassigned to the referenced sector in main store. Data are transferred to a cache sector 16 bytes at a time to form a 64-byte block. The first 16-byte "quad-word" [a 360 word is 4 bytes (32 bits); therefore, a 16-byte block is a "quad-word"] to be transferred into cache store is the one containing the address referenced by the program.

During the transfer of a block, a block validation bit indicates that it is improper to allow a reference to the block. On the model 85 with a four-way interleaved memory and a 16-byte-wide data path, it is possible to achieve a block load in one storage cycle. Figure 8.5 illustrates the organization and functions.

Certainly, the usefulness of this scheme depends upon the degree of

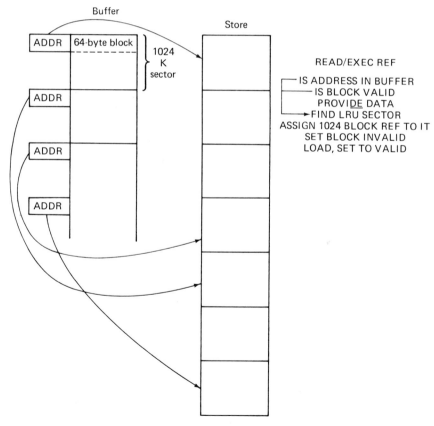

Fig. 8.5. IBM Model 85 Cache.

success experienced in finding referenced data in the cache. To the degree that programs and data are retained, then references to store are minimized and a considerable burden is removed from both I and E units. In a situation where a program and its data are less than cache size, the machine would indeed operate as an 80 ns storage machine.

In cases where, because of the size and reference characteristics of a given program, or because of multiprogramming, there is considerable reassignment of sectors and consequent high incidence of block invalidity, high levels of restoration performance will tend to decline.

No experience has been gained with the approach in an environment having two E units. One might postulate, however, that since the two units would be working on the same problem in the same area, the address generation properties of the program would be similar across the partitions of the cache and that increased invalidation would not occur.

This is a very attractive application of an extended version of the look-behind concept, which we noted in connection with the buffer unit on the I unit. The look-behind is extended to serve not only a recent history of instruction fetches but the recent data experience of the entire system. In an I-E or I multiple E environment all units would share a single cache. Contention problems for sector assignment as well as for physical access would, of course, occur.

8.11. LOCAL E STORAGE

The other approach to the E-S interface is to provide local buffers for the E unit. In order for this to be effective, it is necessary to develop a technique by which some look-ahead is possible to determine what future instructions E might execute and what data it needs for future instructions, and to schedule the fetch of those data while it is executing current instructions. With our current organization it is impossible for an E unit to do this, since only I has any knowledge of operand addresses until it finds a free E and forwards the address to it. E has no way of predicting references to storage.

We have come to a convergence of ideas here. We had previously stated that one way of speeding execution was to increase the parallelism of the E function, and we have done this by adding two E's. Further, we stated that an alternative was to reduce the function of E. This reduction is closely connected to the local operand buffer solution for S degradation of E's operation. In order to explore this area more fully, it is necessary to take a yet somewhat closer look at instruction handling in the I unit and what an I might be asked to do for E.

8.12. BACK TO I AND E: I "EXECUTES" TRANSFERS

The I unit fetches instructions from a buffer and distributes to an op code register at the E interface, which is used synchronously with a selected E unit. Part of the decode function is the forwarding of addresses of operands to the execution unit. We have earlier suggested that it is not always clear where the index register modification capability should be placed, and have suggested the possibility of an asynchronous unit available to I and E.

We have, in the previous sections, made reference to the possibility that E could be relieved of some work and further that certain instruction needs never get to E. Among these instructions we mentioned branching. Surely, if E is doing no more than requesting a transfer address and then shipping this back to I to initiate an instruction request, it makes some sense for I to do this for itself. This can be done since I has the effective address generation capability. Let us determine, then, that in our machine all branches conditional and unconditional will be "executed" by I.

Figure 8.6 shows a functional cycle of an I box, which is capable of executing transfer instructions by itself. If it determines from the op code that this is an instruction to be executed by an E box, it begins to probe for an E box to execute it. This probe may be active or passive. If it is passive, then I will merely search for a signal from an E that it is free. If it is active, I may determine the "freeness" of an E by determining if there is time locally in E to allow the instruction to be sent forward. If the instruction is an I class instruction, I executes it itself.

Since the only I class instructions so far defined are transfers, the only capability I needs to execute is to develop a full transfer address. In order that it do this, the index registers are, of course, required. The response to the request from S is a new instruction word sent to I.

1.   Fetch instruction counter.
2.   Generate full instruction address.
3.   Fetch instruction from buffer (IR) to OP code.
4.   Determine operation class, I or E (I all transfers).
5.   If I, fetch operand address.
6.   Generate full address.
7.   Request transfer to buffer, go to 1.
7a. If unconditional transfer inhibit all instruction advance until buffer filled.
7b. If conditional continue to advance instructions, fill alternate buffer, set tentative mode.
8.   If E probe for free E.
9.   Forward to address register of selected E.
10. Decode.
11. Go to 1.

Fig. 8.6. Functional I cycle.

8.13. I EXECUTES LOAD/STORE

Another class of instructions requires no work from an E unit except address manipulation. These are the data transfer instructions. Since I must have index registers to generate full addresses for sequence transfers, it has the capability of generating addresses for all transfer instructions. In fact, the class of instructions that I executes might be generally defined to be all address manipulation instructions.

However, we require an additional feature to support the I execution of loads and stores. We require an ability for I to cause the transfer of data directly to and from registers and storage. Since paths already exist, the excitation of the activity from I rather than from E causes no particular complication.

A point that should be made is that if I executes its instructions out of the op code register used for E class instructions, we reduce the instruction-passing capability of I. No matter how parallel we make I's activities, it can never pass an instruction to E while an I instruction is being executed. This would mean that no look-ahead could occur at any point beyond a load or a store. We relieve this by providing a separate op and address register set in I for the I instructions. Figure 8.7 shows I with additional registers, the alternate buffer, and access to all index registers. Address-oriented instructions are held in op 2 and E execution instructions in op 2.

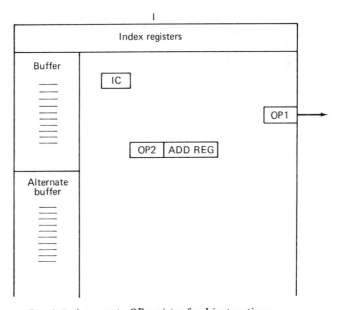

Fig. 8.7. A separate OP register for I instructions.

8.14. OPERAND FORWARDING: I'S CONTRIBUTION TO THE E/S BALANCE

Since we are now executing register loads and stores out of I, we might consider that we have the capability for I to do all data transfers, those imbedded in the two-address form ADD RX, MEM as well as the explicit LOAD RX, MEM. In order to do this we require the local operand buffering of E, and this is exactly the manner in which we will do "look-ahead" on those buffers. Every instruction stabilized by I will be analyzed to determine if data from memory are involved. If they are, the I unit will generate the address and cause the operand to be forwarded to E. One of the requirements for the selection of an E unit would be the availability of a buffer location from the operand to be forwarded.

The difficulty of E's predicting what should be forwarded to an operand buffer is solved by allowing I to make the prediction. The local buffer space in E becomes an invisible repository for operands, so that when E executes an instruction of the form ADD RX, MEM its MEM operand comes from its local store as the result of forwarding from I. E is relieved of S unit delays.

8.15. I NEEDS MORE HELP

One further elaboration is relevant here. Since I will now be executing around 46% of the instructions executed by the system, a serious bottleneck will develop with only one I class op register. If we built a stack of these registers to hold I class instructions as they were encountered in the stream, then it would be possible to pass them off to the stack and get by them to E class instructions. If this were not so, the passing rate of instructions to E would fall off. If an I class instruction were encountered during the execution of an I class, no advance up the instruction stream could be made and any E class instructions would be blocked. The provision of a stack for I's allows the I unit to advance beyond them to locate E class and begin data forward and instruction distribution.

We have now so expanded I that we might be tempted to redraw it as in Fig. 8.8. Here we have added an additional set of buffers. These allow load/store instructions to be passed through op register 2 without waiting for memory response. The buffer stacks up operand stores and fetch requests and allows I to proceed to the next instruction. We are buffering not only the instructions but the memory requests caused by the instructions.

We have now reduced the work load of an E unit in two ways: We have

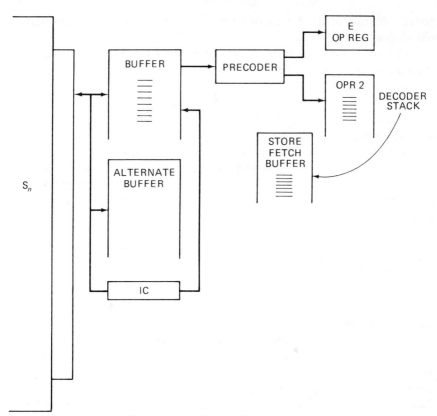

Fig. 8.8. An elaborate I broken open.

reduced the population of instructions that it executes to about 54 percent of the total executed instructions, and we have provided it with a
local buffer store that reduces operand access to a 50 ns buffer reference.
The average instruction time of an E is 325 ns faster than it had previously
been. We have done this without reducing the delivery rate of I, because
instructions forwarded to E are forwarded simultaneously with the execution of I of the I instructions. We have maintained I's delivery rate to itself
by providing operation buffers for it.

8.16. FURTHER GOAL FOR INSTRUCTION PASSING

We have previously reduced our goal for MIPS of the machine from I cycle
to decode time. We shall now reduce it one step further—we wish the
decode time to approach one basic cycle of the machine, where cycle is

defined as the smallest unit of time in which the processor can perform one elemental activity.

The basic cycle of our machine is 50 ns; we would like to drive the machine so that an instruction is decoded every 50 ns. In order to do this, we must have broader parallelism in the I unit. The work of instruction processing must be understood and decomposed so that it requires only one elapsed machine cycle and the work of decoding is redefined and redistributed.

In the face of the fact that we are now moving work back on to I, we must recognize more discrete and therefore distributable decode stages in order to have I achieve the faster rate. We recognize as a limit that no two instructions can be in the same stage at the same time (by definition of a uniprocessor).

8.17. PARALLEL DECODE AND FURTHER ELABORATION OF E

To reduce decode time in I we take one further step; we reduce the function of decode in I to that of instruction passing. We define an instruction set so that from a partial inspection of bits I can decide where to send an instruction and all further encoding is done in the unit to which it is sent. This enlarges our parallelism by providing parallel decode across instruction classes. Further, we split the instruction passing function of I from the execution unit of I, giving rise to a LOAD-STORE-MODIFY-TRANSFER unit (LSMT) and an E unit as asynchronous parallel subprocessors within the processor.

Figure 8.9 is the final form of the machine. We shall eliminate our second E unit and replace it with this functionally specialized system, where LSMT performs some instructions and E other instructions and LSMT does address-forwarding. Since each unit has a local operand store as well as instruction buffer store and decode capability, each unit is truly a processor in its own right. Further, we should be able to add as many of these units as we choose to balance delivery and execution. Finally, since each unit is now a functionally specialized, locally buffered asynchronous subprocessor, some additional work may be done in each unit to speed the performance of its specific functions. We notice that the sequential dependency problem must still be handled with appropriate interlocks between LSMT and E. These must be imposed from I, which releases an instruction only when it can do so without introducing a sequence problem.

Some additional elaborations are possible for the E unit. It is desirable to have an instruction stack in E so that I may have a place to pass

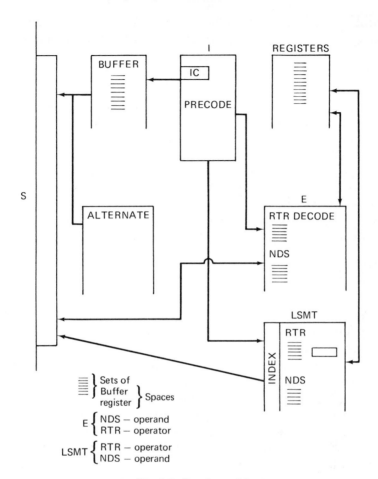

Fig. 8.9. Pseudomachine.

instructions. The argument is the same as for the I stack. I should be able to pass off as many instructions in any period as it can, and finding E unavailable will slow it down, possibly causing delays for I execution and making instruction pass time additive. The limit of how far ahead I should look in the instruction stream becomes the available space in I and E instruction and operand buffers. So long as space is available, I should attempt to decode and distribute.

At any particular time there will be a number of instructions in the system, and the location counter will be pointing some distance ahead of instructions actually being executed. This gives rise to the problem of the "imprecise interrupt" of IBM's 360/91 and 360/195. It is impossible to determine exactly what instruction has caused a signal indicating arithmetic irregularity such as overflow, because the location counter is not set to the instruction or to any precise distance from the instruction causing the overflow.

The E unit itself may be conceived of as a parallel unit capable of performing adds and multiplies in parallel, and the progress of an instruction through E can be broken into parallelizable stages. Alternatively, each function can be thought of as an independent specialized E box with I determining which specialized E will receive the instruction. If this is done, then more discrete op code analysis must be performed in I.

If E units are not buffered and are functionally split apart, the limit of how far I may look ahead becomes less obvious. On finding a referenced E unit busy, the I maintains the instruction. A reasonable limit of how far it will get beyond in looking for an instruction that can find a free E is the number of instructions in an instruction word.

8.18. A PSEUDOMACHINE AT WORK

Figure 8.9 implies some characteristics for the architecture of our machine. Since LSMT has access to index registers and E does not, index registers and arithmetic registers may not be intermixed. An alternative organization is, of course, represented in the Model 91. Further, since E has responsibility for execution of all arithmetic, no distinction is made between fixed- and floating-point registers. Finally, and following IBM in this regard, our instruction set will consist of two instruction forms, register-register and register-memory; op codes will be organized so that the final bit of op code determines which instruction type is involved. Let us take a look at an instruction sequence and see how the units work together. Figure 8.10(a) shows a simple six-instruction sequence, and Fig. 8.10(b) shows the same, coded somewhat differently.

Initially I brings the LD AR1 instruction to the precoder, where it discovers it to be of the class executed by LSMT; it sends the instruction to LSMT, where it is received in the instruction stack. Simultaneously

LD	AR 1, MEM 1 (X)	LD	AR 1, MEM 1 (X)
MPY	AR 1, MEM 2 (X)	LD	E BUF, MEM 2 (X)
ST	AR 1, MEM N (X)	MPY	AR 1, E BUF
LD	AR 2, MEM 3 (X)	ST	AR 1, MEM N (X)
MPY	AR 2, MEM 4 (X)	LD	AR 2, MEM 3 (X)
ST	AR 2, MEM 0 (X)	LD	E BUF, MEM 4 (X)
		MPY	AR 2, E BUF
		ST	AR 2, MEM 0 (X)

(a)   (b)

Fig. 8.10. How it really is.

with the decode of LD by LSMT, I brings MPY AR1 into the precoder and discovers that it is an instruction of the E class. However, it contains a memory reference that must be processed by LSMT.

There is a fundamental difference between LSMT's handling of the two instructions. In the LD the operand from memory may be sent directly to AR1 without passing through an operand buffer of E. This is because the source and destination are explicit in the instruction. We wish, however, for E to find all operands in either registers or its hidden operand buffer. The value in MEM 2(X) must be shipped to such a buffer, so that when E executes the ADD it will be available locally.

What the precoder must do in this case is to decompose the MPY into two instructions, one for E and one for LSMT. The instruction going to E will be in an abbreviated form with the MEM reference deleted. The instruction going to LSMT is received as an LD to E buffer register.

The problem remaining is to coordinate the relationship between asynchronously received instructions and operands in E. At the time of receipt of the LD into the E buffer, LSMT will have enqueued a request for the transfer of MEM 1(X) to R1. In doing this it will have interlocked R1 so that any instruction referencing R1 will be delayed until the interlock is reset. The MPY AR1 in E will find R1 interlocked and will be unable to proceed until the load is complete. LSMT will now enqueue the read to E buffer.

We shall assume a very simple scheme in which the next empty location in E buffer is known by a pointer in E; when an operand arrives, it fills that spot. The local address of that spot must be placed in the MPY instruction when it arrives, or if it has arrived when the data arrive. Since these arrivals are asynchronous and unpredictable, it is necessary for the I unit to have the address of the next available E buffer for placement into the generated LD and MPY. A simple mechanism for providing this can be placed at the interface between E and I.

An interesting possibility for the use of these buffer registers is that in certain cases, where an instruction might be logically independent of a preceding instruction but interlock on an external register reference, it could be possible to execute the instruction by using the buffer registers as pseudoregister storage until the external register was free. The relationship between external registers (the program addressable registers) and these buffer registers is, as was previously noted, not at all fully explored.

After forwarding the generated instructions I will bring up the STR1 and forward it to LSMT. At this point LSMT has enqueued data requests for the real and pseudo load, and E is expecting its operands.

The ST will cause a store request in LSMT to be enqueued. It will be inhibited, however, by the lock on R1 until the completion of the MPY. At the time of arrival of the operands in R1 and buffer 1 the E will execute the instruction. In the time that this execution requires, the I unit will be back around to the next load, and a request for new data for E will have been enqueued so that at the next MPY data will exist in the E buffer. In effect, the I unit serves to modify the code of A so that it resembles the code of B, the intent being to sustain parallel transfer of data into registers and buffers while execution is proceeding.

8.19. HOW WE GOT HERE

It is interesting to observe exactly what we are doing in this chapter and to comment upon it in terms of the examples and analogies and the concept of "maturation."

We have been iteratively investigating relationships and functions, breaking them down into more and more discrete units, in a sense defining our tasks in smaller and smaller elements, and then investigating the possibility for parallel execution between structures specialized to support the tasks. Where this parallelism could be enhanced by adding a local resource, we did so.

We started by separating a processor from I/O, then separated S and I-E, and we then decomposed I. It is this process of perceiving the existence of smaller and smaller functions and then implementing them in functionally specialized units that is the process of "maturation" in computer development.

The reduction of a view of a system from an undifferentiated mass into a collection of discrete functions is the process by which the computers have matured. Much of this has been made possible by technology, but much of it is independent.

It is interesting to note that there is always a three-stage cycle in design advance: (1) Recognize the function. (2) Identify the structure that could perform it. (3) Build more resources into this structure, i.e., elaborate it, until it evolves into a highly intelligent asynchronous subsystem capable of relieving the "main path" of considerable burden.

8.20. TWO REAL ONES

We have, in passing, mentioned the variety of ways in which work can be split once the underlying concepts of local stores, functional specialization, and functional concurrence are recognized and established. High-performance systems currently under design (the successors to the current machines like the IBM 360/91 and CDC 6600) are elaborations of the principle with even more effort invested in balancing the functional units and driving the instruction performance rate to a nominal machine cycle.

Current and past technology has led to the characteristic development of highly specialized functional units for reasons of economics. With the impending development of large-scale integration, where 100's (or 1000's) of circuits can be placed on a chip, there is an economic advantage to the manufacture of identical parts. We might expect to see more functionally identical units of the type of our two E's, rather than the functional splits. This is an interesting turning point.

Up until now, the major concerns for parallel design were the right split and consequent increase in performance at minimum cost. One wished the utilization of the asynchronous components to be high. The industry has always been utilization-oriented because of the cost of systems and subsystems. We own in our normal lives many instruments, some of nontrivial personal expense (like an automobile), whose utilization figures do not directly concern us. There is an overriding consideration of "utility" at time of need that leads us to acquire an automobile even if its total percentage of engine "on" time is less than 70 percent per month.

As computer systems become less expensive and subsystems become small and cheap, we shall worry less about the utilization of units in terms of their actual productive time and consequently may be willing to generalize functions. The n identical E box, as opposed to special-purpose floating-point or store fetch units, may be an outcome of this.

The two currently installed high-performance systems, IBM and CDC's, depend very highly on specialization. The 91 has an instruction unit, a fixed-point execution unit, and a floating-point execution unit. The 6600 has a control unit and 10 specialized functional units: a branch unit that processes all jumps, a Boolean unit that processes logical sums, products, and differences (OR's, AND's), a shift unit, a floating-point add/subtract, two floating-point and fixed multiply units, a divide unit, two short-word fixed-increment units used also for indexing, and a fixed-point add/subtract. Figure 8.11 is a general schematic of the machine.

The short-word add unit (increment units) and branches are the only instructions that may contain direct central memory references. An interesting characteristic of the system is its use of registers for addressing and the means by which it achieves independent operand transfer. The X registers are true 60-bit operand registers addressable by arithmetic and logical instructions. Only the X registers may be addressed by these instructions, so every arithmetic operation depends upon the presence of operands in the X registers. The A (address registers) are 18-bit registers that correspond to X registers. A_0 corresponds to X_0, A_7 to X_7; etc.

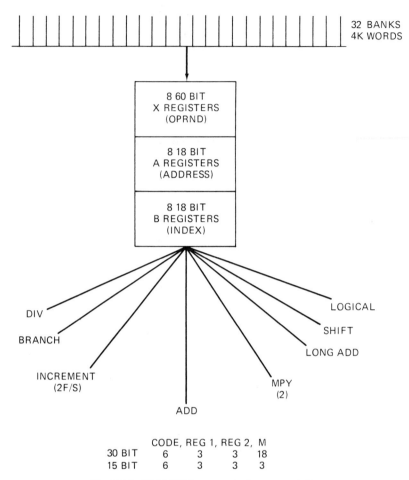

Fig. 8.11. CDC 6600 conceptual representation.

The B registers are true index registers that help to form addresses. They are independently addressable by those instructions that may refer to A registers. The "increment" instructions are allowed reference to A and B. Their function is primarily to form addresses in A registers. Whenever an address changes in an A register, the corresponding X register is fetched or stored from memory. Fetches occur if A_0 or A_5 is referenced, stores if A_6 or A_7. The instruction set has the capability of generating an address in the manner shown in Fig. 8.12. There is quite a variety and flexibility in address formation. These are all relative addresses that are added to a relative location register to form a true machine address. When increment instructions do not refer to A registers as reg 1, then they perform their arithmetic without the implied memory operations.

1. $A_n + I(u) \rightarrow A_n;$ $n < 6$ (A n) is fetch address
                                        $n \geqslant 6$ (A n) is store
2. $B_n + I(u) \rightarrow A_n$
3. $X_{n(18\,bits)} + I(u) \rightarrow A_n$
4. $B_n \pm X_{m(18)} \rightarrow A_n$
    $A_n \pm B_m \rightarrow A_n$
    $B_n \pm B_m \rightarrow A_n$

Fig. 8.12. CDC 6600 basic address functions.

The impressive feature of the 6600 is the breadth and diversity of the parallel asynchronous units, allowing 10 potential independent unrelated instructions to proceed in parallel with the decode of another. The asynchroniy is controlled at the interface of the control unit and the functional units. It is necessary for the control unit to know the status of each register and each functional unit. If the unit required for execution and the registers required are free, the instruction is released to the unit. If not, the instruction is held in control until the unit is free. Successor instructions that are logically independent and whose units are free may be issued by the control unit while previous instructions are held.

The operational efficiency of the unit is quite sensitive to proper instruction sequencing and register assignment, and the compiler for such a system is obliged to pay a good deal of attention to such considerations. The functional units of the 6600 are not locally buffered either for operands or instructions, and consequently each unit may execute only one instruction at a time. Parallelism is achieved across function units. The control unit has an eight-word instruction stack, which may contain up to 32 instructions.

The Model 91 is conceptually similar. Like the 6600, it has multiple programmable registers. The register population, however, is divided into four 64-bit floating-point registers and 16 32-bit general-purpose registers, which also serve as index registers. As in the 6600, there are register to memory and register to register instructions. The register instructions, however, are basically two address instructions.

Memory referencing may occur on arithmetic instructions. The instruction unit of the system basically transforms these memory reference arithmetic instructions into FETCH, FETCH, ARITH sequences in a manner much like our hypothetical machine. Operands are forwarded to the operand buffers of the execution units. There are two of these—a fixed-point execution unit and a floating-point execution unit (itself divided into MPY/DIV and ADD). These are fully buffered for operands and for operations.

There is an implication of a different approach between the two machines. The 6600 achieves its speed by breadth of function; the 91 achieves it by depth. Elaborate instruction sequencing rules are built into the buffered functional units so as to allow them to process more than one instruction at a time within a unit. Various busy register reference bypass techniques, allowing, for example, direct store from a buffer to central memory, are employed to "pipeline" the activity of each functional unit.

The buffering capability of the 91 is rather extensive. Figure 8.13 shows a general schematic of the 91 and the location of its buffers. There

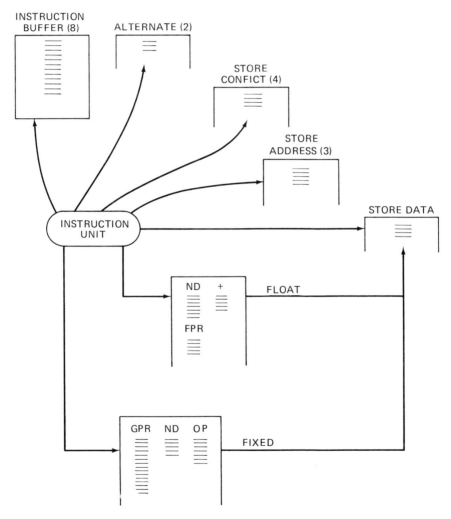

Fig. 8.13. IBM 360/91.

are eight instruction fetch buffers of 64 bytes, each allowing a range of from 16 to 32 instructions (including store to store instructions). There are, in addition, two alternate path buffers. Loop control is provided. In addition, since the instruction unit is responsible for all operand fetching, a family of store-oriented buffers is provided to allow the instruction unit to continue instruction advance and preliminary decoding independent of the storage system responses. The address buffering provided by four storage conflict and three storage address buffers allows the instruction unit to spin off storage requests to busy units and to bypass waits on store references.

The store data buffers allow functional units to free local stores and registers without waiting for storage access, and, of course, the local operation and operand buffers allow functional units to free local stores and registers without waiting for storage access. The local operation and operand buffers allow the instruction unit to continue decoding and fetching independently of execution rate of functional units. The presence of multiple operands and operations in the functional units further allows the multiple execution capability to which we earlier referred.

Particular attention was paid to recognizing the opportunities for parallel execution of instructions referring to the same registers within a functional unit.

8.21. WHY WE WERE HERE

The 91 and 6600 represent the status of implementation of machines of this type. The IBM 360/195 is essentially a speeded-up version of the 91 with the addition of an 85-like cache. The levels of storage associated with the machine are registers, local E unit buffers, and I instruction storage buffers backed by 32,000 bytes of cache storage between local buffers and main store. The machine utilizes both the approaches to the solution of S hang-up which we described.

These machines have a long history and apparently a glorious future. Oncoming are the CDC 7600, the Burroughs 8500, UNIVAC NIKE-X. From the past GAMMA 60, LARC, STRETCH, ATLAS, and IBM 7094 have contributed to the concepts of look-ahead, look-aside, and functional concurrency.

Our interest in these machines is threefold. First, they are, in themselves, important examples of the use of parallelism in design in order to achieve more speed. Second, they form a conceptual basis from which one can build a multiprocessor concept by an extension of I unit membership in the system. Third, the proper sequencing of instructions and allocation of resources present to us the fundamental problems of scheduling and resource allocation for machines at all levels and systems at all levels.

The resource scheduler and allocator for this level of parallel machine is called a "compiler." The next chapter will introduce some basic compiler concepts and attend to the problem of generating good coding for machines of this class.

Chapter 9

INSTRUCTION LEVEL
SCHEDULING

9.1. INTRODUCTION TO SCHEDULING FUNCTION

We shall limit our appreciation in this chapter to the scheduling of that element of work on whose behalf the parallelism of these highly maturated uniprocessors has been devised—the instruction. Scheduling at this level has as its goal the assurance that sequences of instruction are submitted for execution in an order that will maintain the potential breadth of parallel activity inherent in the machine design.

There are commonly three agents responsible for this kind of scheduling. They are the machine itself, the assembly language programmer, and the optimizing phase of a higher-level language compiler. To the extent that any of these agents give concern to maintaining the instruction execution rate of the system and the utilization levels of various independent asynchronous elements, it is making scheduling decisions. They have a great advantage in making such decisions, since they are scheduling elements (instructions) whose performance characteristics are largely known (except for memory interference).

The time for an instruction in a system is known. The resource requirements are known, and the dependencies between instructions are also perfectly known. The difficulty in scheduling at higher levels is the lack of information in these areas and the variations that can occur on each instance of the performance of the task.

9.2. THE PSEUDO 6600

Let us look at a sample instruction stream to determine the effect on performance of a system by this level of decisions. We shall derive our timings and gross machine characteristics from the CDC 6600 in these illustrations. Figure 9.1 shows an encoding of the two arithmetic statements shown above the code. We would like to look at how the machine will perform this simple list of instructions. We perceive that, if each instruction were sequentially performed, the total time on the machine for the entire sequence would be 151 cycles.

We postulate for our machine no local buffering except for an instruction buffer and a distribution of functional units like the 6600. Some of the details of interlock and delays outlined in the CDC 6600 *Reference Manual* are overlooked, but essential considerations for scheduling are maintained. The timing representations, therefore, do not represent exactly the performance of this list on a 6600, but are an approximation on a machine of its general organization.

Each instruction is assumed to have a one-cycle decode time and to be

$$Z = A * B + C/D - E$$
$$V = Z - G + T$$

	Instruction Word		Sub Expression	Elapsed Time
1	FETCH	R1,FIELD A		9
1	FETCH	R2,FIELD B		18
2	MPY	R1,R2,R1	A*B	29
2	FETCH	R2,FIELD C		38
2	NO-OP			38*
3	FETCH	R3,FIELD D		47
3	DIV	R2,R3,R2	C/D	77
3	NO-OP			77
4	ADD	R1,R2,R1	A*B + C/D	82*
4	FETCH	R2,FIELD E		91
4	NO-OP			91*
5	SUB	R1,R2, R6	A*B + C/D−E	96
5	STRE	R6,FIELD Z	Z =	105
5	NO-OP			105*
6	FETCH	R1,FIELD Z		114
6	FETCH	R2,FIELD G		123
7	SUB	R1,R2, R1	Z-G	128
7	FETCH	R2,FIELD T		137
7	NO-OP			137
8	ADD	R1,R2, R6	Z-G + T	142*
8	STRE	R6,FIELD V	V =	151
8	NO-OP			

* BYPASS NOT SHOWN IN TIMINGS

Fig. 9.1. Sequential execution of the list of the code.

spontaneously available from the instruction stack. Not considered is the time to execute a "pass" instruction used as an instruction word filler. Also not considered is the difference in instruction issue time for 15- and 30-bit instructions. Most critically, the CDC 6600's ability to bypass an instruction interlocked on a functional unit or register usage is not shown.

We pay close attention, however, to when the operand register becomes free, when a functional unit becomes free, and what gross system interlocks are. The timings in Fig. 9.1 represent approximate 6600 instruction timings plus a cycle for stabilization plus memory reference time as five cycles.

The cumulative timings of Fig. 9.1 reflect the performance of a machine that does not advance an instruction until instruction end time of a predecessor. That is, the functional unit is considered busy until the result register is set. This leads to an overstatement of consecutive fetch times that are actually performed as in Fig. 9.2 with the fetch unit coming available sooner.

The "stabilize" segment of the FETCH R1, A line is that period of time between the acknowledgment of the instruction by the I unit and the release of the instruction to the fetch unit. As soon as that instruction is passed to the fetch unit, the I unit is free to begin processing of the next instruction.

The interpretation of the FETCH R2, B occurs simultaneously with the fetch unit execution of the first FETCH. In the three-cycle period from time one to time four, the fetch unit is developing the address to be

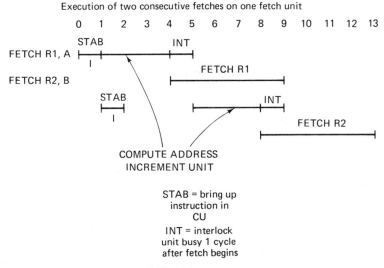

Fig. 9.2. Parallel fetch functions.

fetched by (for example) adding index register value to the passed address. After the development of the address, it is placed in a fetch register (A1) as described in the last chapter. One cycle after this placement the fetch unit becomes free. The change in A1 causes a memory request, which, in the absence of contention delays, will fill R1 by time nine.

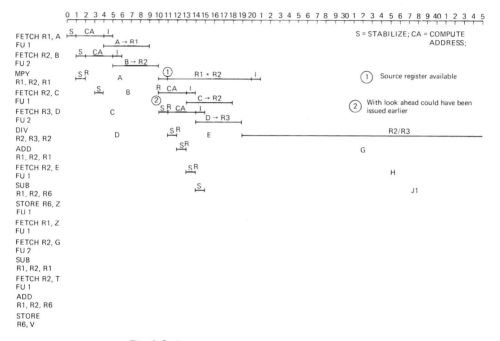

Fig. 9.3. Execution of code on parallel machine.

During this period from one to nine the referenced register is unavailable to other operations using it to hold data but may be available in a controlled way to operations using it as a source of data. In the period one to five the fetch to R2 is held in the I unit, as there is no local buffer in the fetch unit to hold it. Some look-ahead past this instruction might be undertaken (and is in the 6600), but we shall not show that here.

At five time the fetch unit is free and receives the R2 fetch. It computes and places the effective address in A2 while the R1 memory transfer is underway.

The separation of the actual transfer from the computation of the address allows the generation of a memory reference before the completion of a predecessor.

We show the R2 fetch completing by time 13. If it were a reference to the same bank as the R1 fetch, this could not occur, as the memory would not begin to service the second reference until the first reference was complete. Time for two transfers would then be 14, not 13, cycles. In either case the local parallelism in the fetch activity provides for a faster operation than shown in Fig. 9.1, a purely sequential machine.

We shall use the 6600 population of two independent fetch units for

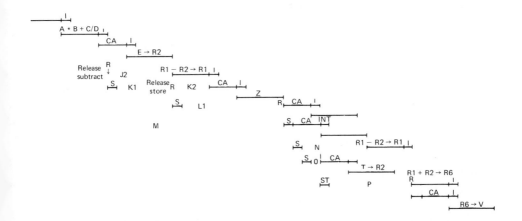

our later example. With two units one can see that the two operands could be loaded into the registers in 10 cycles.

The chart of Fig. 9.2 is called a Gantt chart and is commonly used to represent system flow. Each asynchronous independent unit is listed on the left and has a unique set of line segments associated with it. Time units are laid out horizontally. Those intervals of time which have a line segment for a given functional unit show that the unit was active during that interval. Time units with multiple line segments show parallel activity during the interval. Very basic examples of this type of chart were shown in Chapter 1 to represent the progress of our repairmen.

We follow the sequence of events for the instruction sequence in Fig. 9.3. Initially we notice that the total time for the execution of the 16-instruction sequence is 96 cycles, down from 151 of the sequential example. This is a disappointing result, since our highly sophisticated processor, with potential parallelism between address generation, fetch/store, multiply/divide, and add and with a further two-way parallel potential for fetch/stores and multiply/divide, has achieved only a 55-cycle saving, 38 percent. This is because logical dependency on and contention over resources cause delays in performance and a reduction in the effective parallelism of potentially parallel units.

The resource contention is on functional units and addressable registers, the dependency on the logical relationship between the results of instructions.

Figures 9.4(a) and (b) represent the logical dependencies in the instruction stream. Figure 9.4(a) represents a tree of the kind common in representing arithmetic instruction statements in compiler literature. The bottom nodes represent the tasks of fetching operands; the nodes with arithmetic operations represent the multiply, divide, adds, etc.

The tree has breadth and depth. The breadth is the number of nodes at a given level (L1 through L6), and the depth is the number of levels. A technique for the formation of such trees will be shown later in this chapter. The depth of the tree as represented here has no fixed time connotation. Sequential time is implied by the number of steps and levels shown for the tree. It is impossible to determine from the tree as shown how long the execution of these steps might take.

The function of each node at any level is seen to be dependent upon the completion of the function of nodes that have branches leading into the node. Thus node h is dependent upon nodes a and b but independent of all other nodes.

The breadth of the tree at a given level is a measure of the potential logical parallelism available at that level. Thus logically all seven operands could be fetched at the same time, and the divide and multiply could be executed at the same time. No other logically parallel operations are available, however, since nodes k, l, and m are all dependent on j.

The realization of this potential parallelism rests upon the availability of resources and the relative execution time for each function. Notice that the execution of note j is dependent upon the completion of h and i and that i (the divide) is the constraint, since it takes considerably longer to divide than to multiply. This is not represented on the tree.

The striking similarity between the concepts we are involved with here and the concepts of Fig. 5.2(a) is shown in Fig. 9.4(b), where the tree is redrawn in the form of Fig. 5.2. We had at that time mentioned that the nature of the tasks we were discussing meant nothing to us. We are now using a specific meaning for a task—the execution of a single instruction. Similarly, by a "processor" we mean the functionally specified execution units; and by "resources" we mean the collection of addressable arithmetic registers in the machine.

(a) Tree form of expressions

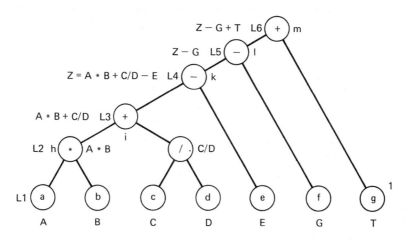

(b) Precedence chart

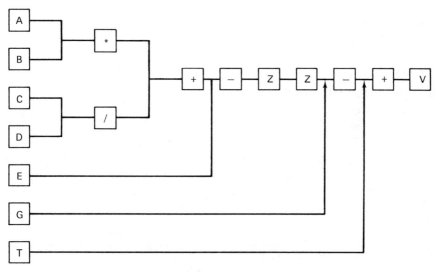

Fig. 9.4.

The phenomenon of parallelism will occur between stages of an instruction and between instructions. The graphs of Fig. 9.4 imply the maximum speed achievable if there were no resource contention. Either an independent resource was always available when required, or with time-sharing no delays ever occurred because of unit unavailability.

If we associate time with this tree, as in Fig. 9.5, we can determine the minimum running time, given no delays. Figure 9.5 is in the form of the graph of Fig. 5.2. It shows that the minimum time for executing the tree (including the store and refetch of Z) is 86 cycles, or 10 cycles less than we accomplished in Fig. 9.3. The differential is due to contention delays.

The 86-cycle minimum, however, is for the execution of the tree as represented. The question must arise as to whether this is an optimum tree for the representation of the two statements. We shall observe later that it is not, and that implies that some better organization at a conceptual level might have improved performance. The programmer's or compiler's concept of the relationship between functions may constrain performance by imposing precedence where it is not truly required. The implication of the operational delays over and above precedence is that there is room for investigation of the resource allocation decision explicit in the coding structure.

We immediately understand from Fig. 9.5 that we may not achieve the 86-cycle time with the given tree because we do not have the four fetch units required to provide operands simultaneously for the start of the MPY and the DIV. To understand the performance of our machine, we shall look very closely at Fig. 9.3, explaining the cause for each delay and developing statements for the rules of our machine.

9.3. THROUGH THE INSTRUCTIONS WITH GUN AND CAMERA

Initially the FETCH R1, A instruction is partially decoded by the I unit (at time 0 through time 1) and recognized to be an instruction to be executed by a fetch unit using R1 for the deposit of the result of the fetch. The I unit determines that a fetch unit is indeed available.

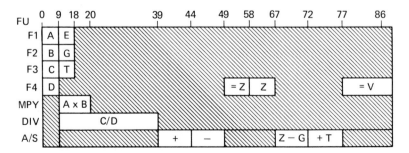

Fig. 9.5. A schedule with 4FU, 1MPY, 1DIV, and 1A/S.

The first rule of the machine is that an execution unit appropriate for the instruction must be available for the instruction to be released by the I unit. The existence of a second fetch unit differentiates Fig. 9.3 from Fig. 9.2. In Fig. 9.2 with one fetch unit the test for a free unit for the R2 load fails until time 5. In Fig. 9.3 the second unit is available at time 2, and the instruction can be immediately issued.

If we assume no look-ahead past a delayed instruction, the impact of delaying on an unavailable unit is to delay not only the current instruction, but the analysis and distribution of all subsequent instructions.

The two register-load instructions reference separate registers as the final targets of the operands transferred from memory. A register designated by an instruction to hold a result (an output of that task) we shall call a sink register. More precisely, we shall recognize the sink condition of a register during the period of time that a referencing instruction is issued from I until the arrival of data to the register. Registers 1 and 2 are in Sink mode at the time that MPY instruction is processed by the I unit. The Sink mode makes a register unavailable to the system for other Sink references. If both fetch instructions had referred to R1, the second instruction would not be released from the I box.

The second rule of our machine is that the referenced sink register must not already be in that status at the time of attempted release. If it is, the instruction is detained in the I box until sink mode is removed.

The reason for this rule is fairly intuitive. In controlling data files we do not allow two programs to have simultaneous update access to the same record because of possibly nonpredictable results. We insist on sequentializing references so that we can at any time determine the final result of the process of update. If both instructions are released, it would not be possible to predict the final contents of R1.

Depending on memory response time (a function of contention) either A or B would wind up as the data in the register. Complete protection against "bugs" is not afforded, however, because the second load of the register will proceed as soon as the first load is completed. When A arrives in R1 the sink mode will be released, and the system will load B into it. What is guaranteed here is that the machine will do what it has been told to do (even if it appears a little unreasonable). To preclude an unpredictable result is one thing; to question the intent of a programmer is another.

The references of multiply are such that it expects an operand (A) in R1 and an operand (B) in R2 and wishes to place a result (A * B) in R1. The register reference to R1 as a place to put a result causes R1 to be designated a sink with regard to the multiply. R1 is also designated as a place from which an operand is to be taken.

In this role as a supplier of operands, R1 is said to be a source register.

R2 is also a source register. The double reference to R1 as a source and a target results in its being regarded as a sink. That sink is the "stronger" or dominating mode is made necessary by the need to inhibit unpredictable results.

This usage is interesting because it reveals an argument for critics of three-address machines. They claim that the incidence of sink/source reference is normally so high that the space taken into the system to represent addresses for two operands is rarely useful. The need to maintain both operands as well as a sum or product is sufficiently infrequent that no meaningful coding efficiency is realized by the provision of space in an instruction to do this. This space is better used for other things such as increased indexing power or for the representation of more instructions in a reduced space frame.

From the tree of Fig. 9.4 we expect MPY to be dependent upon the completion of the two preceding fetches. The dependency comes from the intersection of the outputs of tasks and the inputs of successors. The input of MPY is the output of fetches; MPY can, therefore, not be undertaken before the fetches are complete. There is a precedence relationship recognized by the intersection of register names, R1 as a sink for fetch A, R2 as a sink for fetch B, and both as a source for MPY.

When an instruction references a register in sink mode as a source, the status of that register is changed to sink/source. The source reference instruction is released to the functional unit. This allows the I box to advance to the next instruction.

The equivalent of this is to allow coordinated reading and writing. What is necessary is to assure that the first undertaken operation completes first. The MPY in the multiply unit will be interlocked on register reference until the fetch is complete and the register reverts to source mode. But this interlock will be characteristically brief if it occurs, and it may not occur. If the MPY had used R3 as a source, it would be forwarded to the MPY unit at time 4, and the succeeding fetch would be processed by I at time 5. However, the release of MPY is delayed until time 9 because of the double sink reference on R1. The delay A, therefore, is due to the resource contention on R1 and not the fetch/multiply logical dependency. This observation which nicely shows a distinction between dependency and contention is a result of a design decision in our pseudo 6600—the decision not to allow sink/sink status and to control interlocking from the functional unit.

Multiply is released at time 9 when the sink status of R1 is reset. It is delayed from execution until the availability of both its operands. This occurs when R2 is reset to source at time 10. Finally, for the multiply operation, it is necessary to define the point at which R2 becomes free.

We shall assume a parallel fetch of operands and specify that a source register becomes available one cycle after the activation of the instruction using it. The sink status of R1 continues until the delivery of the product to R1; this is at time 20, one cycle before the availability of the multiply unit.

Any instruction referencing R1 in the interval of the multiply up to time 20 will be delayed in the I box. Because of the late release of MPY, the FETCH R2, C instruction will not be looked at by I until time 9. The delay B is due to this. When ready for issuance I will check and find that the functional unit for fetching is available (in fact both are available). It will then check the status of R2, the sink register of the fetch.

At time 10 this register is still in source mode. It is expecting an active instruction to reference it for an operand. Since it is not in sink mode, it is permissible to release the fetch if it is set to source/sink status. This is equivalent to sink/source, except that the source reference must be made first. If the sink reference is made first, it will be delayed until the source is granted and the register reset to sink mode only. (This is the other possibility for read/write coordination.)

There are other possibilities for the treatment of the FETCH C. If MPY had not been detained on the double R1 sink mode, then FETCH R2, C would have been inspected by 1 at time 4. But at time 4 no fetch unit was available, and the instruction would have been detained in I until at least time 5, the first availability of a fetch unit. At that time, however I would have discovered a double sink reference to R2. The fetching of C would be delayed until the release of R2 as a sink mode register. This would occur at time 10. The preceding multiply instruction, however, would have caused it to revert to source mode in the interval between 10 and 11, so that C FETCH could not access the register until 11 time in any case. The B delay would still have occurred, but it would have occurred with the fetch as the delayed instruction in place of the multiply.

However, the R2 reference for FETCH C is entirely arbitrary. There is no dependency relationship between FETCH B and FETCH C. This is a true contention delay. We could have avoided this delay by simply referring to another register with our fetch.

That decision is equivalent to deciding to allocate and use a further resource (another register) rather than to share a single resource. It is a primitive example of our observation that parallelism is a function of the breadth of available resources.

We notice, however, that even a separate register reference would encounter some delay, since a functional unit (FU1) is not available until cycle 5 and a one-cycle delay would have been encountered on this unit.

The next delay, C, comes about as a result of the long hold of the previous instruction in the unit precluding FETCH R3 from becoming a candidate for issuance until cycle 10. However, this instruction would have been interlocked in any case, waiting on an FU.

Delays D and E, associated with the divide, are due (D) to the inability to bypass the fetch, and (E) to the unavailability of the operand in R3. D is, therefore, a pure resource delay, and E a logical dependency delay due to the requirement that an input of the divide be available. Notice that the divide is issued to the divide unit, and the E delay occurs there. This follows the sink/source issue rule as we described it.

The ADD R1, R2, R1 is stabilized and issued one cycle after the divide. Since ADD specifies its source (R2) to be the sink of the divide, it is interlocked in the add unit, and the G delay is due to the wait for the delivery of C/D to R2. This also follows the source/sink rule we have discussed.

The H delay after the issue of FETCH R2, E is a source/sink delay on the availability of R2. This is a pure resource delay, since there is no logical dependency between the ADD and FETCH.

The J delays on SUBTRACT are interesting because the same delay is caused for two reasons. J1 is the period of time during which SUB must be held in the instruction unit because the add unit is busy with the ADD. This is the period from cycle 15 until cycle 53, when the add functional unit is released. At this point the J2 delay, a dependency delay awaiting the availability of the operand, occurs. The issue of SUB, however, allows the STORE R6, Z to become a candidate for issue.

This involves us in our final rule: No store can be issued while a fetch is in progress, nor can a fetch be issued while a store is in progress, even if a functional unit is available. The reason for this rule comes from our own coding, where Z is stored, then fetched. This is bad code, but it must work. If the fetch of Z were allowed to be issued before the completion of the store, then a wrong contents of Z would be returned to R1. Therefore, it is generally necessary to synchronize and sequentialize fetches and stores. The issue of STORE that would normally occur at cycle 54 is delayed until cycle 55, when FU 2 releases its registers.

The K2 delay after the issuance is a sink/source dependency delay. The L1 issuance delay of the fetch of Z is due to this memory store/fetch synchronization, which assures retrieval of a correct value of Z.

The M delay is due to the retention of the FETCH R1, Z in the instruction unit and is strictly a delay due to sharing and contention. The N, O, and P delays are of the types we have seen; N is sink/source dependency, and O is a resource wait for a functional unit, as is P.

We have now looked at the execution sequence in some detail, inspecting for various delays and determining the logical reasons and rules for assumed logical dependency by the sink/source relationship. We could make further analyses of this stream—for example, the percentage of utilization of all functional units, the MIPS rate, and the percentage of utilization of registers.

9.4. SCHEDULING AND RESOURCE ALLOCATION

Let us go on to make some scheduling and resource allocation statements about our task. From a resource point of view, we see that we have shared resources unnecessarily, and by doing so we have implied logical dependency where it did not exist.

A first improvement in our code is obvious; since we have registers available, let us use them, and let us code so that implied relationships on a register are eliminated wherever possible. Beyond this, however, let us inspect our perception of the fundamental relationships between the subtasks. Given that we wish to retain the result Z, we cannot compress our two statements into one. [There are ALGOL variants where this could be done as $V = (Z = A * B + C/D - E) - G + T.$]

However, by careful scheduling we could force more parallelism out of the machine. We perceive that the tree of Fig. 9.6 is a possibility. This differs from our earlier tree in that the subtraction of the variable E is performed before the end of the division, taking advantage of rules of algebraic equivalence, and further the summation of $G + T$ is perceived to be parallel with other operations. In addition, we eliminate the obviously unnecessary retrieval of Z. Notice that this cannot be expressed directly in the higher-level language.

From the chart of Fig. 9.6(a) we see that, given no resource or processor problems, we could execute this tree in 58 cycles. Our actual execution, from Fig. 9.7, is exactly that, despite the fact that we are sharing functional units, the fetch and add units, across instructions.

Notice the code of Fig. 9.6. We have eliminated FETCH's to the same registers wherever possible and we have scheduled the fetch units as densely as possible without delaying other functions. We have resource delays A and B in Fig. 9.7., but they do not delay the effective start of MPY. The C delay is a true sink/source dependency delay while MPY waits for its operands, but since R1 is also a sink of MPY, it is treated as a sink delay. We have eliminated all apparent dependency by broader distribution of our register references.

One interesting point in connection with this occurs at the use of R2 for G fetch on line 7 of Fig. 9.6(c). No delay is caused here, because the multiply released R2 as a source immediately upon picking up its operand. It is a potential source/sink delay, but MPY has already had its access. A

(a) Precedence schedule

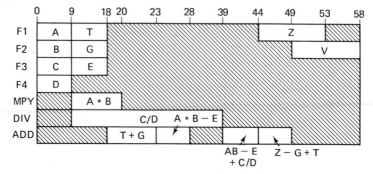

(b) Tree

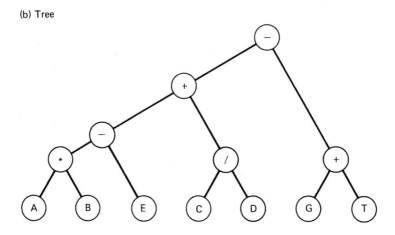

(c) Code

```
FETCH   R1, A
FETCH   R2, B
FETCH   R3, C
FETCH   R4, D
MPY     R1, R2, R1   A * B
FETCH   R5, T
FETCH   R2, G
DIV     R3, R4, R3   C/D
FETCH   R4, E
ADD     R2, R5, R2   T + G
SUB     R1, R4, R1   A * B − E
ADD     R3, R1, R6   A * B − E + C/D
STRE    R6, Z            Z
SUB     R2, R5, R7       Z − T + G
STRE    R7, V            V
```

Fig. 9.6. An alternate form.

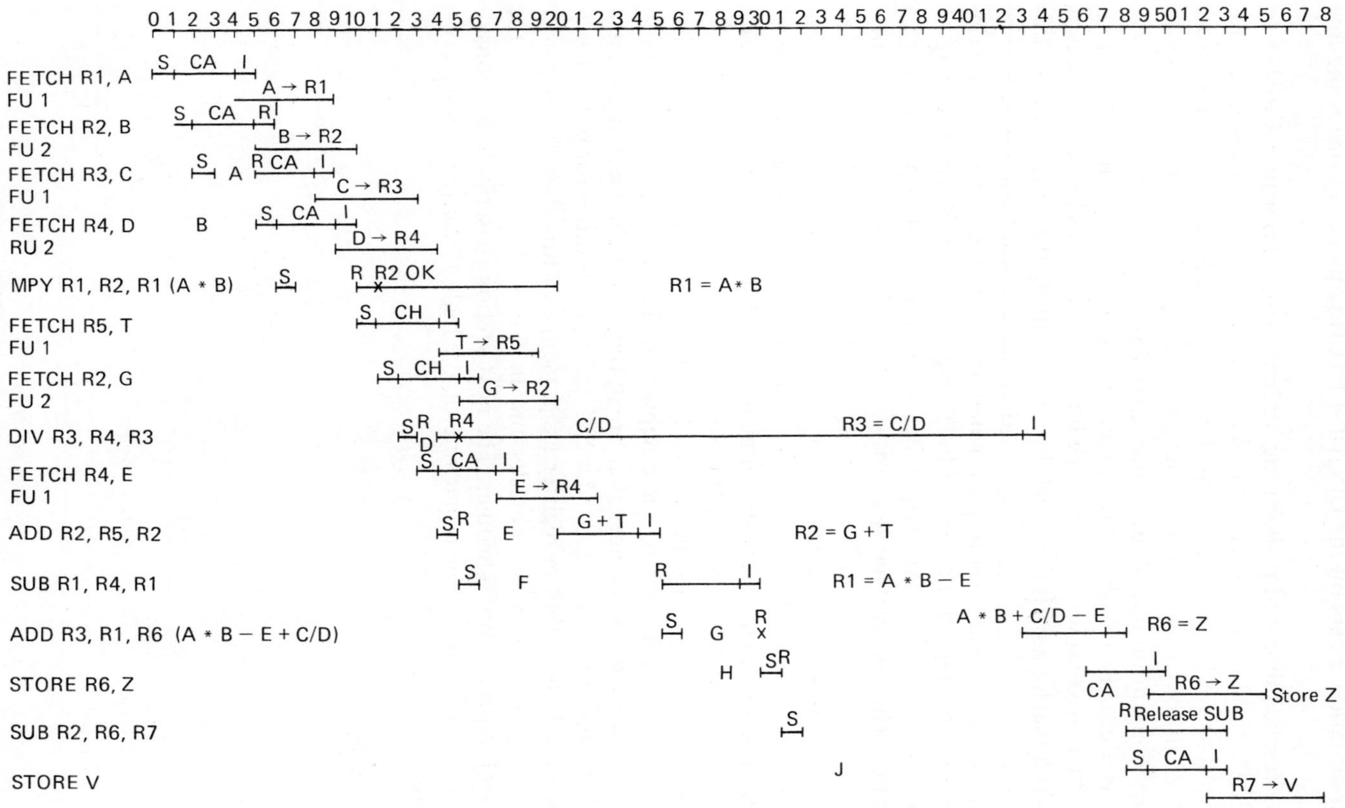

Fig. 9.7. Execution of Fig. 9.6.

similar situation exists on the DIVIDE-FETCH sequence of line 8 and line 9.

By rescheduling and reallocating we have reduced our time from 96 to 58 cycles, from a sequential 151. We have been able to schedule in this way because we knew a great deal about the subtasks with which we were working, although even here we do not have perfect knowledge due to memory contention possibilities from channels.

It is a cruelty to ask a programmer to achieve this form of optimization. There are two ways of relieving him of the responsibility. One way is to build a machine with the capability of executing the code of Fig. 9.1 as if it were the code of Fig. 9.6, or some close approximation to it. The Model 91 has, perhaps, more power than the pseudo 6600 we are describing here in this regard. Because of buffering and register bypass, it may be able to execute instruction sequences in an order more optimum for its hardware. This power has led to some problems in its own right— specifically, the "imprecise" interrupt, the situation where, because of the "chaotic" out-of-sequence execution capabilities of the machine, it is impossible precisely to determine the status of the machine at the time of certain interrupts. However, the fundamental idea of automatic hardware sequence optimization is very attractive.

As an alternative to automatic hardware resequencing or together with this, one must look to the compiler of the higher-level language to generate instruction sequences that are optimum for the machine. It must be able to recognize true dependencies and opportunities for parallel execution as well as optimum register assignments. In this sense the compiler becomes the scheduler of subtasks and in doing so uses very much the same techniques as any scheduler. The next sections describe some considerations for a compiler for a machine of the high-performance uniprocessor class.

Chapter 10

BASIC COMPILER
CONSIDERATIONS

10.1. THE BASIC COMPILER FUNCTION

The fundamental function of a compiler is to generate machine-executable sequences of instructions from the statements in a higher-level language. The variations in compiler design and organization are manifold, but certain basic functions and techniques are general to large numbers of compilers.

Before going further let us review a basic compiler function—the transformation of an arithmetic statement into a form known as Polish postfix and the attendant generation of a statement tree.

The simple arithmetic statement of Fig. 10.1 is expressed in the "infix" form common to most higher-level algorithmic languages. In this form a relationship between two variables is placed between the variable names. The line below represents the same expression in postfix form, where the relationship is placed after the variables.

Notice that the infix form is ambiguous, in that the "true" intent can

$$A = B+C+D+E+F+G \qquad \text{INFIX}$$
$$A \; B \; C+D+E+F+G+ = \qquad \text{POSTFIX}$$

Equivalents

$$A = B+(C+D+E+F+G)$$
$$A = B+C+(D+E+F+G)$$
$$A = (B+C)+(D+E)+(F+G)$$
$$A = ((B+C)+(D+E))+(F+G)$$
$$A = (B+C+D)+(E+F+G)$$

Fig. 10.1. Polish Form.

be interpreted in a number of ways, as shown in Fig. 10.1. These are algebraically equivalent and in most machines numerically equivalent, although numeric equivalency can often not be guaranteed, particularly when multiplications and divisions are involved. The postfix form is absolutely unambiguous in that it can have only one meaning; in this case add B to C; then add D; then E, F, and G, and set the result into A.

The operands of postfix operators grow on the left, as shown in Fig. 10.2. For the class of binary operators—those which require two operands such as add, multiply, divide, and subtract—the two variables on the left are its input. In the first statement of Fig. 10.2, B is operand 1 and C is operand 2 of +. This is easy to see.

$$
\begin{array}{lll}
& \quad\;\; 1\;2 \\
1. & \text{A B C +} \\[4pt]
& \quad\;\;\; 1\quad 2 & \quad 1\;\; 2 \\
2. & \text{A B C + D +} & \text{A R1 D +} \\[4pt]
& \quad\quad\;\; 1\quad 2 & \quad 1\;\; 2 \\
3. & \text{A B C + D + E +} & \text{A R2 E +} \\[4pt]
& \quad\quad\;\; 1\quad\quad 2 & \quad 1\;\; 2 \\
4. & \text{A B C + D + E + F +} & \text{A R3 F +} \\[4pt]
& \quad\quad\quad\;\; 1\quad\quad\; 2 & \quad 1\;\; 2 \\
5. & \text{A B C + D + E + F + G +} & \text{A R4 } G_1 \text{ +} \\[4pt]
& \quad 1\quad\quad\quad\; 2 \\
6. & \text{A B C + D + E + F + G + =} & \text{A R5 =}
\end{array}
$$

Fig. 10.2. Growth of Polish.

The further statements of Fig. 10.2, as the "Polish string" grows, often require an explanation. In Fig. 10.2 (2) the operands of the arrowed plus are the sum of B, C, and the variable D. The entire string $BC+$ is the first operand. This may be alternatively represented as R1, the result of $BC+$, and so on, with all the statements of Fig. 10.2. In (5) the operand 1 is the result of $B + C + D + E + F$, a result 4, and the operand 2 is G.

10.2. A WAY OF GENERATING POLISH

There are a number of ways that are used to generate Polish strings or representations of postfix form that are equivalent to Polish. We shall describe in part a very simple and straightforward technique involving a single left-to-right scan of an expression, and a last in first out list (STACK), which we shall call the operator hold stack, showing only as much as we require for illustrations in parallelism later on. We shall first show the generation of Polish for the simple expression of Fig. 10.1. The

We must initially introduce the concept of operator hierarchy. The basic operators of a high-level algorithmic language like ALGOL, FOR-TRAN, PLI, JOVIAL, and MAD are the +, -, /, *, and **. In addition, there are functions like SIN(X), COS(X), etc. There are relational operators like the FORTRAN .GR. (greater) and .LT. (less than) and logical operators (.AND.).

Each of these operators is assigned a place in a hierarchy which is used to resolve operator precedence in order to avoid ambiguities in expressions. The classical example is Fig. 10.3, where it is not clear whether 1 and 4 represent 2 and 3 or 5 and 6 in intent. In any standard higher language it always represents 3 and 6. This is because * is accorded a higher precedence (is higher in the hierarchy than plus).

In the expression of Fig. 10.1 we have only two operators, = and +. We shall impose the simple hierarchic rule that + always precedes =. Therefore, "plus" will have a hierarchy value of 1 and "equals" a hierarchy value of 0. We further define both operators to be binary operators; that is, they require two operands on the left before they can validly join a Polish string.

We shall also define two modes for our scan procedure, an operator

1. A = B * C + D	4. A = B + C * D
2. A = B * (C + D)	5. A = (B + C) * D
3. A = (B * C) + D	6. A = B + (C * D)
A B C * D +	A B C D * +

Fig. 10.3. Ambiguities.

expected mode and an operand expected mode. These are useful for the detection of bad syntax, for the recognition of function calls, and for the distinction between the unary operator (-), which is identical in form to the binary subtract ($A = -B$).

At the start in Fig. 10.4 we are in operand expected mode, and we have an initial pointer to the A. The routine determines that A is an operand and passes it to the output string. The pointer is advanced to =, and operator expected mode is set; = is encountered and as an operator it is placed in the stack. Since it is the first operator in the stack, it joins without any other action.

The actual process of joining the stack involves a test of the hierarchy value of the candidate operator against the hierarchy value of the operator at the "top" of the stack.

The significant characteristic of LIFO lists is that the "top" of the stack is always the latest entry. The "top" is traditionally indicated by a pointer to the memory location representing the last to join the list. When a new member is placed on the stack, it is "pushed down" in the sense that the top-of-stack pointer is moved to the position of the new entry, which conceptually sits on top of the old. When a member is removed from the list, the stack is "popped up"; the top-of-stack pointer is moved to the member just "beneath" the member being removed.

In Fig. 10.4 the stack empty sentinel $\perp$ is initially at the top, as indicated by the horizontal arrow. Its hierarchy value, -1, is represented here as a field of the word that represents $\perp$ in the stack. When = is encountered in the input string, it is associated with its hierarchy value of 0, and this value is tested against the hierarchy of $\perp$. The rule appropriate here is that a higher hierarchy does not "pop" a lower hierarchy. This is to say that an operator on the stack should not be forwarded to the output string of Polish by an operator having a higher hierarchy value. This is because by definition the higher hierarchy must precede the lower in the string.

Because = has a higher hierarchy than $\perp$, it simply joins the stack. It is not forwarded to the output string, because the necessity for having two operands on the left has not been met.

After = joins the stack, we expect an operand, and in this case we find B, which is passed immediately to the output stream. All operands encountered in the input stream are immediately forwarded to output in the formation of pure Polish postfix. Operator expected mode is set, and the pointer in the input stream is advanced to +.

Plus is associated with its hierarchy value and attempts to join the stack. It does so since its 1 hierarchy value is greater than the 0, = at the "top." Next (6), C is forwarded to the output string.

At this point we encounter another +. Since we now have a situation where the hierarchy value of an operator attempting to join the stack is equal to the hierarchy value of the member at the top, we require a new unstacking rule.

The simplest general rule is that any member on the stack with a hierarchy equal to or greater than that of a candidate must be removed from the stack and forwarded to the output stream before the candidate may become a member.

This is the perfectly rational way of enforcing precedence. The equality rule states basically that operations of the same hierarchy value will be performed in left-to-right order, the sequence in which they are written and encountered in the input expression. The "greater" rule imposes the resolution of ambiguity in sequences like those of Fig. 10.3. In accordance with this rule, the + at the top of the stack (the one which had occurred between B and C) is forwarded to the output stream before the candidate + (between C and D) can join.

After the forwarding the candidate is tested against the new top of stack (0, =) to see if that also must be forwarded. The rule for "popping"

Output	Stack	Expression	Mode
1.	⟶ ⊥-I	A = B + C + D + E + F + G;	OPND
2. A	⟶ ⊥-I	= B + C + D + E + F + G;	OPRTR
3. A	⟶ = o ⊥-I	B + C + D + E + F + G;	OPRND
4. AB	⟶ = o ⊥-I	+ C + D + E + F + G;	OPRTR
5. AB	⟶ + 1 = o ⊥-I	C + D + E + F + G;	OPRND
6. ABC	⟶ + 1 = o ⊥-I	+ D + E + F + G;	OPRTR
7. ABC +	⟶ + 1 = o ⊥-I	D + E + F + G;	OPRND
8. ABC + D	⟶ + 1 = o ⊥-I	+ E + F + G;	OPRTR
9. ABC + D +	⟶ + 1 = o ⊥-I	E + F + G;	OPRND
10. ABC + D + E	⟶ + 1 = o ⊥-I	+ F + G;	OPRTR
11. ABC + D + E +	⟶ + 1 = o ⊥-I	F + G;	OPRND
12. ABC + D + E + F	⟶ + 1 = o ⊥-I	+ G;	OPRTR
13. ABC + D + E + F +	⟶ + 1 = o ⊥-I	G;	OPRND
14. ABC + D + E + F + G	⟶ + 1 = o ⊥-I	;	OPRTR
15. ABC + D + E + F + G + =	⟶ ⊥-I		

Fig. 10.4. Scan of A = B + C + D + E + F + G.

is that all of the members of the stack with higher or equal priority must be forwarded. The = is not forwarded in this case, and the new + joins the stack, giving us 7.

The rest of the expression is decomposed as shown in 8 through 14. After G has been placed on the string, a new operator is expected, but none is encountered or a statement termination punctuation occurs (; in ALGOL and PL/I). Whatever statement end convention is used in the language, its recognition causes all remaining members of the stack to be forwarded to the output string, forming the complete Polish postfix representation of the expression.

The usefulness of this form as compared to infix form may be seen when we wish either to execute the statement interpretively or to compile code. We refer again to Fig. 10.3. To compile machine coding for statement (4) without a transformation into Polish or some similar operation would result in wrong code being generated by a left-to-right scan. The generation would accomplish the intent of (5) and not the conventional intent (6).

Given a Polish representation, the generation procedure consists of passing the Polish string, this time collecting operands and generating when operators are encountered. Figure 10.5 shows the generation of symbolic code for a three-address machine from the Polish developed in Fig. 10.4. Notice that each operand is stacked in a LIFO list; when an operator is encountered, the top two members of the stack represent its operands. These are removed from the stack, and a line of symbolic code is generated. The result of the generated line is placed at the top of the stack as an operand for the next operator. This is the mechanism that supports the concept of Fig. 10.1.

We have not shown a number of associated compiler activities in our simplified description. Variable names are not one character long, and a mechanism is needed to isolate and identify the strings of characters that comprise a variable name. This is traditionally done by collecting the characters between operators, often with the use of a translate and test instruction that provides the address of the first encountered nonalphabetic in a string.

In the so-called table-driven compilers characters are often inspected one by one, and a set of action rules is associated with defined conditions for the string sequences letter letter, letter operator, letter space, etc.

Further, we have not described any activities related to the manipulation of the symbol table or dictionary. This table is kept by compilers so as to associate addresses and attributes with their names. It enables the compiler to determine the correct usage of a variable and note reusage of a variable name so as to use the proper address for it.

Finally, we have not described optimization. Optimization occurs at two levels in a compiler; let us call them conceptual and machine-specific levels. We are interested in both. However, we shall directly concern ourselves only with those optimization techniques specifically directed toward the scheduling of parallel functional units on a machine of the type we have been discussing. All optimization is scheduling, and the techniques used in highly sophisticated optimizers are very like the techniques used in the general scheduling situation.

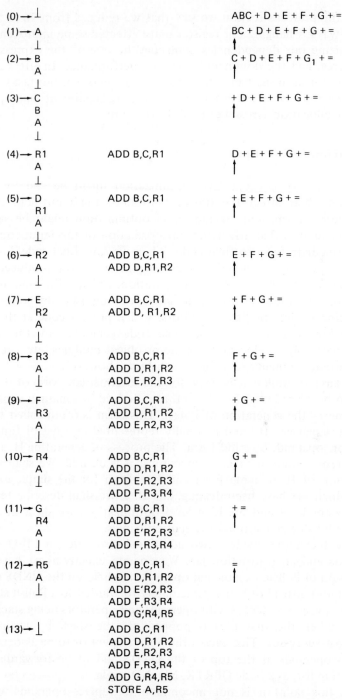

Fig. 10.5. Code from Polish.

In the previous section we saw that we reduced from 96 to 58 cycles in two ways. Changing our register usage effected some improvement, but reorganizing our dependencies, replacing the tree of the expression by a better tree, had the dominant impact on performance. In effect a demonstration that an insight into the relationships between our tasks and their potential ordering and reordering can be more significant than optimizing resource allocation across a given task structure.

10.3. GENERATING TREES

In a rough way concept level optimization might be said to be those techniques that affect the transformation of the input source into an intermediate form, and machine level optimization might be said to be those techniques that affect the transformation of the input source from an intermediate form to final code. When we say "final code," of course, we no longer mean directly executable in the form produced by the compiler. The final stage of code preparation, which is the binding of code to physical locations in store, is no longer accomplished by compilers, but by linking loaders or linkage editors, which process code at the time of their submission for execution. The code generated by the compiler, called, variously, "relocatable" code or object modules, etc., does represent the code sequence schedule in substance, however.

In large compilers it is usual for the intermediate form of the expressions to be represented in a form that is derived in a manner substantially the same as the generation of Polish. This form is often called the intermediate language (IL) and is usually represented as a list of functions in operator, operand, operand form. The process of generating IL allows for a tree representation of an expression if it is desired. We shall show the generation of IL in table form and tree form for the simple expression with which we have been dealing, and then we shall describe techniques for concept level and machine level optimization. We show in Fig. 10.6 the step-by-step generation of execution triples.

The technique involves two additional structures, a LIFO stack for operands and an instruction list. We operate exactly as we have for the generation of Polish, except that operands are held on the ONDS (operand delay stack) instead of being immediately forwarded to a Polish stream.

The process proceeds with operands and operators being stacked until line 7, when the first plus is popped off the stack by its immediate rightmost successor. This causes the binary operator to be associated with its two operands at the top of the ONDS and to be forwarded to the instruction list, as a basic OPRTR, ND, ND triple.

At this point it is not uncommon to replace operands with their symbol table address and the operator with the address of the routine that

"INST." list	ONDS	OHS	
(1)	→ ⊥	→ ⊥	A = B + C + D + E + F + G
(2)	→ A ⊥	→ ⊥	= B + C + D + E + F + G ↑
(3)	→ A ⊥	→ = ⊥	B + C + D + E + F + G ↑
(4)	→ B A ⊥	→ = ⊥	+ C + D + E + F + G ↑
(5)	→ B A ⊥	→ + = ⊥	C + D + E + F + G ↑
(6)	→ C B A ⊥	→ + = ⊥	+ D + E + F + G ↑
(7) 1. + BC	→ 1 A ⊥	→ + = ⊥	D + E + F + G ↑
(8) 1. + BC	→ D 1 A ⊥	→ + = ⊥	+ E + F + G ↑
(9) 1. + BC 2. + 1D	→ 2 A ⊥	→ + = ⊥	E + F + G ↑
(10) 1. + BC 2. + 1D	→ E 2 A ⊥	→ + = ⊥	+ F + G ↑
(11) 1. + BC 2. + 1D 3. + 2E	→ 3 A ⊥	→ + = ⊥	F + G ↑
(12) 1. + BC 2. + 1D 3. + 2E	→ F 3 A ⊥	→ + = ⊥	+ G ↑
(13) 1. + BC 2. + 1D 3. + 2E 4. + 3F	→ 4 A ⊥	→ + = ⊥	G ↑
(14) 1. + BC 2. + 1D 3. + 2E 4. + 3F	→ G 4 A ⊥	→ + = ⊥	
(15) 1. + BC 2. + 1D 3. + 2E 4. + 3F 5. + 4G	→ 5 A ⊥	→ = ⊥	
(16) 1. + BC 2. + 1D 3. + 2E 4. + 3F 5. + 4G 6. = 5A			

Fig. 10.6. Direct to code without Polish.

generates code for that operation. This might be done before or after, however, and is not critical here. The top of the operand stack is filled with the line number representing the sum of B and C, indicating that this sum is an operand (input) to the next operator.

We have "tabled" an instruction in the intermediate language; we may also think of having generated a node on a tree as in Fig. 10.7. The node 1 at lower left represents the + operator with its inputs B and C. Some compilers represent IL in tree form with pointers running from node to node.

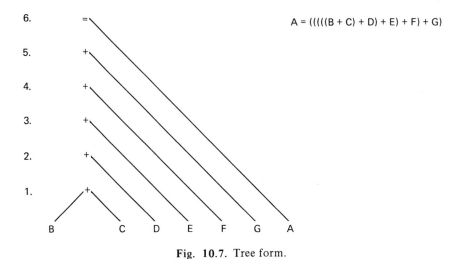

Fig. 10.7. Tree form.

As in our earlier trees, the B, C imply the operations fetch B, fetch C. Each node in the tree is numbered to correspond with a line of code in Fig. 10.6. For both the table and tree representation of the expression we have six levels, of which the last five are dependent upon predecessors for their execution. In this example, as in our earlier treatment of this expression, we are generating a highly logical dependent tree that treats the expression as if it were parenthesized as shown in Fig. 10.6.

But surely we see that the tree of Fig. 10.8 is equivalent algebraically to the tree of Fig. 10.7, and we see that if such a tree could be generated or such a table of IL formed, it would be possible to take considerably greater advantage of potential parallelism in our machine, because we have reduced logical dependency. In Figs. 10.6 and 10.8 logical dependency in the IL table is represented when an operand on the table is the numbered line of a previous operation. In Fig. 10.6 all lines beyond 1 contain line number references for operands and are consequently bound to the out-

A = ((B + C) + (D + E)) + (F + G)

4.

3.

2.

1.

B   C   D   E   F   G   A

1 + BC
2 + DE
3 + 12
4 + FG
5 + 34
6 + A5

1.	→ B A ⊥	→ = 0 ⊥ − 1	+ C) + (D + E)) + (F + G) ↑	COUNT = 2
2.	→ B A ⊥	→ + 3 = 0 ⊥ − 1	C) + (D + E)) + (F + G) ↑	
3.	→ C B A ⊥	→ + 3 = 0 ⊥	) + (D + E)) + (F + G) ↑	COUNT = 1
4.	→ 1 A ⊥	→ + 2 = 0 ⊥	(D + E)) + (F + G) ↑	1 + BC
5.	→ D 1 A ⊥	→ + 2 = ⊥	+ E)) + (F + G) ↑ COUNT = 2	1 + BC
6.	→ D 1 A ⊥	→ + 3 + 2 = ⊥	E)) + (F + G) ↑	1 + BC

Fig. 10.8. An alternative of interest.

put of predecessors; in Fig. 10.8, however, lines (1,2,4) and (3,5) can be executed in parallel.

Whether or not this tree is a "better" tree than the earlier one in actual performance depends on a number of factors. Does the system have the capability of performing in parallel? Can it realize on its resources the functional independence of the nodes? The system that we have been describing could not execute this tree as such, since it has only one add unit; however, a multiple adder system, or a system with multipurpose asynchronous arithmetic units could perform more closely to the form of the tree. Further, our system could take advantage, as we have seen, of the possibility of overlapping fetches. Finally, the register contention would, of course, limit parallel capability.

The question before us is how to develop optimum, or at least improved, trees for systems with parallel capability that rely on the compiler for correct scheduling of instructions. Since the generation of a tree depends largely upon the relationships between operators (the precedence or hierarchic values of operators relative to each other), it is in this area that one could expect a first examination of techniques to take place. If we look at Fig. 10.7 again, we might determine how it is that this particular tree was formed. The answer lies in the explicit parenthesization of the statement.

Parentheses are used in arithmetic statements to resolve ambiguity when the writer of the expression wishes a result different from what he would obtain by relying upon the compiler. This may be done for reasons of machine efficiency, as in Fig. 10.8. Compilers have two basic ways of providing for the handling of parentheses in an expression. One way, more popular, perhaps, is to count the number of left parentheses ("open") and add this count to the hierarchy value of operators as they are encountered. Whenever a right parenthesis is encountered, a subtraction is made from the open parenthesis count.

In Fig. 10.8, line 1, the hierarchy value of the first encountered plus is 3, its normal hierarchy plus the two open parentheses. This is placed on the OHS at line 2. C is then sent to ONDS, and the parenthesis count is reduced by one because of the right parenthesis after C. When the next plus is encountered, the parenthesis count is at 1 and the hierarchy of the operator is 2; therefore at line 4 the +BC is generated and the + is placed on OHS with a hierarchy of 2. In line 5, the next plus has the count again equal to 2. Therefore, it develops a hierarchy of 3. Since the + at the top of OHS has a value of 2, it cannot be popped. Consequently, the generation of +D1 is suppressed, and the candidate plus joins the stack at 6. E joins ONDS and the succeeding right parentheses reduce the open parenthesis count to 0. The plus preceding (F + G), therefore, has a hierarchy value of 1. As such, it will remove both +'s from OHS, forming the independent +DE and the dependent +12. Similarly, +FG will be formed before +34, because the open parenthesis count will increase before the last plus is encountered, giving it a hierarchy of 2 and consequent residence on OHS while G is being collected on ONDS.

The parentheses serve, as we have seen, to impose precedence and ordering on an expression. An alternative way of handling parentheses is to place left parentheses on the stack and to impose two rules: (1) A right parenthesis unstacks all operators down to a left parenthesis and removes the left parenthesis. (2) An operator considers the base of the stack to be the first encountered left parenthesis; therefore, no unstacking may be done past an encountered parenthesis. This has the effect of partitioning the stack within the scope of parentheses. The technique will produce a result identical with what we have shown.

The first technique is interesting because it introduces the idea of a dynamic contextual local precedence for an operator. This is a microcosmic example of a concept of "dynamic running priorities," which will interest us a good deal later in this text. The second technique is to provide a clean method for handling subscripts and function calls of multiple parameters, where it is a convenience to partition the stack.

We might mention in passing that the symbol "(" is truly ambiguous in current usage sometimes representing an expression enclosure, other times a subscript enclosure, and finally the arguments of a function call. A = (A (I + 1) + SIN(X)) is an example. This might more precisely be written A = [A (I + 1) + SIN{X}]. The true meaning of "(" must be contextually inferred in a compiler.

10.4. LOW BROAD TREES

What we wish from the compiler is the generation of low broad (highly parallel trees) as if the proper parenthesization existed when indeed it does not exist. We wish dynamically to modify hierarchy values or to partition the stack on the basis of the patterns of operators that have preceded a given operator in an expression, and by so doing to generate as many lines of code as possible which are not dependent.

A simple first demonstration of this can be given with our same expression. Let us postulate that we wish dynamically to change the priorities of our pluses as if parentheses existed. We search for some way of determining what the hierarchy of a given operator should be. In our simple expression the only quality we observe is the sameness of hierarchy between all operators. We observe instances where we would like to delay the forwarding of an operator and have only sameness of hierarchy value to aid us. We would like to group terms in pairs as much as we can; our goal is to generate a tree of fewest possible levels and greatest breadth. We realize that within trees of the same apparent efficiency there will be differences in performance due to the physical characteristics of the operations as well as to the resource contention phenomenon.

Since we wish to group terms in pairs we search for a means to ensure that no operator may be popped from OHS unless two true terms of the expression are on the top of the stack. We notice from Fig. 10.9 that this can be partially achieved in our expression by delaying every other operator of the same hierarchy value. This can be done merely by providing a switch that has the value ON or OFF.

Whenever an operator removes an operator with the same hierarchy value from the stack, the switch is set to ON if it has been OFF and vice versa. This switch is then used to determine a dynamic relative hierarchy between the two operators of identical formal hierarchy. It is, in a sense, a mechanism for determining dynamic priorities between them.

Let us apply it to our simple expression. Everything proceeds as before, except that we indicate the setting of switch S, until line 5 when we show the switch changing status to ON. This indicates that an operator of equal priority has been taken from the stack. At (7) in Fig. 10.9 the + will find the switch ON and consequently will not pop the stack but will only join it. The switch will be set to OFF, to allow normal stack action for the next operator. At (9) with S = OFF the + will normally unstack. Figure 10.9,(9a), shows the status at the end of the first unstack.

At this point we have a choice. We can continue to unstack, in which case we achieve (9b) in Fig. 10.9. Here we have generated a "node" at a higher level of the tree—that is, the node represented by line 3 of (9b). We would also set our switch to ON. If we do this, we can see that we have generated the exact equivalent of Fig. 10.8. Our switch has, in effect, caused us to treat the expression as if parentheses existed as they do for the expression in Fig. 10.8.

Alternatively, we can limit the unstack to the topmost operator of equal hierarchy and see (9c) in Fig. 10.9. We are delaying the generation of the higher-level node. The final plus will find S ON and will not pop the stack at all, giving us (11) as a final form when EOS is encountered. The final code is equivalent to the alternate parenthesization shown in Fig. 10.9. Whether or not a significant (or any) performance difference occurs depends upon the specific character of the machine. However, the limit on the unstack would seem preferable on principle, since it delays for the longest time the generation of a higher-level node.

We have shown that with a simple switch governing unstacking rules we have managed to simulate the partitioning effects of parentheses even when they are not present and to generate a more parallel tree. We must now, however, go on to look at more complex arithmetic involving mixed operations of different hierarchic levels.

$$A = B + C + D + E + F + G$$
$$A = (B + C) + ((D + E) + (F + G))$$

(1) A ⟶ = o   S = OFF
     ⊥   ⊥-1

(2) B ⟶ = o   S = OFF
     A   ⊥-1
     ⊥

(3) B ⟶ + 1   S = OFF
     A   = o
     ⊥   ⊥-1

(4) C ⟶ + 1   S = OFF   + D + E + F . . .
     B   = o   ↑
     A   ⊥-1
     ⊥

(5) 1 ⟶ + 1   S = ON   1. + BC   D + E + F . . .
     A   = o   ↑
     ⊥   ⊥-1

(6) D ⟶ + 1   S = ON   1. + BC   + E + F + G . . .
     1   = o   ↑
     A   ⊥-1
     ⊥

(7) D ⟶ + 1   S = OFF   1. + BC   E + F + G
     1   + 1   ↑
     A   = o
     ⊥   ⊥-1

(8) E ⟶ + 1   S = OFF   1. + BC   + F + G
     D   + 1   ↑
     1   = o
     A   ⊥-1
     ⊥

(9a) 2 ⟶ + 1   S = OFF ON ENTRY   1. BC +   F + G
     1   + 1   S = ON   2. DE +   ↑
     A   = o
     ⊥   ⊥-1

(9b) 3 ⟶ + 1   S = ON   1. BC +   A = ((B + C) + (D + E)) + (F + G)
     A   = o   2. DE +
     ⊥   ⊥-1   3. 1 2 +

(9c) 2 ⟶ + 1   1. BC +   T only
     1   + 1   S = ON   2. DE +
     A   = o
     ⊥   ⊥-1

(10) F ⟶ + 1   S = ON   1. BC +   T only
     2   = o   2. DE +
     1   ⊥-1
     A
     ⊥

(11) G ⟶ + 1   S = OFF   1. BC +   T only
     F ⟶ + 1   2. DE +
     2   + 1
     1   = o
     A   ⊥-1
     ⊥

(12a) 3 ⟶ + 1   1. BC +   T only off
      2   + 1   2. DE +
      1   = o   3. FG +
      A   ⊥-1   4. 3 2 +
      ⊥

(12b) 4 ⟶ + 1
      1   = o
      A   ⊥-1
      ⊥

(12c) 5   BC +
      A   DE +
      ⊥   FG +
                                   3 2 +
                                   1 4 +

Fig. 10.9. Pseudo parentheses.

Let us look at an expression that contains a succession of operators of various hierarchies, as in Fig. 10.10. The normal Polish and tree for this expression, as well as its effective parenthesization, are as shown. Since we have not looked at the processing of an expression of this complexity in detail, let us do so briefly. We need observe only that at (2) multiply joins the stack, following the normal higher-priority-does-not-unstack rule. It is popped by its rightmost successor multiply, following the conventional equals unstack rule at (3). At (8) and (9) the + unstacks both the divide and the previous plus.

We see by inspection that we would prefer something else from this expression than what we have accomplished. We would, in fact, like the expression in Fig. 10.11. We can begin looking at our scan at the point where we have encountered the second multiply. To here nothing unconventional has occurred. At this point we shall apply our switch rule and, after popping the multiply at the top of OHS, set S = ON. At (2), therefore, the divide will not unstack the multiply and not cause the multiply of AB by C. Instead / joins OHS, setting S to OFF and allowing D to join ONDS as at (3). The plus before F pops divide and multiply as operators of greater precedence to form (4). Notice that we shall now continue, however, since S is OFF, to pop the resident plus and form (5), combining E into the term before the availability of F. Our simple equal hierarchy switch rule has, therefore, achieved some, but not all, of the parallel effect we desired.

We need a further rule. This will be a generalization of the delay we used in Fig. 10.9 to inhibit the unstacking of multiple equal hierarchy operators when S permitted unstacking. We had previously mentioned that we wished "true" terms on ONDS before unstacking an operator. In both instances of our unstack or not unstack decision (Figs. 10.10 and 10.9) we had at the top of the stack an operand that was the result of an operation. We wish to suppress for as long as possible the generation of an instruction which would involve this operand and which would consequently be a second-level node. The rule we applied for the alternative of Fig. 10.9 was basically that if the operator attempting to join the stack has forwarded one operator of equal priority, then it would forward no others before joining the stack. We extend this to include equal or greater priority.

This applies to Fig. 10.11. Let (6) be the successor to (4), following the rule. We shall then get (7). Unhappily, we find that we will generate +3F when we clear the stack and we do not wish this to occur. We need some way of bypassing the 3 on the stack and associating E with F. A straightforward way of doing this is to record on ONDS the source of the operand with the entry, using a code R for result operands and T for

$Z = E + A * B * C/D + F$
$E + (((A * B) * C)/D) + F$

1. $* AB$
2. $* 1C$
3. $/ 2D$
4. $+ 3E$
5. $+ 4F$
6. $= 5Z$

1.	→ Z	→ = 0		
	⊥	⊥ − 1		
2.	→ A	→ * 2	B * C/D + F	
	E	+ 1	↑	
	Z	= 0		
	⊥	⊥ − 1		
3.	→ B	→ * 2	C/D + F	
	A	+ 1	↑	
	E	= 0		
	Z	⊥ − 1		
	⊥			
4.	→ 1	→ * 2	C/D + F	1. $* AB$
	E	+ 1	↑	
	Z	= 0		
	⊥	⊥ − 1		
5.	→ C	→ * 2	/D + F	
	1	+ 1	↑	
	E	= 0		
	Z	⊥ − 1		
	⊥			
6.	→ 2	→ / 2	D + F	1. $* AB$
	E	+ 1	↑	2. $* 1C$
	Z	= 0		
	⊥	⊥ − 1		
7.	→ D	→ / 2	+ F	
	2	+ 1	↑	
	E	= 0		
	Z	⊥ − 1		
	⊥			
8.	3	→ + 1	+ F	1. $* AB$
	E	= 0	↑	2. $* 1C$
	Z	⊥ − 1		3. $/ 2B$
	⊥			
9.	→ 4	→ + 1	F	1. $* AB$
	Z	= 0	↑	2. $* 1C$
	⊥	⊥ − 1		3. $/ 2D$
				4. $+ 3E$

Continue to completion of scan

Fig. 10.10. Multiple operators.

$$Z = (A * B) * (C/D) + (E + F)$$
$$Z = E + A * B * C/D + F$$

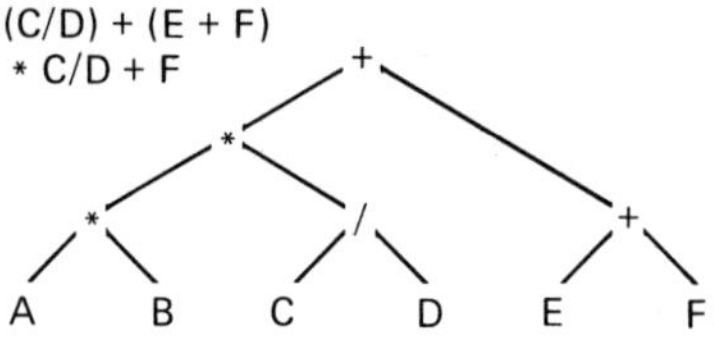

Tree diagram (leaves A, B, C, D, E, F):

```
                    +
              *            +
           *     /       E   F
          A B   C D
```

Operator list:
1. * AB
2. / CD
3. * 12
4. + EF
5. + 34
6. = 5Z

1.	→ B A E Z ⊥	→ * 2 + 1 = 0 ⊥ −1	* C/D + F ↑	S = OFF
2.	→ C 1 E Z ⊥	→ * 2 + 1 = 0 ⊥ −1	/ D + F ↑	S = ON	1. * AB
3.	→ D C 1 E Z ⊥	→ / 2 * 2 + 1 = 0 ⊥ −1	+ F ↑	S = OFF	1. * AB
4.	→ 3 E Z ⊥	→ + 1 = 0 ⊥ −1	+ F ↑	S = OFF	1. * AB 2. / CD 3. * 12
5.	F 4 Z ⊥	→ + 1 = 0 ⊥ −1		S = ON	1. * AB 2. / CD 3. * 12 4. + 3E 5. + 4F 6. = 5Z

6.	→ 3 E Z ⊥	→ + 1 + 1 = 0 ⊥ −1			
7.	F 3 E Z ⊥	→ + 1 + 1 = 0 ⊥ −1		T only	But must suppress + BF
8.	4 3 Z ⊥	→ + 1 = 0 ⊥ −1			1. * AB 2. / CD 3. * 12 4. + EF 5. + 43 6. = Z5

True only when delay = priority operators.

Fig. 10.11. Another expression.

natural or source operands. We might then enter the OPND list to collect the first two T operands for the first stacked +, forming (8), which then gives us what we desire.

We must now assure ourselves, however, that we reject R operands only at the right time. We know that we wish to do this only when we have used our rule to delay equal priority operators. When this rule is invoked, we can set T-only mode. When the operator that caused the invocation of the rule and the setting of T-mode is forwarded, then we set T-only mode to OFF, allowing the normal selection of operands in a LIFO fashion. Refer again to Fig. 10.9. The T-only mode is set from (9c), the point at which our unstack-one-only rule was invoked, until (12), when GF and the + causing the mode setting came off the stack.

We traditionally hold that multiply and divide have the same hierarchy, as do subtract and add. A stronger form of the alternate equal operator rule is always to give divide a higher priority than multiply and to attempt where possible to turn the binary subtract into a unary negative. Similarly, reciprocal multiples are sometimes used for divides. Although this has no parallel impact, it can speed operation by replacing a characteristically slower instruction with a faster one.

Let us look at a few statements and see what useful rearrangements of trees come out of the basic rules we have given. These rules are summarized in Fig. 10.12. The example expressions come from an article by T. V. Milne.

The simple expression of Fig. 10.13 shows the application of rules 1, 3, and 4. (1) of this figure shows the highly dependent tree of normal parsing. Notice at (2) the divide higher rule prohibits the passing of multi-

1. Divide always has a higher priority than multiply.

2. Never allow an operator to unstack more than one operator of the same priority.

3. Never allow an operator to unstack an operator of the same priority if he has previously unstacked operators of higher priority.

4. When an operator is on the stack because of 2, 3, set a mode so that result representations of the stack are considered unavailable for use.

5. Never allow an operator to unstack an operator of the same priority if this had occurred on a previous unstacking.

6. Parentheses may be overridden in a limited way. Terms collected in parentheses may not be combined with terms outside before the parenthesized expression is formed, however the group in parentheses may be moved relative to other operations.

Fig. 10.12. Stacking rules.

Rules 1, 3, 4

$$Z = A * B / C * D$$
$$((A * B)/ C) * D$$

1.

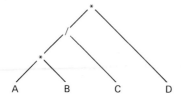

1. * AB
2. / 1C
3. * 2D

2.   → B   → /   C * D   Rule 1: DIVIDE > MPY HOLD MPY ON STACK
      A   *   ↑
      Z   =
      ⊥   ⊥

3.   → C   → /   * D
      B   *   ↑
      A   =
      Z   ⊥
      ⊥

4.   → 1   *   D   1. / BC   S = OFF, T = ON
      A   *   ↑   Rule 3: HOLDS OFF A * BC. * DOESN'T
      Z   =   POP * BECAUSE POPPED /. INHIBIT CAUSES T (TRUE) ON
      ⊥   ⊥   S OFF BECAUSE = HIER NOT OCCUR

5.   aD ← ⌊*⌋   1. / BC   Rule 4: BECAUSE T = ON (OPERATOR IN STACK
      1   *   2. * DA   BECAUSE OF RULE 2) 1 IS REJECTED
      A   =   T = OFF
      Z   ⊥
      ⊥

      b2   *   1. / BC
      1   =   2. * DA
      Z   ⊥   3. * 12
      ⊥   4. = 3Z

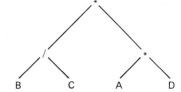

Fig. 10.13. An effect of new stacking rules.

ply against A and B and allowing the residence of C at the top of ONDS. The arrival [at (4)] of the second multiply causes the BC division. By rule 3 we do not unstack the OHS resident multiply and consequently do not form the product A*B/C. The inhibition of the unstacking causes the T (result exclusion) mode to be set so that at (5) when a multiply is finally unstacked it rejects result 1 as an operand, forming instead the product *DA. As this is done, T is reset, so that the next multiply may accept results 1 and 2 as operands.

The expression of Fig. 10.14 shows the application of rules 3 and 4 only. Notice that after divide is formed T is set ON, the unstacking of the resident plus is prohibited, and consequently result 1 is rejected to form the sum A + D.

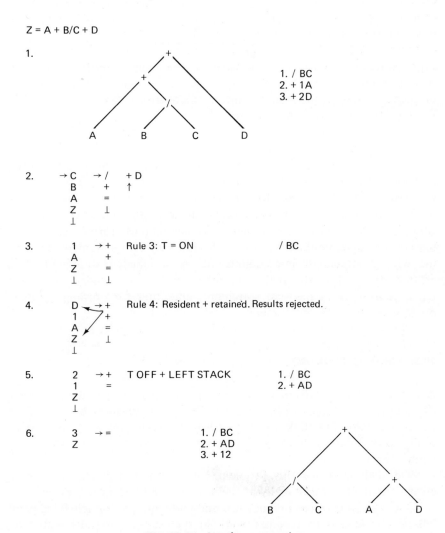

Fig. 10.14. Another regrouping.

Figure 10.15 shows an extension of rules 3 and 4 to effect (6). The interesting action occurs at (4). Here the right parenthesis unstacks the add of D to B * C. Normally, it would unstack the left parenthesis to expose the minus at the top of the stack. When the final minus came to the stack, the resident minus would be popped, with no reason for T to be on; the logically dependent difference of A and the expression in parenthesis would be formed, excluding the desired independent (A - E). If we apply our rules, however, the right parenthesis is inhibited from unstacking the left parenthesis. This sets T to ON and protects the resident minus. Consequently, the A - E subtraction takes place, moving as if it were the parenthesized expression relative to the other terms.

What we have basically shown are some techniques for representing expressions in such a way as to maximize their logical independence. We shall go on in a later section to show what can be done at the next level to prepare a final schedule from the intermediate tree or table.

One final remark is needed here. We have shown all our work at the expression level, optimizing for potential parallelism locally. It is more and more common for compilers to attempt regional or global optimizations in which rearrangements, compressions, etc. are undertaken on segments of code that are greater than an expression. As we saw from our very first example of coding for a machine of this class in this chapter, it is useful to be able to work with a larger code grouping. One frequently used concept is the concept of a region. In its simplest form a region is a sequence of statements beginning with an assignment statement and ending with a statement immediately before a labelled statement. It is assumed that within this region no entry can be made other than sequential execution from the first statement, and, therefore, arrangements of operations can be liberally made.

10.5. WHAT IS OPTIMUM?

A problem that occurs repeatedly in various guises in parallel systems is the question of "local optimization." Consideration must be given as to whether a particular task to be performed in a complex of tasks necessarily contributes most to system performance when it itself is in "optimum" form.

We shall later see this problem considered in a multiprogramming environment. It occurs equally at the level of compiler optimization.

The principle of redundancy is a known technique for achieving parallelism. The idea is to organize work in such a way that although more

Z = A − (B * C + D) − E

1.

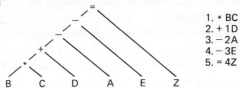

1. * BC
2. + 1D
3. − 2A
4. − 3E
5. = 4Z

2. → C   → *   + D) − E
       B   (↑
       A   −
       Z   =
       ⊥   ⊥

3. → 1   → +   D) − E   1. * BC
       A   (↑
       Z   −
       ⊥   ⊥

4. → D   → +   ) − E
       1   (↑
       A   −
       Z   =
       ⊥   ⊥

5. → 2   (− E   T = ON   1. * BC   Rule 3: Don't uncover
       A   −   ↑   2. + 1D   (because) popped +
       Z   =

6. → E   → −   T = ON   Rule 4: 2 not eligible for −
       2   (
       A   −
       Z   =
       ⊥   ⊥

7.   3   → −   1. * BC
       2   =   2. + 1D
       Z   ⊥   3. − AE
       ⊥

8.   4   → =   1. * BC
       Z   ⊥   2. + 1D
       ⊥   3. − AE
                                     4. − 32

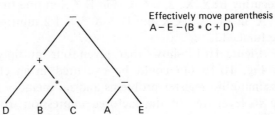

Effectively move parenthesis
A − E − (B * C + D)

Fig. 10.15. "Bypassing" parenthesis.

work is done to accomplish a goal, enough of it is done in parallel that elapsed time is reduced.

Any asynchronous element performs most effectively when there is a minimum dependency on other elements. Consider that the elimination of common subexpressions is an undesirable undertaking for a parallel machine of any sophistication. The elimination of multiple independent points of computation of a value causes there to be a general dependency of all successor activities on the completion of the single occurrence of the computation. The work saved by eliminating excessive computation may cause the sequentialization of activities that might otherwise be undertaken in parallel.

All "parsing" achieves flexibility of the type we have described by reliance upon the algebraic equivalence of forms of arithmetic statements. They assume a total rearrangeability within algebraic laws. Numerical analysts and others concerned with precision point out that in many cases algebraic equivalents are not numeric equivalents because of rounding and normalization in multiply and divide. The "cavalier" rearrangement of expressions written with problems of precision in mind does not much contribute to successful results. A respect for parenthesization existing in an expression is a minimal courtesy claimed by this group. Ambitious optimizing compilers may not respect these parentheses.

The techniques further depend upon the expression's occurring in a directly manipulable form. Consider Fig. 10.16. The example comes from Hellerman's article in the bibliography. The initial expression of line (1) transforms into the final factored expression of line (2). Notice the trees of the two equivalent expressions. The tree of the initial expression (1) allows for a high degree of parallelism. It is usually avoided in computation, however, because exponentiation is characteristically a subroutine with attendant linkage burden and because, as shown in its alternate, the number of multiplies a characteristically expensive instruction, is twice the number in tree (2). However, given a sufficiently large number of multiply units, we could execute (1) in two multiply and two add times, executing $B*X, X\uparrow 2, C*X$, and $D*X$ at one time, absorbing the add $A + B*C$ in the $X*CX$ and $D*X*X\uparrow 2$ multiplies and then completing the final adds.

Figure 10.17 shows that, given four multiply units, the four-level tree of Fig. 10.16 (1) could be executed in an elapsed time of 30 cycles (assuming no register problems and no memory delays). The corresponding six-level tree of the factored expression would require 45 cycles to execute. Notice further that the execution on a 2 MPY unit system would require 48 cycles, only three more than the factored tree. This is accomplished even without taking advantage of the availability of $X\uparrow 2$.

1. $Z = A + B * X + C * X \uparrow 2 + D * X \uparrow 3$

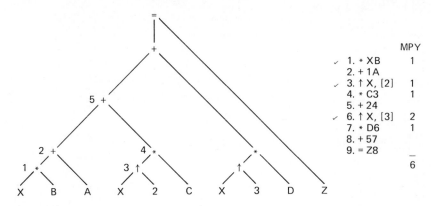

	MPY
✓ 1. * XB	1
2. + 1A	
✓ 3. ↑ X, [2]	1
4. * C3	1
5. + 24	
✓ 6. ↑ X, [3]	2
7. * D6	1
8. + 57	
9. = Z8	—
	6

2. $Z = A + X * (B + X * (C + X * D))$

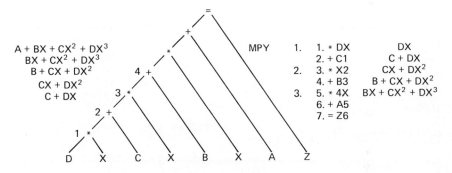

$A + BX + CX^2 + DX^3$
$BX + CX^2 + DX^3$
$B + CX + DX^2$
$CX + DX^2$
$C + DX$

		MPY
1.	1. * DX	DX
	2. + C1	C + DX
2.	3. * X2	$CX + DX^2$
	4. + B3	$B + CX + DX^2$
3.	5. * 4X	$BX + CX^2 + DX^3$
	6. + A5	
	7. = Z6	

Fig. 10.16. What is optimum?

The recognition would have allowed D * X to occur at level 2, D * X * X ↑ 2 at level 3, and the final add to occur at level 4, reducing elapsed time to 37 cycles. Here the elimination of a multiply helps us because we can eliminate a level.

Notice that the system has accomplished 78 cycles of work in 30 cycles in Fig. 10.17. This is, at this point, not surprising for us, but it introduces two points which we see here for the first time and which reoccur when we talk about parallelism at higher levels for tasks in an operating system sense in a multiprocessor environment. The fundamental point is that we shall often find that total system time spent on a task may well increase as we decrease elapsed time. The second point, very much related to the first, is that it is often legitimate to force this increase

(1) Unfactored: 4 MPY units, 11 CYCLES
                  1 ADD unit,   4 CYCLES

Elapsed Time			Processor Time
L1	11	11*4	44
L2	11	11*2 + 4	26
L3	4	4*1	4
L4	4	4*1	4
	30		78

(2) Factored

Elapsed time = Processor time	
L1	11
L2	4
L3	11
L4	4
L5	11
L6	4
	45

(3) 2 MPY units unfactored, no common X ↑ 2

Elapsed Time			Processor Time
L1	11	2*11	22
L2	11	2*11 + 4	26
L3	11	11 + 4	15
L4	11	11	11
L5	4	4	4
	48		78

(4) 2 MPY, common X ↑ 2

Elapsed			Processor
L1	11	2*11	22
L2	11	2*11 + 4	26
L3	11	11 + 4	15
L4	4		4
	37		67

Fig. 10.17. Factored vs. unfactored.

in order to achieve elapsed time benefit. The means of forcing this is often to allow redundancy where that redundancy will reduce logical dependency and the need for synchronization. The multiple derivation of X ↑ 2 is such a redundancy.

Whether it is better to factor or not on any given parallel system is a point depending, like all scheduling decisions, on the task and the resource pool. The point that must be stressed here is that no current compiler will undertake to expand (2) of Fig. 10.16 into (1) of Fig. 10.16 to investigate opportunities for parallel execution, no more than any optimizing compiler would have undertaken to factor (1) into (2) for the benefit of a sequential machine. Such string manipulations are not impossible, and an increased interest in true algebraic manipulation by compilers is becoming current in the field, but at this stage compilers are very much limited in such exercises and do not go beyond the application of the basic commutative arithmetic laws in expression analysis. The burden of efficient expression is, therefore, even with sophisticated compilers for highly sophisticated machines, not removed from the programmer.

10.6. BASIC CODE GENERATION

The conceptual rearrangement of individual subtasks has left us with a list of operations to be performed which have, because of a minimization of dependency, a maximum flexibility of ordering with regard to scheduling of specific functions.

Let us take the IL form of one of the expressions we have been considering and expand it so that it more closely approximates a true internal form. We shall in Fig. 10.18 show an expanded form. We are merely expanding our triples to show that the fetches of operands from storage to registers is a schedulable independent operation. The expanded

1.	* BC	1. FETCH B	1
2.	+ ID	2. FETCH C	1
3.	− AE	3. MPY 1,2	1
4.	− 32	4. FETCH D	2
5.	= 4Z	5. ADD 3,4	2
		6. FETCH A	3
		7. FETCH E	3
		8. SUB 6,7	3
		9. SUB 5,8	4
		10. STRE 9,Z	5

	Cycles	Register in Use
1. FETCH B	9	1
2. FETCH C	9	2
3. MPY 1,2	11	1
4. FETCH D	9	2
5. ADD 3,4	5	1
6. FETCH A	9	2
7. FETCH E	9	3
8. SUB 6,7	5	2
9. SUB 5,8	5	1
10. STR 9,Z	9	0
	80	

Fig. 10.18. Expansion of macro using pseudocode.

representation shows the dependency of arithmetic on load/store operations already represented in the tree.

The expanded form of Fig. 10.18 still does not show register usage. Let us take a first look at usage by counting the number of registers that are used at the end of each operation. Each fetch is assumed to require one register; the multiply on line (3) deposits a result in a single register, reducing usage to one, and similarly down the line. We see that we never run out of registers, and in this case we eliminate the need for generating temporary storage locations and stores and fetches for the storage and retrieval of intermediate results. Any implementation of a compiler would, of course, find it necessary to provide for this possibility, and further to provide a means for reordering fetches and arithmetic so as to minimize the use of intermediate storage and references to it. We may consider our final list of Fig. 10.18 to be the result of such optimization if we choose.

We should now like to assign registers to our code to produce a first feasible schedule. We initially generate Fig. 10.19 using simple rules. Since the machine uses R1-R5 for fetches and R6 and R7 for stores (reverting to a 6600 convention), we shall assign all fetches cyclically R1-R5 and all arithmetic results to R6 and R7. If result registers are not available, we shall use the highest-numbered fetch register. If a fetch register is not available, we shall move an operand to a store register. If an instruction refers to a result in a result register as an operand, the result is returned to that register. (This is a simplification discussed further below.)

We may do this freely, since we are assured that there is a total sufficient number of registers by our previous register usage counts. We

	1	2	3	4	5	6	7
1 FETCH R1,B	SN	–	–	–	–	–	–
2 FETCH R2,C	SN	SN	–	–	–	–	–
3 MPY R1,R2,R6	S/S	S/S				SN	–
4 FETCH R3,D	–	–	SN				
5 ADD R3,R6,R6	–	–	S/S			SN	–
6 FETCH R4,A	–	–	–	SN		SN	–
7 FETCH R5, E	–	–	–	SN	SN	SN	–
8 SUB R4, R5, R7	–	–	–	S/S	S/S	SN	SN
9 SUB R6, R7, R7	–	–	–	–	–	S/S	SN
10 STRE R7, Z						–	S/S

SN = SINK;   S/S = SINK/SOURCE

Fig. 10.19. Register status.

have passed over register optimization very lightly here, because we are interested in it only to the extent that we wish to suppress resource conflicts. It is a highly subtle art, however, with a vast literature of its own. The simple scheme shown here is only to allocate across the broadest base of registers in order to minimize those times when sink/sink or sink/source delays occur because of assignment and not logical necessity.

Line 1 of Fig. 10.19 assigns R1 as a sink register for the fetch of B. On a register availability table shown to the right of the instruction list, R1 is marked off as being in sink mode. A test to determine that an assignment for a fetch register was necessary indicated this was to be a sink assignment, and the next available register (R1) was assigned to FETCH R1, B.

On line 2 of Fig. 10.19 we assign R2 for the same reasons, now putting R1 and R2 in sink status. The MPY instruction requires a sink register for its result. We assign the first available result register (R6) and put it into sink mode. We must now make an association between the source registers of MPY and previous sink registers. In Fig. 10.18 the operand fields of MPY refer to the line numbers on which it is dependent, 1 and 2, the two FETCH's of B and C. We pick up the register assignment made for sinks of the FETCH's and place them as source registers in the register fields of MPY. The status of these registers now changes, since a reference to their contents has been made. In a true implementation this would be slightly more complex, involving perhaps a counter associated with the lines using a register as a sink; each time a subsequent instruction referred to this line, the reference counter would be incremented. At register assignment time this reference counter would be associated with the register assigned to the line and decremented as source references where generated, becoming available when the reference count went to 0. In our simple example we have no multiple references (only MPY refers to the FETCH's) and set a register available after a source reference is made. We indicate it in sink/source mode until a successor instruction is processed so as to indicate its potential unavailability and consequent potential delay. It is actually available one line before—at the source reference line.

The generation of line 4 is accomplished merely by assigning the next available register R3, even though R1 becomes available again after the MPY reference. Line 5 receives the next available result register, R6, free because of its own reference, and is linked to the FETCH and MPY provided operands by its pointers in Fig. 10.18.

Lines 6 and 7 are assigned next available registers. Line 8 is linked to its source registers by the mechanism we have seen. Its sink register is R7. After the reference in R6 by ADD, it remained unavailable because of

ADD's result reference. R7 is the currently "up" register due to the cycling of R6 and R7. Line 9 is linked in the usual way to its source registers and assigned R7 as a sink because R7 is an operand reference. R7 is truly free because of the line 9 operand reference.

From Fig. 10.20 we see that the code compiled as it is would execute in 40 cycles with a high degree of functional overlap achieved. The sequential execution time of 80 cycles is reduced because of the inherent parallelism in the machine. We notice from Fig. 10.20, however, a number of delays that might be avoidable if we had a different instruction sequence.

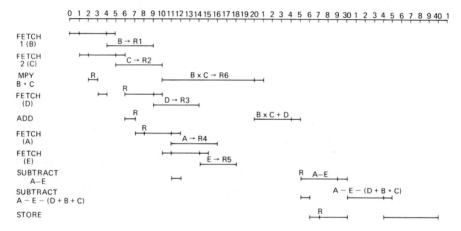

Fig. 10.20. Execution of code.

10.7. CODE REARRANGEMENT

We are in a splendid position to reorder our instruction list as we see fit. We know all subtasks; we know the precedence relations; we know the execution time of each task; we know the processors and resources available. It is an ideal scheduling situation. Let us first develop a statement of finishing and starting times for all subtasks in order to achieve a no-delay execution of the sequence of instructions.

Figure 10.21 shows the tree of the expression reflecting the necessary starting and finishing time. We derive this by working backwards on our tree, starting with the root node (STORE) and assigning it a finish time of 0. All relative times will be negative to this starting time. Since we charge 9 cycles for the store, its derived starting time must be -9. In order for this

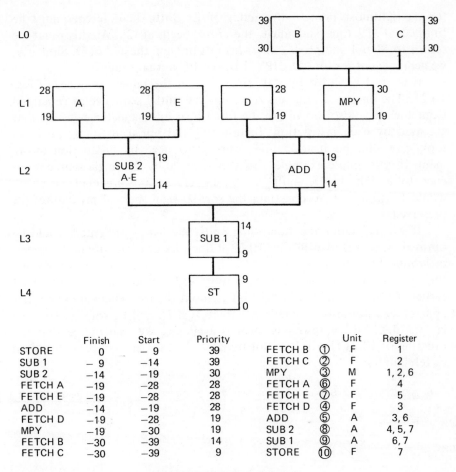

	Finish	Start	Priority			Unit	Register
STORE	0	− 9	39	FETCH B	①	F	1
SUB 1	− 9	−14	39	FETCH C	②	F	2
SUB 2	−14	−19	30	MPY	③	M	1, 2, 6
FETCH A	−19	−28	28	FETCH A	⑥	F	4
FETCH E	−19	−28	28	FETCH E	⑦	F	5
ADD	−14	−19	28	FETCH D	④	F	3
FETCH D	−19	−28	19	ADD	⑤	A	3, 6
MPY	−19	−30	19	SUB 2	⑧	A	4, 5, 7
FETCH B	−30	−39	14	SUB 1	⑨	A	6, 7
FETCH C	−30	−39	9	STORE	⑩	F	7

Fig. 10.21. Rescheduling code.

to occur the (A - E) - (B * C + D) must finish at -9 and, therefore, must execute at 14. The subtraction here depends upon A - B and (B * C) + D. Therefore, the A - B must end at -14 and begin at -19. Similarly, the add of D must be complete by 14 time and begin at -19.

We are traversing the tree of the expression going from a node to its predecessors to determine the necessary start and finish times. The method of traversal may vary. We here shall go to the leftmost ancestor, following the chain always to the left. When we are finished, we shall return and do rights at various levels. Therefore, we have not yet investigated the D add (the right branch of SUB 1), and it does not appear on the table. The next entry on the table is the fetch of A, the leftmost ancestor of SUB 2. Since this is the end of the line, we return to SUB 2 for its rightmost branch, the fetch of E. Both these fetches must be finished at -19 time and must, therefore, begin at -28. At this point we return to SUB 1 and process its rightmost branch, the add of D. Similarly, we generate entries for FD, MPY, FB, and FC in that order.

If we order the list by start times we get the tabled sequence of Fig. 10.21. The encircled numbers represent the initial sequence as generated from the tree. Also shown are the functional units and the registers that are used by each instruction. In resorting our initial instruction list, we took care that no irrationalities were introduced by instruction movement. In our simple example, this was assumed by the single source reference to a sink. Where multiple source references were made, a check would be necessary to constrain the sort so that operand procedure was preserved.

If we had unlimited functional units, the list as presented would be optimal, and as shown in Fig. 10.22 it does, indeed, execute in 34 cycles, or around 85 percent of the original list. It accomplishes this by reducing the operand wait time for the A, E subtract because A and E are fetched earlier. We notice, however, two things about the list: There are many ties apparently arbitrarily ordered, and there is at L1 and L2 (of Fig. 10.21) a greater demand for resources than is available. We also notice that the ordering of FETCH A,E,D is not trivial in the face of resource contention of fetch units.

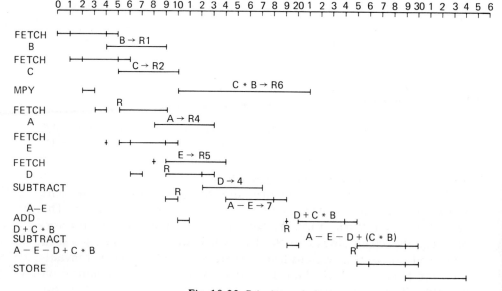

Fig. 10.22. Priority ordering.

Two known techniques for further scheduling do not attend to all problems and make no change in the performance of this sequence. One technique is to float fetches and sink stores. This would give us the list of Fig. 10.23. Although not effective for our example, it tends in general to reduce operand wait time. It is particularly useful over large regions, because it has the effect of considerably earlier fetch of late operands and the consequent true interleaving of statements.

A similar technique involves the generation of a skeletal FETCH list with a dynamic clock running to represent current region time and finish time for each fetch. Arithmetic is scheduled into a completed instruction list at the time of the readiness of operands. Initially the instruction list preparation would link all fetches together into a fetch list. The first fetch is assigned to fetch unit at time 0 and completes at 9; since another unit is free, the second fetch is submitted for completion at 10. The earliest start time for MPY is 10, but at the time the instruction is looked at region time is 2—the time for submission of two fetches. We consequently have an eight-cycle time possibility of doing something else before scheduling

```
FETCH     R1, B
FETCH     R2, C
FETCH     R3, D
FETCH     R4, A
FETCH     R5, E
MPY       R1, R2, R6
ADD       R3, R6, R6
SUB       R4, R5, R7
SUB       R6, R7, R7
ST        R7, Z
```

Fig. 10.23. Floating.

the MPY. We do not break our chain to the fetch of A. This instruction, however, requires a fetch unit, and none will be available at 2 time. One will be available, however, at 5 time, so we can squeeze in our fetch, since the time between 5 and 10 time (by 2 time) is greater than the time to push the fetch through the instruction counter. We submit the fetch, updating counters to region time = 5, FU I free time to 9.

We had previously listed the MPY on a delay list, with its line number and its start time. Whenever instructions are on the delay list, their start time is compared to current region time. On this compare we find a five-cycle time opportunity to do something more. We look at the next sequential instruction, the E fetch, and compute that the FU 2 will be available in one cycle, or four cycles earlier than MPY start time; consequently we have time for another fetch. Register arrival times are computed, region time is increased, and functional unit times are increased. Notice that the increases in region times are adjusted to reflect the interlock time in which the issuance of fetches was delayed due to waiting for the functional unit to be free. (We remember that instructions are held in the instruction unit on a unit busy condition.)

At the end of issuance of E fetch it is 7 time. We have three cycles until MPY start time. We look at the next instruction, the D fetch. We determine that an FU will be available in two cycles, and, therefore, we may undertake this fetch. This will take the clock to 9. We have only one cycle until MPY must start. Therefore, we initiate multiply at this time, removing it from the delay stack and taking a pointer from the last fetch to line 3; we update all clocks appropriately at line 3.

It is now 10 time at the end of MPY submission. We go back to the next sequential instruction, the ADD; we find it dependent upon R6, which will not be ready until 20 time. Therefore, we delay the add, placing it on the delay list, and look at the next instruction, the A-E subtract. This instruction cannot occur until 15 time, however, the time of availability of E in R5. We delay it, placing it in sort on the delay list, and advance to the next instruction, the final subtract. We find, however, a contingency in 20 time here also and so must delay. Further, we have sink/source problems that we are not showing here. The store is also contingent on the time of R7 and is delayed, since nothing is scheduled for 7.

We now have a delay list in time order and in initial order, where times are equal. We have no choice but to submit the subtract first, and we do so, removing it from the delay list and linking MPY to it. We then abandon the list.

This is an important improvement over the float fetch only approach—we have enabled our independent A-E subtract to get ahead of the dependent ADD. In calculating status at the end of issue of SUB, we see that the SUB will be issued by 11 time, but that an operand delay will hold until 15 time; therefore, the unit will not be free until 20 time, but the result will be available at 19 time in R7. If we had a sink conflict, it would be necessary to reassign registers. Since we are exchanging with ADD, we could reassign SNK registers to the two instructions. It would have been possible to delay register assignments until this ordering occurred. What we are doing now is remembering the cyclic nature of register assignment and behaving as if SUB had occurred first.

The effect of this technique and its essential quality is to simulate the status of the machine and assign instructions to reflect the status, squeezing in arithmetic at the earliest time its operands are available.

Neither the simulation nor the float fetch breaks priority ties or optimizes the order of fetches. It is not uncommon that scheduling techniques tend to be arbitrary when two tasks of equal priority, both meeting precedence requirements, compete for a resource.

10.8. FINAL REMARKS

In ordering a task list there are two basic strategies that may be employed. The techniques we are discussing reflect them both. The original reordering of the list by descending start time, working up the tree, reflects a "latest possible assignment" strategy in which assignments are made on the basis of the last possible time that an activity may be started in order to avoid delaying a successor. The effect of this in a broad resource system is to balance resource utilization by staging usage within an upper time constraint. It is a preferable technique, since it does achieve balance and increases utilization across a smaller resource framework. An alternate technique is the "earliest possible" strategy, where activities are undertaken as soon as they are precedent unbound. This tends for large queues on limited resources and for very irregular utilization of resources in very broadly equipped systems. The float fetch technique is of this general nature, in effect scheduling fetches at the earliest possible time in anticipation that they may cause bottlenecks. The clocking simulation is a mixed technique that starts from earliest possible and adjusts backwards, squeezing in as much as possible before causing delays. It is essentially, however, a forward scheduling technique.

To a large extent, compiling for machines of this class lies at an intersection between compiling for true multiprocessors and compiling for sequential machines. The essential differences between multiprocessor compilation and what we are discussing is the level at which attempts are made to recognize parallelism implicit in coding. The essential difference between a good compiler for our machine and any good compiler for any machine is very small; one expects to find good optimization techniques in eliminating redundant code in both. What we have seen in this chapter are some specific considerations for these machines which are added to the set of optimization rules.

A wide variety of opinion exists as to exactly how much should be done in the transformation to IL and in the manipulation of IL for all optimization techniques and particularly in this area. The general level of prerogative a compiler should have in rearranging program text presents no trivial problem to the industry. The extent of recognizability of a program by a programmer is a very real issue in the expensive process of debugging and in the serious area of programmer, as opposed to process, productivity.

PART TWO:
SOURCES AND FURTHER READINGS

Allard, R. W., Wolf, K. A. and Zemlin, R. A., "Some Effects of the CDC 6600 Computer on Language Structures," *Communications of ACM* (February, 1964).

Amdahl, G. M., "New Concepts in Computing Systems Design," *Proceedings of IRE,* 50 (May, 1962).

Amdahl, G. M., "Validity of Single Processor Approach," *AFIPS Proceedings,* 30 *Spring Joint Computer Conference* (1967).

Amdahl, G. M., "The Model 92 as a Member of System/360 Family," *AFIPS Proceedings,* Fall Joint Computer Conference (1964).

Anderson, D. W., Sparacio, F. J., and Tomasulo, R. M., "Machine Philosophy and Instruction Handling," *IBM Journal of Research and Development* (January, 1967).

Baer, J. L., and Bovet, D. P. "Compilation of Arithmetic Expressions For Parallel Computations," *IFIP Congress Proceedings,* Booklet B (1968).

Boland, L. J., Granito, G. D., Marcotte, A. U., Messina, B. U., and Smith, J. W., "IBM System/360 Model 91: Storage System," *IBM Journal of Research and Development* (January, 1967).

Chen, T. C., "The Overlap Design of the IBM System/360 Model 92 Central Processing Unit, " *AFIPS Proceedings,* Fall Joint Computer Conference (1964).

Conti, C. J., "System Aspect: System/360 Model 92," *AFIPS Proceedings,* Fall Joint Computer Conference (1964).

Conti, C. J., Gibson, D. H., Pitowsky, S. H., "Structural Aspects of the System/360 Model 85—General Organization," *IBM Systems Journal,* Vol. 7, No. I (1968).

Control Data Corporation, *Control Data 6400/6600 Computer Systems Reference Manual,* Publication number 60100000 (September, 1966).

Dreyfuss, Phillippe, "System Design of the Gamma 60," *Proceedings of Eastern Joint Computer Conference* (1958).

Dreyfuss, Philippe, "Programming Design Features of Gamma 60," *Proceedings of Eastern Joint Computer Conference* (1958).

Eckert, J. P., Chu, J. C., Tonik, A. B., and Schmitt, W. F., "Design of UNIVAC LARC System: I," *Proceedings of Eastern Joint Computer Conference* (1959).

Flynn, M., "Very High Speed Computing Systems," *Proceedings of IEEE,* 54 (December, 1966).

Flynn, M., and Low, P. R., "IBM System/360 Model 91: Some Remarks on System Development," *IBM Journal of Research and Development* (January, 1967).

Hellerman, H., "Parallel Processing of Algebraic Expressions," *Transactions of IEEE,* Vol. EC15 (February, 1966).

International Business Machines Corporation, "IBM System/360 Model 195 Functional Characteristics," Form A22-6943 (August, 1969).

Liptay, J. S., "Structure Aspects of the System/360 Model 85—The Cache," *IBM Systems Journal,* Vol. 7, No. 1 (1968).

Pirtle, M., "Intercommunication of Processors and Memory," *AFIPS Proceedings,* Fall Joint Computer Conference (1967).

Seelye, M. A., and Lorin, H., "Proposal for A NICOL Compiler," Service Bureau Corporation Document (September, 1966).

Sperry Rand Corporation, "UNIVAC 1108 Multi-Processor System–System Description," UP-4046 (1968).

Squire, J. S., "A Translation Algorithm for a Multiple Processor Computer," *Proceedings 18th ACM Conference* (1963).

Stone, H. S., "One Pass Compilations of Arithmetic Expressions for a Parallel Processor," *Communications of ACM* (April, 1967).

Thorlin, J. F., "Code Generation for PIE Computers," *AFIPS Proceedings,* Spring Joint Computer Conference (1967).

Thornton, J. E., "Parallel Operation in CDC 6600," *AFIPS Proceedings,* Fall Joint Computer Conference (1964).

Tomasulo, R. M., "An Efficient Algorithm For Exploiting Multiple Arithmetic Units," *IBM Journal of Research and Development* (January, 1967).

3

Multiple Machines

Chapter 11

MULTICOMPUTER
CONFIGURATIONS

11.1. INTRODUCTION

In the previous chapters we have discussed parallelism within the framework of a single processing system. Despite the high degree of asynchronous, concurrent operations, we were investigating the behavior of a "single
thread," a single collection of highly related subtasks constituting a single
program. Our fundamental motive for increasing the parallel capabilities
of the machine was to achieve a greater speed of task execution for the
task. We were, in effect, investigating how we could organize the resources
and work habits of a single repairman in order to increase his productivity.

In the course of this investigation we saw a number of ways of achieving concurrent activity. The duplication of identical resources to reduce
bottlenecks, the functional specialization of semi-independent units, the
use of local queuing space, and the distribution of "intelligence" all contributed to achieving two apparently different but truly identical types of
parallelism: the parallelism between two different instructions and parallelism within stages of instructions.

These are truly the same thing if we define our processor subtasks in
units of sufficiently discrete size and abandon our concept of instruction
execution as a fundamental operation. We watched a "uniprocessor" ma-
turate and evolve into a machine whose sequential nature was very much
undermined and which awaited just a few more elaborations before emerging as a multiprocessor.

In this chapter we shall be interested in the other pole, as it were, of a
continuum that converges at its center on the generalized multiprocessor

as we understand it today. A very rough picture of the development from both ends may be seen in Fig. 11.1. We see the development from the uniprocessor coming down in stages with a side development to multi-compute due to the development of greater and greater capability in the I/O side of the CPU-I/O functional split. In the upward line we see the closer and closer couplings of stand-alone machines to form both master-

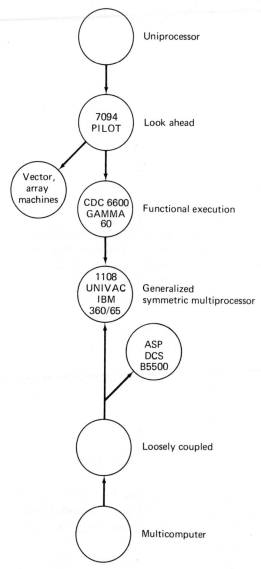

Fig. 11.1. Convergence on multiprocessor by merger and growth.

slave and generalized multiprocessing systems. A master-slave system generally embodies the concept of functional specialization, whereas the generalized multiprocessor's outstanding attribute is the presence of identical processing units. Within any system, of course, functionally specialized units, inside the processors or to support I/O, will exist.

11.2. HISTORICAL DEVELOPMENT AND PROJECTION

The development of multicomputer systems has two fundamental sources Primary, perhaps, is the association of computers in on-line command and control systems. The underlying motivation here was the necessity for achieving reliability within some reasonable cost constraint and with some potential usability of the system for non-on-line support-oriented activity.

A second source of the system design was very large computing installations with a complex set of independently installed computer systems. Low-level (device and channel) associations of systems were seen as a way for smoothing operational procedures in transferring data from one system to another. The relationships at these interfaces have now been so elaborated and investigated through software and hardware that quite sophisticated implementations of asymmetric dual-processors appear in generalized hardware-software packages.

In both the reliability-oriented military systems and the operationally oriented systems there was in the formative stages quite a high degree of system specificity. This is not other than expected in local solutions to local environments and local needs. From these specific solutions, however, general principles were observed and formed the basis of contemporary computer hookups.

In the future we may expect a large interest in the association of geographically dispersed computing systems through communication lines. Current interest in time-sharing systems—that is, access by a typewriterlike or cathode-tube device to a computational capability some distance from the "terminal"—will mature into an interest in intercomputer hookups as technology expands the capability of low-cost terminals to include some local processing (such as line-editing or, going further, programming language syntax checking).

Interest in the so-called remote job entry systems, where a low-capability, nonresponsive, local device transmits card-, paper tape-, or even tape-and-disc-based representations of jobs to a central computer and receives output for local printing, already forces attention on the physical and functional relationships between geographically dispersed computers. Message-switching and data base applications as they mature will force more and more interest and attention in this area.

Notice the hypothetical system of Fig. 11.2. At location A there exists a highly capable computer system with a large data base. The organization of the system at A includes four types of processors: a generalized processor that has two incarnations (one might visualize this as a machine of the UNIVAC 1108II class), a specialized telecommunications central processor (a UNIVAC 418III ?) which handles all data line transmission and data-line-oriented functions, a special I/O processor to handle all local I/O such as tape, local printers, card-oriented devices, and a specialized storage hierarchy control processor, which controls the movement of data "forward" and "backward" in the storage hierarchy of the data base.

The system is associated with a family of local terminals through the teleprocessing computers, as well as with another teleprocessing computer at B, which has local I/O devices associated directly with it. It is also associated with the center C, an intermediate size center with a multiprocessor IBM 360/65, which supports a communication computer at D as well as local terminals. Here the I/O computer and telecommunications computer functions are performed by the two processors on a time-shared basis.

The network shown here raises very profound questions about local capability, load levelling, levels of access, transmission of large amounts of data, allowed degrees of planning, etc. A number of points arise from this. First, all of the scheduling and optimizing questions occur in double strength in these systems; second, our interest in multicomputer operations is not purely or even primarily historical. To what extent geographical separation and the interposition of communications lines will change or influence the approaches to multicomputer design remains largely to be seen.

11.3. DIFFICULTY IN DEFINITION

We should undertake a definition of a multicomputer that enables us to see the distinctions between it and other parallel capable systems. A very limiting definition that is sometimes used constrains a multicomputer system to be a connection between two computer systems, each of which was designed to operate as a stand-alone system primarily, but which have been interfaced so as to allow some coordination of activities. This would exclude systems like PILOT and LARC, which were inherently designed as dual systems, and would consider them multiprocessors.

We also wish to distinguish between a processor with a highly intelligent channel and a two-processor system. We would admit a UNIVAC

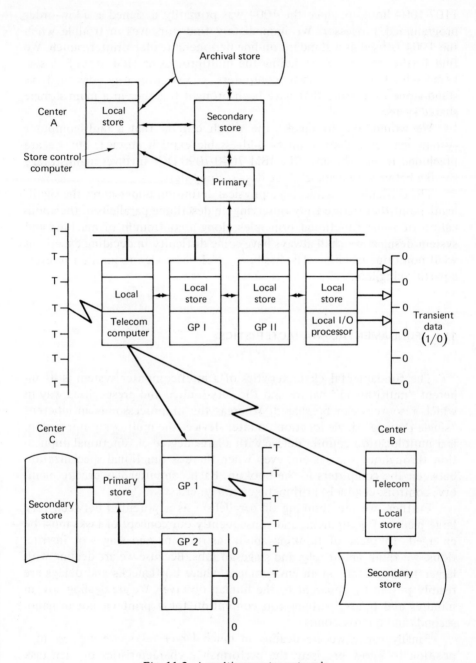

Fig. 11.2. A multicomputer network.

1107-1004 hookup, since the 1004 was primarily designed as a low-order programmed processor. We immediately find ourselves in trouble when the 1004 is used as a standard on-line peripheral reader/printer/punch. We find further trouble in excluding the multiprocessing IBM 360/65, a system which involves two computers (65's) primarily designed as stand-alone processors that have been brought together in a main storage shared system.

We would like to invoke the rough criteria that a multicomputer system may not share primary addressable explicit storage, but we are precluded from this, since the IBM 7040-7090 DCS involves main storage sharing between two stand-alone design computers.

The difficulty in achieving a physical definition underscores the significant point that we are truly observing, in describing parallelism, the application of some functional principles along a continuum of machine and system designs; we shall always have some difficulty in deciding exactly at what point one machine class has become close enough to another to draw a partitioning line.

11.4. FUNDAMENTAL CHARACTERISTICS

The fundamental characteristics of a multicomputer system is its inherent "multithread" nature and the very limited and prespecified way in which resources can be shared. Whereas the uniprocessor is an inherent "single-thread," single-location counter device, the multicomputer system is a multilocation counter device with a large degree of functional duplication throughout the system, even when there is functional specialization between the computers in the hookup. Each computer has private memory, controls, channels, arithmetic circuitry, etc.

Further, we are thinking of parallelism as a potential between very large pieces of equipment, and consequently our concept of tasks must be enlarged. In place of thinking about instructions and stages of instructions, we think about jobs and stages of jobs. Because we are dealing with larger tasks, we are in an environment where bottlenecks and delays are readily physically apparent to the human observer. We are dealing now in minutes and hours, perhaps (on contention for a printer), not in nanoseconds and microseconds.

Finally, since we are dealing in much larger tasks, we are less in a position to know precisely the performance characteristics of each task each time it operates.

11.5. CONCEPTUAL INDENTITY OF
TWO ASYMMETRIC SYSTEMS

Figure 11.3 attempts to show the fundamental similarity that can be seen at the polar levels of parallelism, if a very particular relationship is assumed to exist between the multicomputers of the multicomputer system. We have forced the correspondence here to show the conceptual similarity, but we have not fundamentally distorted processes or relationships.

Consider the functions of an instruction unit in an elaborate uniprocessor (IU in Fig. 11.3). We shall postulate a machine in which store/fetch operations are done through the IU (not unlike a model 91). Consider the multicomputer system represented beneath it. We postulate a system here where there is a support computer (SC) and a computational computer (CC). All scheduling is performed by the SC in the system.

The equivalent of generating an address in the IU is the SC selection of a job on a job queue to be run next. The SC maintains on a tape or disc a list of all jobs submitted to the system. At the end of each job the SC selects a new job to be run. Surely the selection criteria and selection process are more complicated and time-consuming, even given the simplest disciplines, but this activity of an SC is functionally identical to stabilizing a new address for a next instruction in IU.

Uniprocessor, Highly Evolved

```
IU   STABILIZE INSTRUCTION 1 ─────────────────┐
IU   FETCH INSTRUCTION 1            STABILIZE INSTRUCTION 2
IU   DECODE                         FETCH INSTRUCTION 2
IU   GET OPERANDS                   DECODE
EU   EXECUTE                        GET OPERANDS
IU   STORE RESULTS                  EXECUTE
                                    STORE RESULTS
IU = instruction unit;
EU = execution unit;
SC = service computer;
CC = computation computer.
```

Multicomputer

```
SC   SCHEDULE PGM 1 ───────────────────────┐
SC   GET JCL                        SCHED 2
SC   ANALYZE JCL                    GET JCL
SC   COLLECT DATA AND RESULTS       ANALYZE JCL
CC   EXECUTE 1                      COLLECT DATA
SC   PRINT OUTPUT                   EXECUTE 2
                                    PRINT RESULTS
```

Fig. 11.3. Functional equivalence.

When a job is selected, the description of work associated with it must be brought to the SC. This is analogous to fetching an instruction. Initial decoding in an IU to determine what operands are needed and who is to execute is equivalent in an SC to an analysis of the work description to determine what data are to be brought to the system, to allocate devices to data, deliver mounting messages, etc.

The actual performance of these set up activities is equivalent to getting the operands for an instruction. This setup may be done on the SC, in which case data are intermittently forwarded to the CC from devices associated with SC, or directly on CC, which then refers to its own devices for data. In any case, operands are forwarded in a sense through the activity of the SC, either indirectly or directly.

The actual execution of the job, like the execution of the instruction, is performed on CC. In this system, whenever an output record is formed, it is sent to SC to be enqueued for printing or writing on tape. In the same way, contents of registers that hold results of E-unit activities may be forwarded to core through storage buffers in an I unit, as on the 91. The intent is the same in both places, to create a queueing mechanism that will separate the result generators (E or CC) from the speed of core or output devices. From this example we can see that the multithread capability of a multisystem computer can be used to support a functional equivalent of a mature single processor. The possibility of other organizations is, of course, present.

11.6. VERY LOOSE COUPLING

The loosest possible coupling of computers is that which associates them both to a common input stream, and where this common input stream is entirely self-scheduling so that there is no selection or scheduling function performed by the computer system.

A fundamental example of this is the connection of two computers to a single real-time communication channel. Messages arrive at the computer interface at a rate determined by a sending agency (or agencies), and the function of a computer is to receive the message, process it, and transmit a response. In order to perform processing a bank of background information must be available to the computer, as well as a collection of programs which constitute the "application." In a very loose couple this collection of programs and data may be visualized as being duplicated on both computer systems. The two computers are in "primary" or "back-up" status. The primary computer always is "on-line" to the channel; the back-up computer is always in a standby state, either idle or performing non-on-line functions. At scheduled times or on the occasion of failure of the primary system, the real-time channel is operator-switched from the malfunctioning to the standby machine.

Figure 11.4 is a representation of a machine hookup of this type. We notice that only the remote input channel is in any sense shared between the computer systems, each of which has a private collection of devices and of data and programs. It is a perfectly symmetric multithread system with any processor capable of anonymously performing any task in the system. In an application where resource data are not changed over time, the switch from one computer to another involves merely the physical

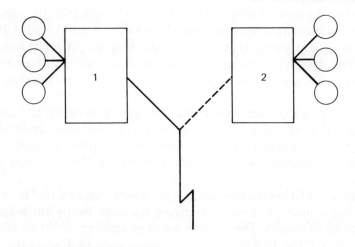

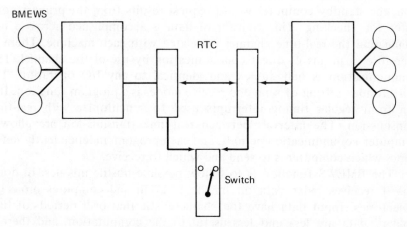

Fig. 11.4. Basic connections.

switching of the channel. There is no direct communication between computers. The motive for such a system is, obviously, pure reliability.

It might be possible to achieve reliability of the same order by replication of parts at a lower level. Early systems, especially the UNIVAC I, invested a great deal of equipment in achieving reliability through duplication of internal registers and arithmetic circuitry. This has the disadvantage, however, that it greatly increases the number of lower-level components and their interconnections; this increase in connections may itself introduce a reliability problem. In the case of malfunction it is also potentially necessary to retire the entire system in order to repair.

With current technology, protective partitioning on lower-level functional units might be achieved and full processor replication need not be undertaken. In the technology of the late 1950's, however, the association of multiple standard computers was a better way to achieve reliability and a "fail-safe" capability. It is not yet clear that we yet understand "fail-safe" fully. An obvious potential advantage in associating full computers is the potential capability to make productive use of the additional hardware.

In many applications in real-time, changes to resident data are made or results depend on accumulations of input over time. In these cases more effort is required to switch from one computer to another. It is necessary to establish a current environment for the computer coming on-line.

One way of keeping the standby computer "up to date" is to have it follow the primary computer, receiving the same inputs but suppressing the sending of results. The operational programming would be cognizant of the status of the machine and would adjust itself to this status.

The BMEWS system used an approach very much like this. In addition, the standby computer would request results from the primary computer for checking. This transfer of data is accomplished across a line connecting the real-time channel associated with each machine. There is, therefore, a means of direct communication by use of the channel. This communication is half duplex and identical to any I/O operation. The synchronous setting of send and receive modes is a program function. It is run from regular timing interrupts used to synchronize with real-time transmission. The intervals between real-time transmission are allowed computer communication periods, and each program independently determines which computer is to send and which to receive.

The BMEWS function is to follow possible hostile missiles. In doing this it receives radar data on missile location and computes projected trajectories. Input data have the characteristic that over periods of time "older" data are less and less useful to the computation, and there is indeed a cut-off point at which such data are useless. This enables the standby computer to catch up with the primary computer after it has been started and to achieve parallel operation after a given time. This catch-up feature is critical when the computer coming on to standby has been "down" for some period of time.

The fundamental advantage of the connection seems to be in the capability of a continuous check of results by the standby machine. Another use for the connection is that a path does exist by which limited data can be transferred from one machine to another. In an application where the switching of status between computers could be facilitated by a transfer of some data from the retiring to the oncoming primary computer, this transfer is facilitated through the channel connection. We notice that it is not possible for one BMEWS machine to shoulder-tap (interrupt) another.

11.7. TIGHTER COUPLING

The BMEWS approach might be quite costly in another applications environment. The dynamic maintenance of a data base of any size would probably find the private duplication of files for each machine quite a prohibitive cost. One approach to the problem is the provision of switches that allow devices to be switched from one computer to another as required. This device switch capability is a first step in a continuum of sharing arrangements that proceeds through programmable switching of channels from each processor to the device and finally to true data sharing, where not only is the device dynamically accessible but there may be common access to logical data in the same data set. The switching of devices is also very common in large computational centers. Sometimes the switching capability can be rather extensive.

See, for example, Fig. 11.5. This is a large computer center with perhaps hundreds of peripheral devices and of the order of ten or more computers of various types. We show here two types, I, a large-scale scientific machine with up to three channels, and a smaller machine, II, with up to two data channels. These might be IBM 7090's and 7040's or CDC 1604's and 160-A's, for example. The general arrangement of Fig. 11.5 has been developed for both systems. The IBM version is slightly altered so that only disc files and high-speed tapes are available through the switch, and conventional tapes and printers, punches, and readers form private pools. A further significant difference between Fig. 11.5 and a described 7090/40 hookup is the presence of the scheduling monitor computer (SMC) in this configuration.

Regardless of the variations, the motives are the same: to increase the

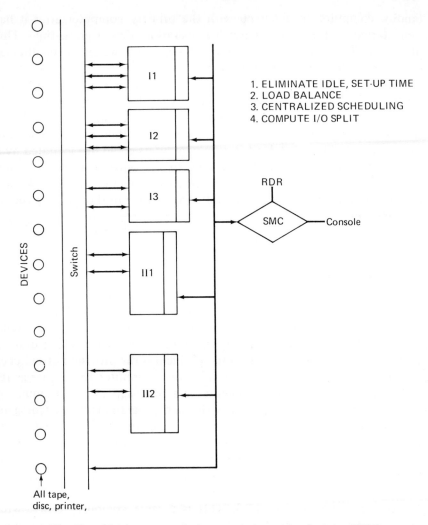

Fig. 11.5. Multicomputer device association with scheduler (SMC).

productivity and utilization of equipment by smoothing the operational
interfaces between jobs and job stages, to reduce setup time, and to take
advantage of the load-balancing potential in scheduling from a very
broadly defined dynamically assignable pool of I/O devices.

The complex is essentially a loosely coupled asymmetrical multicomputer system containing a degree of specialization of function between computers of different types. One natural split would be between
I/O and computational computers. The basic function of the I/O computers would be to unburden the computational computers from the

I/O-bounded tasks of a job. Initial system input would be streamed onto discs or tape by the I/O computers, and these devices would then be switched to the computational computers. Output written by the computational computers on disc or tape would then be made available to an I/O computer for printing or punching.

In the presence of the monitor, computer scheduling is centralized. A central queue of work is maintained on a device private to the SMC. A specialized scheduler is executed only by SMC. Each individual operational computer is under control of the operating or executive systems (or a modified version) that would be operative if it were a stand-alone.

All device allocation/deallocation decisions are made by the scheduler, which maintains contact with the operators in the machine room. The actual switching of units may be logically accomplished by machine or operator, although it is highly desirable to have SMC equipped with a switch selection mechanism, as shown in Fig. 11.5. SMC maintains at all times a directory of equipment and file status. In a disc-oriented system, one might visualize shared discs, of course. A communication capability between SMC and other computers is necessary to allow SMC to direct the undertaking of a job for a selected machine and to allow a machine to report completion of a job.

It is not necessary to restrict one's image of functional split to I/O and computation. In the same environment one can visualize I/O- and computer-bound tasks being assigned to preferred computers, or any number of diverse criteria used to select a computer type. The capability of preemptive scheduling exists with an SMC that has knowledge of job characteristics and priority for all jobs running on the system.

If the functional split is conceived of as being computation and I/O, there would be two basic ways of providing support. We mentioned this in passing at the beginning of this chapter. One approach is dynamically to interpose the I/O computer between the computational computer and each I/O reference, the memory of the I/O computer becoming essentially a dynamic buffer for the I/O of the computational computer. The alternative is for the I/O computer to collect all introduced data for a job onto a shared device, where it is quickly accessible by the computational computer. In the absence of an SMC (supervisor computer) each computational computer would schedule itself when it required work. I/O computers would pool job descriptions and requests on a common file. As a computational computer needed work, it would select a job from the common work queue. Premounting messages and device allocation might still be accomplished by the I/O computers.

The parallelism of such systems as these comes from the asynchronous operations of many computers on many jobs. In a non-real-time system, reliability may not be dominating, and specialized fail-safe, fail-soft components may not be designed into the system. The flexibility of pooled resources provides reliability as well as efficiency by offering many alternative devices. If a computer goes down in an SMC system, this failure need only be reported to the SMC so that it will not attempt to schedule it. This is a sort of easily achieved fail-soft. The failure of SMC, of course, requires a redistribution of its functions, some to human operators and some to the local executive systems in order to maintain any system performance.

Quite considerable performance and cost performance advantages have been realized with systems of this type. Reports of as much as 8 to 12 times throughput are made for the 7040/7090 hookup under simulation. A very critical decision area, of course, exists in determining how many computers of what type and how many peripherals of what type should be associated with a system. The decisions here are very similar to the determination of how to balance subprocessors in the uniprocessors.

A number of asymmetric multicomputer systems have been developed as generalized offerings by various vendors; of these the IBM 360 ASP (Attached Support Processor), the IBM 704X-709X Direct Couple System, and the CDC 6600 are perhaps best known and representative of a continuum of intercomputer connections. Each system has at its underlying concept the definition and removal of system overhead and I/O functions from a large computational processor.

11.8. ASP—THE CHANNEL-TO-CHANNEL ADAPTER

This system is the loosest association of computers and is truly a multicomputer hookup. In its initial version a support processor (an IBM 360 model 40 or model 50) was associated with a main processor (an IBM 360 model 50 or 65) through a device called the channel-to-channel adapter (C to C). There was no other interconnection between the computers. Figure 11.6 shows an overview of an ASP configuration, showing a model 50 attached to a model 65. The attachment is made by associating one of the three high-speed channels of a model 50 with the adapter and on the other side, one of the up to six high-speed channels of a model 65.

The channel capabilities of these two machines are differently implemented with regard to integration with processors, with the 65 being the first machine to have truly independent channel device attachments, but this difference need not concern us here.

The C to C actually physically attaches to one of the channels. For both associated channels the C to C occupies a control unit position and appears to be a control unit to the channels that it connects. In effect, it makes the other computer appear to be a device associated with a control

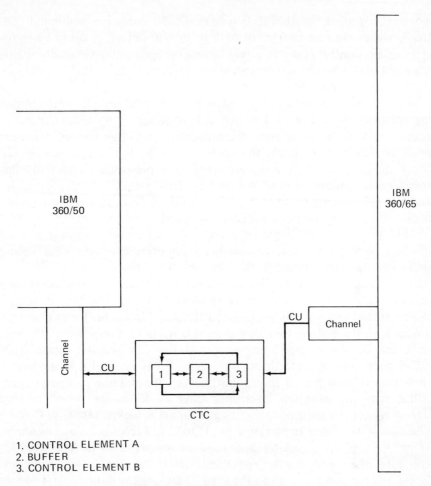

1. CONTROL ELEMENT A
2. BUFFER
3. CONTROL ELEMENT B

Fig. 11.6. ASP interconnection.

unit on a channel. Computer-to-computer operations across the C to C, therefore, are basically I/O operations identical to I/O operations in normal control units and devices. (An interesting alternative use of C to C is on two channels of the same computer, where it is used to move data from main storage to other points in main storage over the channels, freeing the CPU from support of block transfer functions.)

A synchronization activity must be present between the two channels in order to effect the transfer of data. Reads and writes must be matched on the connected channels, and it is this synchronization that is the basic function of the C to C unit.

Reads and writes are presented to C to C as normal channel command words (CCW's) of the IBM 360 system. CCW's contain a command code (read, write, etc.) an address of data to be transferred, a count of bytes, and certain control codes to effect channel programming (basically scatter read and gather write, accomplished by indicating chains of data addresses and chains of command).

The C to C receives and decodes the command code from the initiating channels. If it is a read or write, it is necessary for a corresponding read or write to be issued from the other channel before the operation can proceed. The C to C holds the command in its buffer and notifies the other channel that there is activity. That channel issues a command that transfers the command in C to C buffer. This command is then examined and a synchronizing command is issued to the C to C, allowing the performance of the I/O command initially requested.

Let us take an example of a computer wishing to transfer a block of data over its channel to another computer. The computer issuing the write forwards through the channel the write command. If the C to C is busy with a previous command, a busy and immediate disconnect is sent to the requesting channel. If C to C is free, it accepts the command and determines if a read is already in the buffer of C to C. If a read is in the buffer, then this write is considered a response to the read initiation of the reading computer, and the data transfer immediately takes place byte by byte through the adapter. If no read is in the buffer, then the write is placed in the buffer, and the reading computer is signalled with an attention interrupt. After an indefinite period of time during which the requesting (writing) channel is locked into C to C and is unavailable for other operations, the C to C will receive a sense command from the reading computer and forward the write to it. The receiving computer then analyzes the command, determines that it is a write, and issues a corresponding read to the C to C, causing the data transfer to take place. It is, of course, possible to avoid the transmission of the sense and issue an immediate read, but this could be accomplished only if the receiving computer had a way of predicting exactly what command was waiting in the C to C when it was interrupted.

Notice that the speed of the complete operation is limited by the responsiveness of the receiving computer and that during the interval from acceptance of the write by C to C and corresponding issuance of sense and read by the other computer, the channel of the initiating machine is "locked" in to C to C. This feature causes there to be some considerable disparity between channel utilization measurements when investigation is made of the usage of C to C and associated channels.

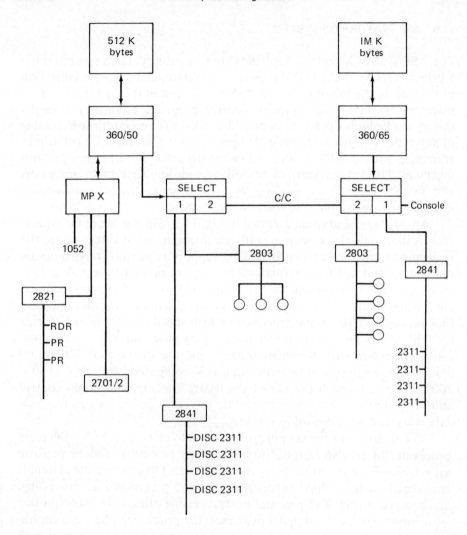

Fig. 11.7. Typical ASP.

Since the system truly perceives the support processor as a responsive slave, all action for data transfer is initiated by the main processor and the C to C associated channel of the main processor spends a good deal of time awaiting response from a rather busy support processor. The use of C to C on a channel does not preclude other control units and their devices from being associated with a channel on the main processor.

11.9. ASP—NATURE OF SYSTEM

The ASP system is built around the C to C capability. Each computer has a private version if the OS/360 operating system, and a private collection of devices. In the initial release of ASP the version of OS/360 used in both machines was the basic primary control program, essentially a single-thread stacked-job operating system that allows for no multiprogramming of any type. This has since been changed to allow OS/360 MVT (multiprogramming with variable number of tasks—the full OS/360) to run on both machines. Other later variants of ASP include multiple main processors and shared direct access devices. We shall describe here the underlying qualities and philosophies of the basic ASP system.

An ASP configuration is shown in Fig. 11.7. Since it is not infrequent that ASP installations operate both machines in stand-alone mode, the main processor is often configured to support minimum stand-alone requirements and not for cost/performance optimization of the ASP configuration. (Various usage approaches to the ASP system in various processing environments will be discussed later on.) Characteristically, one may find printers on the main processor, a high-speed drum resident storage for OS/360. A growing usage of the support processor is as a communications interface with terminals through the use of the IBM 2702 communications control unit or with other IBM computers, 360/20's, 1130's, 360/30's, etc., through the use of the IBM 2701 communications control unit. This usage is in support of remote job entry and low-speed terminal data entry and job submission systems.

The function of the support processor is to enqueue jobs for the main processor, to receive output from the main processor, and to perform asynchronous utility routines as required. Each job entering the system is considered to have three fundamental stages: preprocessing, processing, and postprocessing. The pre- and postprocessing phases are accomplished asynchronously by the support processor; the processing phase is a coordinated activity between both processors. For certain utility operations such as tape to print, a card to tape, or tape to tape for reformatting, the processing stage is performed on the support processor. These utilities are operator-scheduled.

Figure 11.8 shows the active relationship between the processors. The support processor is receiving the job A, presented to it in the normal manner that any job is presented to OS/360, a collection of job control language (JCL) statements containing a full description of the job structure in terms of steps and required devices, a program (perhaps in FORTRAN), and some associated data. The support processor creates an internal representation of the job on a storage device and creates a scheduling element for entry onto a queue of waiting jobs. Additions to the

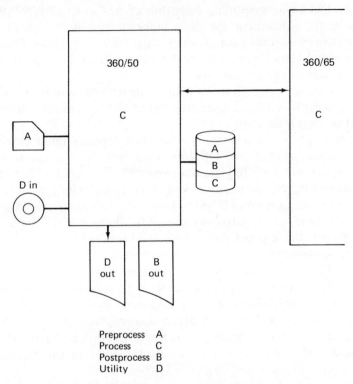

Preprocess   A
Process   C
Postprocess B
Utility   D

Fig. 11.8. Work flow—ASP.

normal OS/360 JCL allow for the user to describe what mounting requirements exist for the job. The support processor sends messages to the
operator directing mounting of units. Allocation of main processor devices
is, therefore, a support processor function. The intent here, of course, is to
minimize the time that the main processor must wait for data mounting
when a job is selected for execution. A completes the preprocessing stage
and is ready for execution selection before its devices have been mounted.
If, when it is selected for processing, mounting is not complete, the job is
delayed.

   Job B in Fig. 11.8 is in the postprocessing phase. This means that it
has been serviced by the main processor and that an output file has been
collected for B which requires printing or punching or some other locally
defined utility operation. The support processor undertakes this function
asynchronously, reading records for the disc device onto the printer.

   The processing of job C in the "execution" or processing stage directly involves the use of the C to C. At the termination of B the main
processor had requested a new job from support. The selection of the new
job, C, involved the running of a scheduling algorithm on the support

computer. The basic scheduling algorithm of ASP is an external priority driven with the adjustment for the temporary bypass of higher-priority jobs that require mounting for lower-priority "nonsetup" jobs. This "non-setup padding" allows new work to flow through to the main processor while setup jobs are being mounted.

Various locally specialized scheduling algorithms have been provided for ASP to meet specific requirements, but in all installations scheduling is a support processor function.

The concept of setup and nonsetup jobs depends upon two funda-mental concepts of ASP and OS/360. These are the concept of SYSIN (system input) and SYSOUT (system output). The SYSIN "data set" is the stream of data presented to the system. If all the information required for the execution of a job, including necessary data input, is in the input stream, then no input mounting is required for that job.

Essentially, the support processor acts like a card reader for the main processor, forwarding through its core over the C to C images of the SYSIN "data set," as requested by the main processor. The stream is pool-buffered in the support processor so as to achieve a maximum flow across C to C. This SYSIN stream was originally placed on disc storage by the support processor during the preprocessing phase of the job. Part of the requirement for a nonsetup job is, therefore, that all input data come from the SYSIN stream—what is generically called in the industry the primary input stream.

If the output of a program requires only one file and the user wishes to use the convention, then the SYSOUT file is used for this output. In a three-stage compile-load-go, the printed output of the compiler, all system messages, and the output of the executed program would appear on SYSOUT unless the user specified a private file to separate his execution output.

If all input comes from SYSIN and all output goes to SYSOUT, a job is a "nonsetup" job. SYSIN is represented on the disc device associated with the support processor and is forwarded record by record on request of the main processor. SYSOUT is represented on the disc device and is forwarded record by record by the main processor, through the memory of the support processor to the device.

The primary use of the C to C is to support the transfer of informa-tion associated with SYSIN and SYSOUT between the machines. It is possible for other files to be associated with C to C by applications pro-grammers writing for the main processor, but the underlying system con-cept of ASP is for all files other than SYSIN and SYSOUT to be files private to the main processor or to the support processor.

When additional input files are required on the main processor, and/or when private output files are required, a job is a "setup" job in the sense that volumes must be mounted on the main processor before the job may proceed. The function of C to C during the cooperative execution of job C (from Fig. 11.8) is to provide a path for the quick submission of SYSIN to the main processor, and the quick removal of SYSOUT records from the main processor.In order to support this function, large amounts of support processor core are organized into a buffer pool dynamically assignable to input or output records and with a possibility, through the dynamic core acquisition capability of OS/360 on the support processor, of acquiring more space for buffers when needed.

The final activity, D, of Fig. 11.8 is the independent running of a tape to print service routine, a utility function entirely under control of the support processor.

11.10. ASP—ORGANIZATION OF SOFTWARE

It is obvious from our descriptions that, although both processors are running under the PCP (Primary Control Program) version of OS/360, the support processor is running in a multiprogrammed mode. In this hookup multiprogramming can now be achieved by running the machine under an intermediate version of OS/360 called MFT (multiprogramming fixed number of tasks) and by running ASP as a task along with other tasks. This new feature, however, does not provide for the multiprogramming inherent within the ASP functions themselves.

Time-sharing of ASP activities is accomplished by a lower-level monitor within the ASP task called the multifunction monitor. This ASP element switches the attention of the system between active functions. It maintains a list of all active functions (the Function Control Table) and, when entered, selects the highest-priority active function that can operate in order to cede control of the support processor.

All functions in ASP are provided by dynamic support programs (DSP's), whose basic functions are shown in Fig. 11.9. Notice that 1 to 7 are associated with main processor support, and 8 to 13 with independent utility functions. Additional DSP's may be provided by an installation for servicing of unique processing needs. The user may designate which of these DSP's are to be considered resident and which transient. Further, each DSP may have a locally assigned priority and may run parallel incarnations of its function, e.g., multiple printer operation.

CPU switching occurs entirely on a "politeness" basis. When a function cannot proceed, it calls upon the multifunction monitor with a system macro (AWAIT) to assign it to a waiting status until it can proceed.

1. Main service	Coordinate SC/CC
2. Input service	Read input stream
3. System setup	Allocate devices, issue messages
4. System breakdown	Return devices
5. Print service	Print data sets
6. Punch service	Punch data sets
7. Purge	Remove job, return disc allocation
8. Card/tape	
9. Card/print	
10. Card/card	
11. Tape/print	
12. Tape/tape	
13. Tape/card	

Fig. 11.9. Basic ASP DSP's.

This characteristically occurs when an I/O request must be completed. The function is put into a waiting state until its request is fulfilled, at which time it becomes a candidate for CPU time.

In addition to the DSP's, ASP consists of a job segment scheduler (JSS), a console service program, a DISK I/O handler, and various other support programs considered part of the ASP nucleus. The function of the job segment scheduler is to maintain the number of functions that can be multiprogrammed (limited by the size of the Function Control Table active by selecting qualified jobs whenever entries in Function Control Table are available. The JSS and other resident functions of ASP appear to the multifunction monitor as essentially high-priority DSP's.

11.11. ASP—PERFORMANCE

As with any functionally specialized system, the efficiency of ASP depends upon the matching of a system's capability of performing a function and the requirement for performance.

Since we are dealing with two substantial pieces of equipment, we are interested in showing that cost/performance, turnaround, and throughput are enhanced by the hookup as opposed to running two independent systems. The reduction of equipment costs in an ASP system depends upon the extent to which peripheral and I/O devices can be removed from the main processor. This is a function in part of a center's dedication to ASP, the degree to which the installation is required to run stand-alone jobs (a condition that may occur because of a requirement to run under a variant operating system, or because of a requirement to provide special condition services for certain users), which might require minimum equipment levels, and in part to the feasibility of using C to C considerably more broadly than as a SYSIN/SYSOUT transfer mechanism.

Since considerable potential penalty may be experienced due to the channel "lock-in" of the requesting processor, which may result in idle CPU time and would certainly result in deeper bufferings of C to C-oriented requests in the main processor, there seems to be a true constraint on the breadth of C to C usage, and consequently on the reduction of equipment on the main processor. Cost/performance is not characteristically improved with this system because of equipment reduction.

The question must arise, therefore, as to what is an "ideal" environment for an ASP system. Clearly, one wishes very high CPU utilization on the main processor and very high channel utilization and channel overlap on the support processor. To the extent that the work of an installation consists of short compute bound jobs with a high degree of single-input and single-output streaming possible and with voluminous output produced by each short job and further with a significant amount of utility processing required by the installation, ASP will be an effective performer.

The support processor will unburden the main processor from those support-like activities that it can perform equally well. High CPU utilization will be achieved because of the fast I/O response time across the C to C, and a maximum number of devices can be removed from main processor. The requirement for allowing the main or support processor to operate in a stand-alone mode can never, however, be entirely dismissed, even in a pure nonsetup environment, because of reliability considerations. The core of the main processor may be effectively used for expanded procedure space, since deep buffering on the main processor side is not purposeful.

Consider, however, an environment in which there are few rather long jobs characteristically I/O bound and where there are very extensive and complex I/O device requirements. In this situation the main processor will have relatively low CPU utilization and will receive trivial support from the support processor. Since jobs are long, the look-ahead setup capability becomes minimally significant.

If we further postulate that a minimum of human usable output is produced (there is a great deal of sorting, for example) the channels of the support processor will tend to dry. What, indeed will the support processor do during a six-hour main processor run? We shall, therefore, experience both low support utilization and low main CPU utilization.

Various approaches have been taken by various installations to proper scheduling environments for ASP. These involve gross scheduling decisions and can be categorized as being of two major types:

1. *Deferred bad times.* This strategy depends upon the availability of sufficient short jobs to fill a prime shift. Maximums are imposed by the support processor on estimates of CPU time, lines of print, and punching

for a job. Jobs which exceed these limits actually, despite low estimates, are aborted by action of the multifunction monitor. The scheduling and I/O support functions of the support processor are maximized with whatever utilizes that space allows for run at a low priority. Large independent runs are deferred to a second shift, where the support processor may be run stand-alone.

2. *Preferred Long Runs.* The long independent runs, like six-hour linear programming jobs, are admitted to the system; support processing is utilized to perform a high level of necessary utility runs; short jobs are processed in priority order overnight if necessary. Since many ASP installations have a 709X prehistory and run 709X Emulation on the main processor, there is usually a large utility burden on the support processor in connection with 7094-oriented files. "Emulation" is a feature of IBM's that allows certain 360's directly to execute most programs written for IBM 1400 or 700/7000 line computers.

11.12. DIRECT COUPLE SYSTEM

The direct couple system is a chronological predecessor of ASP which, however, involves a potential for considerably closer coupling of the main and support machines. Particularly, it provides a hardware basis for a shift in the balance of master/slave in the direction of the smaller computer and tends further to reduce the larger computer to a computational slave by providing complete I/O stripping of the large machine. The system includes any member of the IBM 704X family (IBM 7040, 7044) as the small computer and any member of the IBM 709X family (IBM 7090, 7094, 7094 II) as the large computer.

The physical interconnection between the units is, as in ASP, a feature that makes the small computer appear as a channel to the large one, providing a path for interrupts. Additional logical facilities are present in the system, particularly the ability of each processor to cause an interrupt (trap) in the other independent of I/O and the ability of the 704X to cause direct transfer from/to its memory from/to the memory of the 709X.

Communication between the processors is, therefore, not limited to an I/O-type command. The interruption of one processor by another is implemented by specific trapping instructions; particularly, the 704X has specific instructions to enable it to interrupt the 709X whether the 709X is running with or without interrupts enabled.

The system may operate in two basic modes, settable by the 704X. One mode called "HIP" (halt on I/O) causes the 709X to halt on all I/O instructions and to interrupt the 704X. The 704X executes I/O and re-starts the 9X when it is complete. This mode, also called "compatibility mode" is used to support programs that violate the conventions of the DCS system.

The system may also be in "direct" mode. In this mode the 704X is trapped and transmits the I/O request data to its core storage, the 709X continuing as if the request were a normal asynchronous channel request.

The fundamental difference between the modes is the specific cause for excitation of a 704X interrupt. In direct mode the interrupt is caused by the execution of a "HEY" instruction interrupting the 704X and caus-ing it to find and interpret the I/O request. In compatibility mode the interrupt is caused by an attempt of the 709X to execute an I/O com-mand resident in its program.

In both modes it is assumed that all 709X I/O equipment is being simulated by the 704X. Data flow from the core of the 704X to the core of the 709X, which is directly addressable by the smaller machine. The interface at the data level is at the block level, so that item level handling is accomplished by the 709X. This means that any buffering system devel-oped in the program running in the 709X will operate in DCS. The effect of this is to somewhat bias performance improvements to systems being put on DCS which previously ran stand-alone without buffers vis-à-vis previously buffered programs. Since a basic design goal of the system is to provide I/O service at memory speeds, the improvement in performance of well buffered, well balanced programs would indeed not be as great as for seriously I/O bound programs previously unbuffered.

The ability of the 704X to access core directly is implemented through the DC channel, which is capable of transmitting a word in 16 μs from one memory to another. As an additional feature of DCS, it was also possible to retain some direct channel control on the 709X, this feature being provided to accommodate devices special to a 709X system, as, for example, the IBM high-speed ("hypter") tapes.

The operating systems environment for the DCS systems, as for ASP, basically involves an operating system in each machine. The 704X system (DCMVP) is designed specifically for the hookup, unlike ASP, where the connection support is implemented as a task under the normal operating system. The 709X operating system is a modified version of IB SYS, the 7090/7094 operating system.

Modifications are in the loader/allocator and the input/output services (DC-IOEX) to support the simulation of 709X I/O by the 704X. Device allocation is performed by the 704X, and communication is provided through common core to map 709X I/O addresses onto 704X allocated devices. A map of relationships between 7040 file/unit assignments and

7090 units is made available on request by the 7090 supervisor at job initiation time, and the 7090 does a "shadow" allocation, mapping 704X symbolic assignments onto its control tables. Scratch tapes and blanks for output are simulated by disc allocation on the 704X; only specifically named SETUP units are assigned 7040 tapes.

The 709X operating system sees all control streams as transmitted to 704X, operating essentially, except for I/O requests, as if it were a stand-alone machine. If it recognizes a request for execution requiring compatibility mode, it issues a request to be placed in compatibility mode to the 704X. The loading of the monitor or of specific system processors is accomplished for the 709X by the support machine. It is possible to transmit directly to 9X memory from disc under control of the 704X.

Programs written for the stand-alone 709X IBSYS system could run unaltered in direct mode under DCS, since only the I/O library was modified to support these programs. Data tapes, however, which did not adhere to DCS format conventions, required streaming onto disc by a routine that edited them into DCS format during the transfer. It is also possible to use a system tape-to-tape utility ot create DCS format tape. All tape input to the 709X must be in this format.

In its general conceptual operation the system is much like ASP, splitting post-, pre- and execution phases across the two machines in an ASP-like way. In a sense, one feels that a minimum realization of the potential of the hookup was realized because of the constraints of compatibility.

The great difficulty with both ASP and DCS and particularly with the latter is the determination of how to deal with already stable software-oriented stand-alone operations. To what extent the loader should run on the large machine, to what extent the compilation function should be moved forward, and to what extent all tapes should be prestreamed to disc.

Each advance in the separation of function implies a cost obsolescence in existing software. Surely DCS has been more ambitious than ASP and more willing to modify existing large machine software. Whereas modifications to OS/360 have been minimal in ASP, rather extensive modifications appear to have been undertaken for DCS to facilitate core-oriented I/O buffering.

Both these systems faced the problems of being late concepts in the environments with which they were associated. The next system we discuss is a further step in the realization of asymmetric processing in that initial design envisioned the functional separation. It is still limited, however, in the realization of the distribution of functions across machines.

11.13.   CDC 6600

We have earlier discussed a machine that very much resembles the "large machine," the CPU of the CDC 6600. What we have not described is the system environment in which this CPU runs. The main frame is isolated from I/O operations entirely.

Associated with the system are ten apparent "peripheral processors," each with a private memory of 4K 12-bit words and an instruction repertoire different from the repertoire of the "main frame." Basic logical, arithmetic, and decision instructions are included in the command list, making these processors true fully realized small systems.

Each processor has an apparent path to all I/O channels in the system and an ability to transfer to and from main memory. Each has an ability to interrupt the main processor and cause it to perform an exchange jump. This instruction basically provides the peripheral processors with control over the execution of the main frame. The exchange jump causes the CPU to "trade" one program for another, picking up the program whose register status and starting location is at an address specified by the peripheral processor, and storing the dynamic data of the program it was processing prior to the EXN interrupt.

The most interesting aspect of the peripheral processors lies in their implementation. Although they operate as, and appear to be, entirely independent processors, they are indeed time-sharing decode circuitry and all control and arithmetic capabilities.

On a fixed time division basis, 1/10 of a full cycle (a major cycle of 1 μs), a peripheral processor gains control of the control circuitry to execute, or attempt to execute, an instruction. There are ten positions in a hardware list of the instruction counters, arithmetic registers, etc., of each processor. When a processor's time comes, it gains control for one minor cycle (100 ns) and executes. At the end of the cycle the next processor gains its turn.

All processor-dependent information is retained in the "slot" associated with the "processor," so that the control and arithmetic circuitry is truly a time-shared, interinstruction independent processor that cares nothing for instruction sequences or dependencies. It accepts a flow of instructions from the set of "peripheral" processors in much the same way that a tube tester would receive a flow of tubes to be tested without regard to the set from which they came.

This process-independent operation is possible because the processor slots retain the dynamic information necessary to show instruction effect for a given instruction sequence. Although a "perhipheral" slot does not have control, it is still capable of transmitting and receiving information from channels or main memory into its private core.

Here is a form of parallelism almost the reverse of what we saw in the main frame. There a single thread was distributed to functional units; here independent threads are collected by a general-purpose unit. The scheme is not unlike the original "multiprogramming" implementation on the Honeywell 800, where eight program statuses were time-division rotated, each program receiving the attention of the CPU as it came "up" in rotation.

The operating system for the CDC 6600 defines one of the peripheral processors as the control processor. This is a software specialization. All other peripheral processors may be "on-line" to provide I/O support for the main processor program or may be privately occupied. Each processor is specialized as a tape, disc, or card processor. This specialization derives from the limited memory of each peripheral processor and the efficiency derived from holding stable resident device handlers in the core of each.

The control processor performs the functions of the support processors in the ASP and DCS systems. It monitors job stacks, assigns resources, dispatches I/O requests, and receives reports of I/O completions. Control by these control processors over the CPU is through the exchange jump; control over other processors is through communications regions in central memory.

Each peripheral processor is directed to undertake an operation when the control places a request in a central memory location, which the peripheral processor scans on a time fixed basis to determine whether there is anything for it to do. The control processor scans the communication region to determine the status of requested operations.

The central processor contains routines that provide an interface with the control processor. Basically the function of these routines is to provide inputs to loading queues which are scanned by control and the requested function undertaken. Loading and relocation are accomplished cooperatively with a peripheral processor program, LDR, performing actual relocation and the CPU cooperative LOADER resolving linkages and resolutions. Control periodically scans I/O request queries, distributing functions to the various peripheral processors (PP's) through the central memory communications region. When control observes that a program cannot continue, it switches the attention of the CPU through exchange jump.

The main frame is truly a computational slave of the control processor. This philosophy is modified somewhat in the presence of extended core storage (ECS), because this ECS, used as a system control and staging area, is directly readable only to the CPU. Consequently, there is a pull

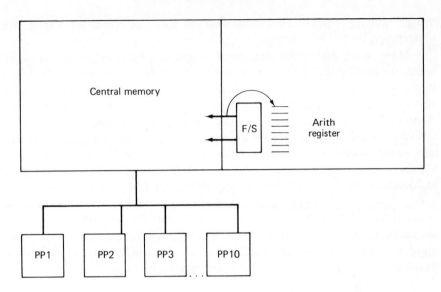

Fig. 11.10. 6600 CPU/PP.

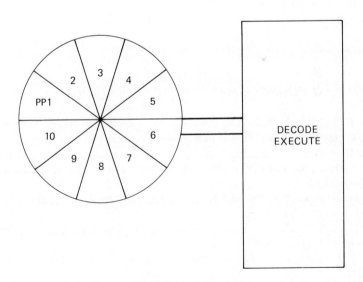

Fixed time multiplexing

Fig. 11.11. 6600 "barrel."

toward moving control functions back to CPU, although all I/O is still undertaken by the PP community.

There is an enormous degree of functional specialization and parallelism in this system. Yet the approach is fairly conservative, the functional split still being the traditional I/O-CPU split in the presence of a control processor. This is a very serious elaboration of the idea of asynchronous intelligent channels, but perhaps the full potential of asynchronous specialization is not yet tested. In a fully realized asymmetric system one could visualize a class of processors that do all the compiling, another for sorting, a third for matrix inversion, etc. Perhaps machine architecture could be dynamically adjusted so that, depending on load and need, different processors could take on the form which best suits them for a given function. Microprogramming, the ability to write the logic of instructions in special fast stores in a very low-level language and implement the instruction set of a machine as macros, might provide a key for a dynamic asymmetric system that would relieve the user from committing himself to specific system capabilities and balances in the face of varying requirements.

As we showed in connection with ASP, there is a general problem of optimizing an asymmetric system. This complicates attempts to build general-purpose systems, which must assume certain operational environments.

PART THREE:
SOURCES AND FURTHER READINGS

Baldwin, F. R., Gibson, W. B., and Poland, C. B., "A Multiprocessing Approach to a Large Computer System," *IBM Systems Journal* (September, 1962).

Clayton, B. B., Dorff, E. K., and Fagen, R. E., "An Operating System and Programming Systems for the CDC 6600," *Proceedings AFIPS*, Spring Joint Computer Conference (1964).

Control Data Corporation, "SCOPE Reference Manual," Publication 60189400 (April, 1965).

Dennis, J. B., "A Position Paper on Computing and Communications," *Communications of ACM* (May, 1968).

Dow, J., "Programming a Duplex Computer System," *Communications of ACM* (1961).

Gosden, J., "Operations Control Center Multi-Computer Operating System," *Proceedings National ACM Conference* (1964).

International Business Machines Corporation, "IBM System/360 Attached Support Processor System (ASP) System Description," Form H20-0223.

International Business Machines Corporation, "IBM System/360 Attached Support Processing System (ASP) System Programmers Manual," Form H20-0323-4.

International Business Machines Corporation, "IBM System/360 Special Feature Channel to Channel Adapter," Form A22-6892.

International Business Machine Corporation, "IBM System/360 Model 50 Functional Characteristics," Form A22-6898.

International Business Machines Corporation, "IBM System/360 Model 65 Functional Characteristics," Form A22-6884.

International Business Machines Corporation, "IBM System/360 Operating System Job Control Language," Form C28-6539.

Leiner, A. L., Notz, W. A., Smith, J. L., Weinberger, A., "Pilot A New Multiple Computer System," *Journal of ACM* (July, 1959).

Roberts, L. G., "Multiple Computer Networks and Inter Computer Communications," *ACM Symposium Operating System Principles* (1967).

Schmitt, W. F., and Tonic, A. R., "Sympathetically Programmed Computers," International Conference on Information Processing (1959).

Smith, E. C., "Directly Coupled Multiprocessor System," *IBM Systems Journal* (September, 1963).

4

Input/Output, Multiprogramming, and Operating Systems

Chapter 12

CONTEMPORARY
I/O SYSTEMS

12.1. REVIEW OF I/O DEVELOPMENT

In an earlier chapter (Chapter 6) we discussed the separation of the CPU
from the total control of the I/O operation in order to introduce the
concepts of distributed intelligence and localized function on a hardware
basis. This separation is especially interesting, however, for its own sake
and for its impact on programming and system techniques. Considerations
of utilization, performance, and efficiency of resource use, such as we
have demonstrated for the arithmetic operations of high-performance
CPU's, apply directly to CPU/I/O usage with perhaps considerably greater
impact on the utilization and productivity of systems. Without the concept of parallelizable I/O, two generations of software concepts would not
have developed, and the current state of the programming art would not
have been achieved.

The fundamental transformation is the step from the sequential read
only, compute only, write only machine to the read/write/compute machine. In the sequential machine, since all operations, read, write, or arithmetic, preclude any other operation, the total time to run any program is
the sum of all write time, read time, and compute time. The concepts of
I/O limited or CPU limited do not exist, since no conceptual separation
between I/O time and compute time is available. Regardless of the distribution of devices in the system, each record read or written contributes a
period of time to a single line representing elapsed time on the system.

12.2. VARIATIONS IN OVERLAP

Given a WRITE function, a COMPUTE function, and a READ function, we can see from Fig. 12.1 that there are various possibilities for
overlap. The purely sequential machine, compute or read or write, is our
initial condition. We have enabled the second case (compute and read or
write). A third possibility, compute or read and write, has also been
implemented in extensions of the IBM 705. In many situations this capability would make considerably more sense. This is because it would allow
us to parallel the more significant paths of read and write when they were
both significantly longer than CPU time.

1. Read or Write or Compute
2. (Read or Write) and Compute
3. (Read and Write) or Compute
4. Read, Write, and Compute
5. Read/Read, Write/Write, and Compute

Fig. 12.1. Overlap

In the calculation of elapsed time in the READ/WRITE or compute
machine, we select the longer of READ or WRITE time to add to internal
time for total elapsed time.

Finally there is the READ/WRITE/COMPUTE machine capable of
doing all three at the same time. The elapsed time represents the longest
of the three parallel paths. This longest line identifies the limiting function, and the device performing this function is the limiting device. Under
no circumstances can a run complete in less time than that required to
perform the limiting function.

The basic R/W/C machine is capable of a single read and a single write
with compute. In its contemporary and generalized form it is capable of
various patterns of multiple parallel reads and writes. In intermediate
forms some limited capability for READ/READ or WRITE/WRITE could
be provided. With multiple read and write machines the limiting function
is either the CPU or the time to read or write the data that takes longest in
the system.

We may call this limit the "hypothetical optimum" run time for the
program. It is a goal to be achieved, and the difficulties in achieving it for
any program or for any population of programs on a given machine are
impressive indeed. As an aside, this particular optimum implies nothing
about the quality of the programming or coding; it merely indicates that
as written what the minimum running time could be. If the "critical path"
was the CPU, an attempt at recoding for more efficiency might well be in
order.

We remember our manipulation of utilization for I and E boxes and the relationship between "throughput" and utilization. Since we are going to take, on a R/W/C machine, only the time of longest path to complete and since we are not changing the amount of work each functional unit must perform, we have increased the utilization of the CPU and each device. We have, by allowing the simultaneity, reduced the elapsed time and consequently the sparseness of usage.

In effect we have created an opportunity for the CPU and the I/O devices to work harder, since they are no longer inherent structural bottlenecks. They are even more independent than an I and an E. But of course they do represent limits on each other, intermittently bottlenecking in much the same way that an instruction processor and an execution unit do.

The concept of the limiting resource is also useful, because it identifies that unit whose utilization must be kept at a critical load for the duration of the run in order to achieve the hypothetical optimum time. It is the mutual task of a programmer and a system to sustain this usage.

12.3. HARDWARE BUFFERING—TWO EARLY MACHINES

There are various techniques for achieving overlap, and various tradeoffs have been made in order to achieve it. The buffered computers referred to in this chapter and generally described earlier paid a price of fixed physical record sizes. When the interfacing device was essentially a fixed record size device, this represented no additional penalty for the system. The 650 and UNIVAC Solid-State buffered card images on their drums. When tape was introduced on the 650, a core buffer was added to the interface between tape and the CPU. This core buffer was addressable by the programmer; particularly, he had the capability of setting tape block sizes with an instruction that defined the location and extent of the input or output block. Variable physical block size was possible up to the size of the core buffer unit. Since there was only one buffer and this had to support all devices other than card equipment, traffic became a little heavy in this unit. Essentially the machine performed as the compute and read, compute and write system that we introduced above.

Early UNIVAC equipment used two fixed-size, nonaddressable buffers, RI and RO, allowing a simultaneous read/write/compute. The system had the backward read feature on tape from the beginning and performed a reversal so that on a backward read data words arrived in the buffer in proper order. The basic tape read command called for a transfer from tape to the buffer unit of a fixed-size block. The processor was released at a point after startup, and information came into RI asynchronously. Since RI was a private memory for input, no cycle stealing or processor delay of any sort occurred during the operation.

If an input instruction was given during the activity of a prior instruction, the processor "interlocked" on the completion. An interesting feature of the structure was the combined buffer transfer to memory, unit transfer to buffer. When the "I tank" (RI) was filled by an initial read, all subsequent reads could simultaneously read out of the buffer and into the buffer. The effect is very much like the two-doored elevators on the Eiffel Tower or the London subways. When a stop is made, the doors on both sides open; those in the elevator go out one door and those waiting to get on the car go in the other simultaneously. The scheme was basically the same for the card unit buffering on the 650, where a read instruction transferred from a buffer into fixed positions on the drum and from the ready to the buffer simultaneously. These buffering schemes are inherent in I/O interfacing. Control units characteristically have small buffers in which they collect characters or words of data in order to smooth the rates of speed of delivery from a unit and acceptance by a memory. The approach of large independent hardware buffers to achieve simultaneity became unpopular in the late 1950's for a number of reasons. The fixing of the block size was considered to be a disadvantage because of the inflexibility of data organization that was imposed.

Second, the early system imposed real limitations on what could be accomplished. A parallel read and a parallel write were not always truly significant. One can easily postulate a program with considerably more reading than writing with the input sets distributed across multiple units where parallel read/read might be significantly more use than read/write. This is a limitation on the read and compute or write and compute machine also. The assumption of a balance between input and output is not valid over such a significant population of programs that the inflexible buffering schemes are not adequate. It is possible, of course, to provide a buffer for each unit on the system, but this is expensive and does not solve the fixed record size problem.

Lately some concepts have developed that might cause a redevelopment along these lines. The fixed record size is coming into some greater favor, because of the desire of proponents of virtual memory/paging machines to extend the concept of paging and the transparency of data locations to the I/O area. The data associated with a program are conceived to be part of an enormous conceptual space that is addressed like real core. Reference to data would be supported by the mechanisms described later (Chapter 34). Since current paging systems are organized around fixed-size pages, supporters of the concept would like to see I/O "blocking" conform to page size.

Another development is the increase of interest in so elaborating I/O subsystems that they in effect become an I/O processor with a local memory acting as data buffers. For the moment, however, external device buffering seems unlikely to replace cycle-stealing, where, in effect, memory acts as the ultimate buffer for I/O subsystem devices.

12.4. CHANNELS AND PATHS

The interest shifts from providing a separate residence to providing paths and in determining what distribution of intelligence should be made along the path from a CPU to a device and back to memory. Basically, a channel provides a path for data and control signals to be passed. Channels vary by characteristics of speed (how many units per second of data they can pass), by transfer width (how many bits are passed in parallel), and in functional capability. Simplex channels pass data in one direction only; half-duplex channels pass data in either direction, but not at the same time; full duplex channels pass data simultaneously in both directions.

Channels are limited in their association with devices by the transfer rate of the devices; a low-speed channel is obviously unable to handle a device too fast for it because the delivery rate of data from the device is higher than the rate at which the channel can accept it, causing either overlayed or lost data. In many systems there is a mixture of channels by speed. General Electric, for example, offers two separate speed channels in its configuration; IBM fundamentally does this but goes farther in the elaboration of channels and channel logic to offer what amounts to general-service high-speed channels and specialized low-speed channels for card, tape, and printer.

Specialized channels are a feature of many systems where, in addition to general-purpose channels, there are specialized printer channels, card reader channels, etc., where some special logic accruing to the device exists at the channel interface. To the extent that special logic exists in a channel, of course, the flexibility of the I/O system is retarded.

At the CPU interface one finds a variation in the amount of support that the CPU provides for I/O operation, the extent of true independence of channels from the CPU and from each other, the degree of integration of the hardware supporting channel operations, and the hardware supporting the CPU. IBM's 360 line, for example, while preserving apparent logical identity, varies its I/O implementation seriously for each machine in the line, with true independent "external" channels appearing only at the 360/65 level.

On the UNIVAC 1108, control words are kept in memory, a memory access is needed for each word transferred to memory to acquire and modify the "access control word" to reflect the count of the number of words transferred. In the presence of an I/O controller (IOC) this information is sent to the IOC, and the channel interference for control word update is eliminated.

In System 360/65 and higher, all of this is accomplished in the channel. The channel in the IBM sense is a small processor with a set of registers and an instruction set that it performs. It contains a pointer (sent to it by the processor) to a list of memory locations containing a command list for it to execute; it forms and passes interrupt words from the devices to memory; it controls words and record counts.

Perhaps no more elaborate and intelligent channel has been built than the 7904 channel for the IBM 7094. Even the predecessor and more popular 7607 channel could be programmed to accomplish unexpected things, and "channel programming" reached a high point of fashion with this machine. The basic capability of an intelligent channel is data or command chaining, and in an unusually intelligent channel conditional transfers around an instruction list and even dynamic relocation.

The I/O pattern is fairly common at this point. There is a CPU instruction START I/O or INITIATE which specifies a device. The address is commonly decomposed to channel address, control unit address, device address by the system. The CPU execution causes a word to be fetched by the channel from a fixed location. This word contains the address of the beginning of I/O coding. I/O code is represented by control words containing a function (READ, WRITE, etc.), an address, and a count. In 360 these are called channel command words, on 1108 access control words, on other systems other things, but the function is the same. This is almost an industry standard, and it is popular because it works well, providing a flexible form for command structure and allowing quite a variation in the expense of the I/O subsystem required to support it.

Those systems that allow the definition of an I/O record to be read in segments of blocks are now rather common. In its most elaborate form an I/O command structure can be fundamentally independent of physical representation on devices. The capability comes from "chaining" of individual I/O commands. If a program wishes to define an input record that consists of words 17–33 of an initial 60-word block followed by words 6–9 and 47–52 of a second block, he may do so. He may form a chain of I/O commands associated with the instruction to perform a read such that the first words of block one are skipped, words 17–33 are read, the final words of block one are skipped, initial words of block 2 skipped, etc.

The device will cross the interrecord gap at full speed. The chaining capability is a feature on an "intelligent" channel. On nonchaining machines there also often exists the possibility of crossing the gap by submitting an I/O command to a device within a given amount of time of a previous completion. For example, on a Univac 1050, the time might be 700 μs. In some earlier systems a limited form of chaining, which permitted definition of a logical record within a physical block, was available. This feature, "gather read" and "scatter write" on Univac III, allowed for the distribution of words on tape to noncontiguous locations in core or the collection from core of noncontiguous locations into continuous blocks on tape.

The "chaining" consists of relating "command" words to each other. In the IBM 360 there are two ways of doing this. One is to chain "data." This is basically scatter/gather I/O, where segments of a block may be collected or distributed. The other is to chain "commands," where new functions can be processed against data structures. A bit in the channel command word (CCW) determines if chaining is occurring and whether it is command- or data-chaining.

At the other interfaces a channel has devices or control units. It is common to specialize whatever is unique at a channel/device interface into a control unit. The control unit (CU—also called synchronizer) is a specialized unit that contains logic and capability required by a given unit—a tape or disc at a channel. Rather than specialize the channel, a CU is developed that interfaces with the channel. This permits the channel to interface with other type CU's as well, keeping the flexible generality of channel operations. On the other side, the CU allows considerable economy in device development, allowing the time-sharing of logic by tapes or discs which would otherwise have to be duplicated in each device.

The question of balancing and machine efficiency that has been traditional at the CPU/I/O interface is no less critical now. There are two apparent schools of thought, which might seem to polarize around the 360 and the 1108.

The 1108 system attempts to balance its CPU with a large number of relatively unintelligent channels. The 360 appears to attempt balance with relatively few more intelligent channels. We doubt at this point that any IBM technologist would deny that the high end of the 360 line is under-channeled, that the number of bits per second required to maintain CPU activity are just not available from the channels available with the system. This is partially because of a problem at the interface of channels and disc control units, which requires that a channel devote additional attention to certain I/O during times when it might be thought to be free. But it is particularly because there is just not enough transfer in/out power available to support the incredible speed of a 195. More channels might surely be helpful for the high end of this line.

On the other hand, the development of IOC by UNIVAC and similar devices by GE indicates that they feel that they suffer in some cases because of the stupidity of the basic channel. One must put in a sentimental word for the 7094 with its highly intelligent channel and with the impressive flexibility a user had in determining how many channels he should have on his system, choosing from one to ten of the highly evolved channel units.

12.5. INTERRUPTS—MINIMAL SUPPORT

Of particular importance to a multiprogramming machine is the interrupt. In a single program the presence or absence of an interrupt system as opposed to buffer testing or interlocking will make little difference in performance. However, a full description of the multichannel asynchronous computer is impossible without some comments on interrupt. This is a concept central to asynchronous control.

The principle of the interrupt is commonly known and is rather simple. Five elements are involved: a signal, which causes the phenomenon of interruption; a place in core associated with the signal from which a next instruction is to be taken (or the address of the next instruction); a place in core to place the location of the instruction normally next to be executed if the interrupt had not occurred; a mechanism for precluding other interrupts for some interval of time when an interrupt has occurred; and a place in core for reflecting the exact cause of the interrupt and which identifies the source of interrupt and the status of the action related to the interrupt.

Various systems differ in the organization of these five elements. In the most primitive systems, for example, it is possible to have a single fixed address in low core which holds the address of the next instructions to be executed for all interrupts. There would similarly be a single fixed address in which the location counter was placed for all interrupts and a single fixed address for the interrupt status word. When a signal was developed which caused an interrupt, the location counter would be placed in low core and the new instruction address would be brought to the location counter, which would bring up the instructions to the instruction register.

Alternatively, the instruction itself could be held in low core and immediately brought to the instruction register. This could be accomplished by placing the fixed location of the instruction to be executed at an interrupt in the location counter as part of the basic hardware interrupt function.

The function of the first postinterrupt instruction must be to disable other interrupts. Instructions of this type exist on a number of machines, including UNIVAC III and the 1108. The interrupt handling routine now performs an analysis of the interrupt by investigating the interrupt status word. This word contains a pattern of bits that indicate the type of interrupt that has occurred. There are two basic interrupt types in a simple interrupt system, I/O completion and processor error. A code representing each type is defined for the system and identified by program action.

When the interrupt type is determined, two things can occur. The system may execute an instruction that permits interrupts of other types than the one being processed to occur. This would involve the transfer of the interrupt status word from its fixed location to another word in core and with it the transfer of the location counter stored at the time of the interrupt. A rather complex software support is implied, because it is now necessary to provide a software mechanism to "walk back" through the interrupts. In most multiprograming systems the low core locations are transferred into various system areas that are associated with the program that has been interrupted. Second, a further analysis of the interrupt within its type is undertaken. If the interrupt is determined to an I/O interrupt, it is necessary to determine what channel caused it to occur, what device on that channel, whether it is a successful completion, and what error has occurred if it is not a successful completion.

After determining the source and nature of the interrupt, the interrupt handler may call upon a number of service routines to respond to the interrupt. In the case of successful I/O completion, one of the things that might be done is to determine if there exists an unqueued request for I/O on the now free device.

When all interrupt-related functions are complete, the interrupt routine returns control to the interrupted program (or to some mechanism to determine to where control should be granted). In a system with a disable interrupt instruction this may be accomplished by a "jump indirect" and "enable interrupt" instruction. The location counter value of the interrupted program resides in low core; the indirect jump specifies that address in low core as holding the address of the next instruction to execute.

This is a feature worth dwelling upon, since many contemporary programmers are unfamiliar with it. The IBM 360 does not have indirect addressing in the generally accepted meaning of the word. When an instruction references an address directly, it is the contents of that address with which we wish ultimately to deal. A BRANCH X designated that X is the location we wish to execute from next. A BRANCH X* (using the SLEUTH II convention for representing an indirect address) designates that at X there is an address and that that address is the place from which we wish to execute the next instruction. If X contains Y, the next instruction will be executed from Y.

There are many variations in the implementation of indirect addressing, usually concerning what instructions can and cannot specify indirection and whether there is more than one level of indirection and how the indirection is finally stopped. It is a powerful feature for a machine. UNIVAC III, UNIVAC 1107/8, IBM 7094, and other machines had this feature in some form. In the IBM 360 it is approximated in a sense by the load address instruction.

If there is no indirect addressing on the machine, then a load location counter instruction is needed in order to place the return point in the location counter. With this logic a separate ENABLE interrupt instruction is required.

12.6. MORE ELABORATE INTERRUPT SUPPORT

This minimum system is fairly inadequate, because the amount of analysis required to determine what has happened is much too long and the work required to enable any other interrupts during the analysis is too cumbersome. Consequently, we usually find more elaborate interrupt systems.

The fundamental elaboration is to provide separate interrupt locations for the basic types of interrupts allowed on the machine. I/O interrupts, processor error interrupts, and operator interrupts each have different location counter store and next instruction fetch locations on the UNIVAC III. When an interrupt occurs, it is no longer necessary to determine which basic type it is, since the routine that is entered to handle interrupt is entered because an interrupt of its type has occurred.

When this separation is made, the overhead of allowing other types of interrupt to occur is much reduced, since there are independent location counter storage, first instruction, and interrupt status words for each interrupt class.

The concept of relative interrupt priority is formed here. Each interrupt type is associated with a priority such that any interrupt of lower or equal priority cannot occur during processing of a type, but any interrupt of higher priority may occur. In order to allow interrupts of lower or equal priority, the "enable" instruction must be used. The effect of separation interrupts in this fashion is, therefore, both to reduce analysis time and to provide hardware support for more flexible interruption.

Beyond elaborating positions for class, some systems (UNIVAC 490, for example) provide separate interrupt locations for each channel in the system, eliminating the need for determining by analysis which I/O channel has caused the interrupt to occur. A common approach found in the

1108 and 360 is to elaborate the locations for instructions but to have multiple interrupts share status words. This is an attempt to provide for quicker analysis but to reduce the amount of core required.

12.7. IBM 360 INTERRUPT SUPPORT

The IBM 360 interrupt system is fairly elaborate. New classes of interrupts have been defined for the machine. This is of particular interest, since their involvement in the support of multiprogramming is fairly direct. In addition to processor error (machine check) interrupts and I/O interrupts, there are program and supervisor call (SVC) interrupts on the machine. The supervisor call interrupt is executed when the program executes an SVC instruction. This instruction causes the CPU to enter supervisor state. It is necessary to be in this state to execute certain instructions on the machine. All I/O instructions and all instructions having to do with machine status and storage protection, particularly those which can effect the enabling or disabling of interrupts, must be executed in this state.

The system does not have a direct enable or disable instruction, but enables and disables by "privileged" manipulation of an extended location counter described shortly. The privileges of the supervisor state versus program state is a fundamental characteristic of contemporary multiprogramming machines. The basic intent is to prevent a user program from undertaking such actions as to endanger the integrity of another program or to undermine the intent of the control software of the machine.

In the UNIVAC III the instructions to enable or disable interrupts could be used by anyone as they could in the 7044. In the 360, instructions with this effect may be executed only in supervisor state. This is true of all contemporary equipment. In addition, a program interrupt will occur whenever a reference is made outside of a program's own memory space and when an attempt is made to execute a privileged instruction.

The IBM 360 provides low core instruction store areas for external program, supervisor call, machine check, and I/O interrupts. Any I/O interrupt, therefore, goes to the same location. There is one I/O interrupt status word (channel status word, or CSW). Actually, the CSW as defined for 360 is more than an interrupt status word. It contains all the information required to provide a complete picture of channel condition at any point. For this discussion, however, we shall be interrupted only in the STATUS field of the CSW, that is, in that portion of it which is an interrupt status word. Any malfunction of the I/O subsystem by device, channel, or control unit is indicated; the busy or free condition of the subsystem element referenced is indicated.

The contents of the channel status word are not sufficient for com plete interrupt analysis and control. A 360 feature which involves itself here is the PSW, or program status word. Whenever a program has control of a 360, a considerable extension of a location counter is associated with it. In addition to the location of the next instruction, the PSW represents the current value of the condition codes (set as a result of compare and arithmetic instructions), the bit pattern that must be associated with a memory area to make it legitimately addressable by this program, the supervisor or problem program (privileged or not privileged) status of the machine, two mask fields, and an interrupt code field. These last two are of particular interest.

The two mask fields are the system and the program mask. Their use is in the enabling and disabling of interrupts. The system mask contains eight bits; the program mask contains four. Each bit corresponds to a particular interrupt source. Bits 0 through 6 of the system mask represent channels of the system. The bits of the program mask represent possible sources of arithmetic interrupts (overflow, underflow, significance). When the bit is "on" (set to one), the interrupt associated with it can occur (it is enabled). When the bit is off, the interrupt is inhibited. Instructions "set system mask" and "set program mask" exist to alter the status of interrupt allowability by storing the location of designated fields into the "current" (active) PSW. Since these are privileged instructions, only a supervisor state routine can effect the masks. This is the direct equivalent of the enabling and disabling instructions of other systems.

The interrupt code field of the PSW is used to identify the sources of interrupts. Each interrupt places an identifying code in this field. On an I/O interrupt each channel is identified in the field of the PSW. Analysis proceeds, using the PSW for source identification and the CSW for status.

It is, of course, the human condition that each design advance, when accepted, generates a new set of objections. It is currently true that the elaboration of interrupts and the burden that interrupts place upon a process are beginning to generate some criticism, and alternate mechanisms that allow more CPU discretion are coming back into discussion (reliance on memory-based flag lists, time-driven status inspection). However, it is surely sane to predict that the next ten years will see more reliance upon interruption and more elaboration of interrupt mechanisms.

12.8. SOFTWARE RESPONSE TO I/O DEVELOPMENT

With the development of the asynchronous channel-control unit-device, interrupt-oriented I/O system, we may begin to look at some of the software responses to such systems.

Programmers responded in two ways—first with a realization that the design of the I/O interface had become sufficiently complex so as to represent a burden to a user or programmer who wished to get on with the business of solving his problem. The golden era of "Read a block from Tape 3 into 500" or "Punch Card" of the early commercial machines was gone. The feeling developed that the true I/O interface had to be "covered" in some way, made "cleaner" to the user. There were just too many things to do to write or read. A basic I/O support package developed which ran with a user program and was called upon to do I/O and to handle interrupts.

The support of I/O's is available in modern operating systems at three general levels. The basic level, the "device level," provides support for handling the eccentricities of the device; the user maintains control over all auxiliary functions such as core location assignment, block assembly, and disassembly.

In OS/360 this is the "EXCP" macro level. The interface between the user program and the device is through the device handler, by way of the macro expanded in the user's program. A second level, the READ or WRITE level, provides for system control of all activities necessary to present a block of logical records to the user, a single physical block. The third level, the GET, PUT level, provides the user on request with a logical record, a true item of his file. More and more functions have been added to the I/O support area until it has been transformed into "DATA MAN-AGEMENT." This involves the maintenance of catalogues indicating the location of data files on line to the system, and perhaps basic file maintenance functions, such as file update. Another characteristic of the support of modern I/O architecture was to bring to the industry an idea from the basic R/W/C machine which had already been highly developed by UNI-VAC I people, the concepts of buffering, buffer control, buffer pooling. The importance of the idea of buffering we shall explore presently.

Chapter 13

PERFORMANCE OF
PROGRAMS
IN A
UNIPROGRAM ENVIRONMENT

13.1. INTRODUCTION

Notice Fig. 13.1. What is shown is a simple tape file update program
reading from a master tape and a change tape and writing out the master
on a new master tape. We shall look very closely at this program in order
to derive some observations on the performance of a program on a machine. We see that elapsed time on an unoverlapped machine is the cumu-
lation of all functional times associated with the processor and with I/O
component performance. For our primitive program example we shall
assume some file characteristics, as shown in Fig. 13.1. The expected
compute time relates to an assumed 30 percent activity in the master file.
Seventy percent of the time processing will be required only to inspect
and advance past a record; 30 percent of the time more work must be
done.

The expected record processing time by the CPU is the relative frequency of active records times their processing time plus the relative
frequency of inactive records times their processing time.

Notice our I/O characteristics. Associated with the input/output device are some characteristic hardware stages, each with an associated time.
To start a tape drive from a stilled status requires a certain amount of time
until the tape is moving at full speed. This is a number that has varied
wildly from system to system over time, being as high as 600 ms and as
low as 5 ms. Figure 13.1 shows a 10 ms start time for a postulated type
drive for use in our application.

Another interesting number is shown in Fig. 13.1, "Gap pass time."

Input file: Tape "1": 50 word records, 10,000
Change file: Tape "2": 30 word records, 3,000
Output file: Tape "3": 50 word records, 10,000

ASSUMPTIONS
1.   No blocking
2.   Process time
      A. Active record: 50 milliseconds
      B. Passive record: 10 milliseconds
      C. 30% activity
      D. Expected process time = .70 x 10 + .30 x 50 = 22ms
3.   Hardware times
      A. Start time (tape inactive to pass data status): 10 milliseconds (read or write)
      B. Gap traversal at full speed   5 milliseconds
      C. Time read or write one word   1 millisecond
4.   Program time, Read + write + compute
      A. 3,000 reads of tape "2" at 40 milliseconds per read = 120,000 milliseconds
      B. 10,000 reads of tape "1" at 60 milliseconds per read = 600,000 milliseconds
      C. 10,000 writes of tape "3" at 60 milliseconds per write = 600,000 milliseconds
      D. 10,000 computes at E (comp time) of 22 milliseconds = <u>220,000</u> milliseconds

                                     1,540,000ms = 1,540 seconds

      R2 R1 C W R1 C W   .   .   .

Fig. 13.1. A classical sequential update on a nonoverlap machine.

There is a space between records on tape commonly called the interblock gap or the interrecord gap. The size of this gap is a function of the construction of the unit and is related to the speed at which a tape drive can be brought to a stop after performing a read or write operation and the speed at which it can be brought back to speed. After completing an I/O operation, a device must be brought to a halt; that is, there is an interval of time after the completion of the transfer of data during which the device is still operative and requires monitoring and control.

In a purely sequential system this time would be added to the CPU, since it would have the responsibility of bringing the unit to a halt. The distinction between data transfer and time and unit free time is common to contemporary systems. In the 1108, for example, two interrupts are available, one at the time that the data transfer specified by the operation is complete, and one when the device is completely settled after the completion of the operation. The "gap passing time" is the time that is required for a drive at full speed to pass over the space between records. In some systems it is possible to submit I/O requests such that more than one block is read in order to fulfill the request. In other systems there is a period of time after the end of data transfer and before device release where the submission of a new I/O function to the device will inhibit its final "shutdown" and avoid the necessity for a full restart.

Our example will assume start/stop mode; consequently, with an assumed 32-bit word and a capability of reading one every millisecond, reads and writes of the master file each will require 60 ms (including startup and making no charge for shutdown. Reads of the change file will require 40 ms.

The complete time for this run is shown at the end of Fig. 13.2 to be 1540 seconds. If, for some reason, this number is not acceptable, either because the output is required less than 1540 sec (26 min) from the availability of the input, or because of a general need for more work to go through the system in a given interval (a shift), a number of things can be done. One thing, of course, is to get a faster machine.

But faster in what sense? By "faster" one can mean that the characteristic time it takes to do things is shortened by shortening the "length of the line," as it were. We see easily that if tapes were doubled in speed of the processor, or both, we would certainly decrease the time of the program. By "faster" we may also mean what we have meant up to this point: a reorganization of functions so as to shorten the line by removing segments from it and placing them alongside.

There are solutions other than a faster machine. One is to replan the work. In the middle 1950's, when IBM's 704 and 705 had no parallel capability, it was frequently recommended that one organize update systems so that only the changed records were written. In our example this would eliminate 7000 writes or 420 sec, or nearly 30 percent of elapsed time. This would, however, involve some end-of-cycle additional work to form a complete updated master. Also, of course, in many applications it was necessary to read the "change" tapes during the update run. The UNIVAC I and II were read/write/compute machines at this time (as was IBM's 650) which characteristically avoided any necessity for deleting writes.

We shall at this point look at our update program and all the impact on it of the multichannel machine.

CPU
   1. CPU executing non I/O: 220 sec, utilization = 14%
   2. CPU executing I/O (WAIT) 1320 sec, idle = 86%

I/O
   1. Tape "1" utilization: 600 sec, utilization = 39%
   2. Tape "2" utilization: 600 sec, utilization = 39%
   3. Tape "3" utilization: 120 sec, utilization = 10%
   4. Total I/O utilization: 1320 sec, utilization = 86%

   Maximum speed-up if CPU/I/O 100% overlap = 1320 sec

Fig. 13.2. Utilization of system during 1540 seconds of elapsed time of update run.

From Fig. 13.3 we see that we can allocate one input tape (the master) to one channel, another input tape (the change) to another channel, and the updated output tape to a third channel. This enables us to run three channels in parallel and reduce elapsed time to 600 sec.

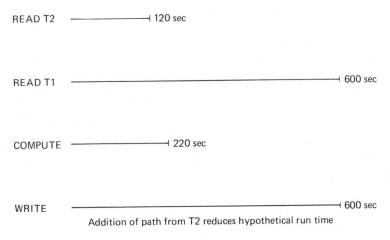

READ T2 ——————————⊣ 120 sec

READ T1 —————————————————————————⊣ 600 sec

COMPUTE ————————————⊣ 220 sec

WRITE —————————————————————————⊣ 600 sec

Addition of path from T2 reduces hypothetical run time

Fig. 13.3. Total elapsed time with parallel paths of R/R/W/C/ system.

13.2. BLOCKING

We have at our disposal a very popular and effective technique for speeding the run beyond this, and it might very well pay us to apply it to our limiting path(s) in order to shorten them.

We have a time, "start time," which represents for our 50-word records one-sixth of the total time necessary to read or to write. We can achieve a distribution of this startup time over multiple records, however, if we employ the technique of blocking.

Blocking effectively increases the rated speed of a device by increasing the amount of information available in memory from it during any time interval. The effect of blocking 10 logical records to a physical block is to distribute the 10 ms start time across 10 records, giving an average record read time of 50 ms plus 1 ms start time, a total of 51 ms per logical record.

Time to read or write 10,000 records becomes 510 sec, and since total elapsed time on the critical resources is reduced, then the actual run elapsed time is hypothetically reduced. Now, of course, the limit here is the 500 sec that would represent the time to read 10,000 records if the tape were started before the run (to absorb initial start time) and read continuously at full speed one uninterrupted record of 500,000 words. This is neither commonly achievable or a desirable limit to press for. It is excluded by placing a reasonable limit on core size available for a block.

There are other limits to block size. In a sense, the formation of a block is the development of a queue. In a sequential tape situation there is 100 percent predictability that every element of the queue will be processed and that no loss will occur due to the commitment of system resource to the transfer of data irrelevant to the process.

Blocking is just a form of "prepaging," and the penalties for bad guessing can be high. In a disc situation there is a "startup" time that is considerably more significant than tape startup time. This is "seek time." Certainly, one would like to acquire as much information as possible from an individual seek; that is, one would like to maximize block size on disc as well as on tape. With a disc-oriented system, however, there is a certain possibility that the next record that is desired is not the ordered successor of any referenced record.

Sequential processing on disc is an approach to system design, but it is not the one being most explored with this device, and a high degree of "surprise" is associated with disc-oriented systems. Serious expansion of block size in such an environment tends to increase the possibility that multiple records will be processed in the same blocks, but it also increases the probability that some records will be transferred into scarce core over precious channel time and that time and space given over to this will be a total loss.

The size of a block is limited by its competition for core with other users. Other input files, output files, and coding compete for core space that can be portioned between them. Demands for larger block space involve tradeoffs which should be based upon appreciations of critical performance factors, but which characteristically are not.

An interesting element in the relationship between CPU and I/O timing is that blocking, which has a tendency to lower I/O time, has a tendency to increase CPU time, since it is now necessary to provide record advance and block control routines that increase the amount of processing for each record in the processor.

A strongly CPU-bound run could very well, depending on the I/O processor synchronization pattern, run longer in the presence of dense blocking than without it, particularly if the space for the large block precluded more use of core for opening loops, etc., which would result in reduction of processing time.

These three elements—CPU burden to control a feature aimed at providing system efficiency, the contention for memory space, and the optimization of I/O—are the three legs of the stool of performance, and consequently of multiprogramming and by extension multiprocessing design.

13.3. BUFFERING — PERFORMANCE WITHOUT IT

The underlying idea of the buffer is to improve the interface between asynchronous functions so as to allow asynchroniety. Consider our example without the allocation of buffer space. In Fig. 13-4 we show a profile reminiscent of our charts for the high-performance processor. We are assuming here that we shall block all input and output, but we shall make no charge to CPU for block level processing.

The hypothetical minimum run time for this run is now at 510 sec, as we have before shown. In order to achieve this, however, it is necessary that a read is given whenever the input channel can accept one and that a write is given whenever the writing channel can accept one for the read and write of the master tapes. That is, it is necessary that there be no gaps in the elapsed time line of tapes 1 and 3. There may be gaps in CPU and change tape utilization, of course there will be, without a necessary delay in completion time, but it is possible for certain patterns to form that will preclude minimum running time, and this is shown by Fig. 13.4.

With ten records committed at a time in a single block from both the change and master, there is an initial delay to the CPU that cannot be avoided, but the time of this delay does not offset our run time once we are not pressing for utilization of the CPU in order to finish "on time." At the time of delivery of the block, at 510 ms after initiation, CPU processing can start.

We will assume for this example a very rigid 30 percent activity, so that for every ten records there will be exactly three active, and CPU processing time for each block will be a constant 220 ms. We can say in this instance that this run is I/O bound with some assurance, but the looseness of this concept will be revealed when we relax that distribution rule.

Processing of the first record does not begin until 510 ms after a read. During the time that a block is being processed, no read can be given to the input master, because there is no place to put an entire new block. Consequently, there is effectively no overlap between processing and input.

We can see that there is almost perfect overlap between read and write master but that a problem develops between the master and the change file. At the find of the tenth changed record it is necessary to read a new change block. During the read nothing else can happen, because there is truly nothing else to do. The regularity or probability of this cannot be known previously, of course, because the distribution of changes will not be as shown here; on some running of the file the additional time for unoverlapped change reads may be better, and on some worse, than what we have shown here.

What we are observing in general here is that in the absence of buf-

206

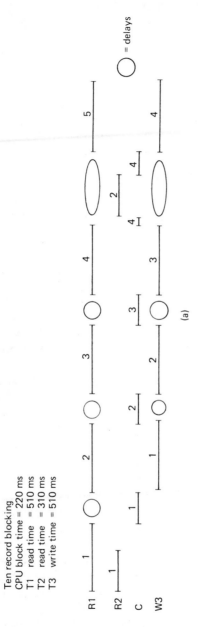

Ten record blocking
CPU block time = 220 ms
T1 read time = 510 ms
T2 read time = 310 ms
T3 write time = 510 ms

(a)

Five record blocking
CPU block time = 90 ms (1 active) or 130 ms (2 active)
T1 read time = 260 ms
T2 read time = 160 ms
T3 write time = 260 ms

(b)

 = delays

Fig. 13.4. Performance with various record block sizes.

fering asynchronous machines tend to degrade into sequential machines. Since there is no CPU overlap with I/O and since we show a constant 220 ms gap between reads and writes, we compute the "effective" time of the limiting path to be 730 ms per operation, or a total run time of 730 sec without counting the further delays caused by unoverlapped reading of the change tape, which one could estimate to be around 80 additional seconds, showing a total run time of 810 sec. This is achieved on a machine using R/R/W/C with a hypothetical minimum of 510 sec. The true performance is less than the potential for a R/W/C machine. (Certainly it leads one to observe that T_1 and T_2 could be put on the same channel with very little further decay in performance.)

13.4. IMPACT OF BUFFERING ON PROGRAM PERFORMANCE

The reason for this decay is (indirectly) core. Given a core which could hold only 1300 words of input/output work space in the presence of our process, we decided to block into 500-, 500-, and 300-word blocks, using a single buffer area. Did we make a right choice? We might look at the alternative of blocking the change and master tapes at five records rather than at 10 records each. This would then provide across 1300 words two areas for the input master and new master and two areas for the change file.

What we are doing is investigating the tradeoff in decreasing the efficiency of a particular element in the system so as to increase the degree to which it may be overlapped.

At five records per block, average record read time goes up to 52 ms and the limiting path becomes 520 records. The processor begins to operate at 260 ms (after a read of one new block), and as it operates on the records from the first block, a second block from the master is being read into core, as is a second block from the change tape. If we assume that the processor will find two active items on the first block, it will have formed an output block 130 ms after processing begins.

In order to complete this run in minimum time it is necessary to have an output block to write every 260 ms, so that there is no gap in the write line. Whether this can be done is a function of the distribution of changes. If we impose the assumption that we will find exactly two changes in the first block and one change in the next and subsequently two in every odd-numbered block and one in every even-numbered block, we can see whether we can do it under these conditions.

If the first block of master is available at 260 ms, then the first output block will be ready at 390 ms, since it requires 130 ms to process a two-change block. The write of that block will complete at 650 ms, and a new block must be ready at that time. From 390 ms time until the delivery of the second master block the CPU will have nothing to do (except handle the interrupt for the completion of the second change block).

At 520 ms the second master block is in. If one change is found, then a second output block can be ready at 610 ms; at 650 ms time the write operation will report in and, finding a write request enqueued for it already, will proceed without delay. At 780 ms time the third block from the master will occur; at 910 ms time the second write will complete. There is only, therefore, a 130 ms interval to process the third block of master input.

Now if we have a two-change block we can just do it, delivering the completed record at 910 ms time. Since the delivery time of a given block trails the write start time for that block by 390 ms and since maximum processing time is assumed at 130 ms and read time at 260 ms, we shall continue to deliver output blocks to the write channel, and we shall complete in theoretical minimum time unless some interference can develop because of the reads of the change tape.

Change tape reading can delay the processing and as a consequent delay of processing perhaps even delay a read of the master tape. The change file is also buffered and blocked. It will have delivered two blocks, ten items, by 320 ms after run initiation. Although the channel is free at this point, no activity can be undertaken, because there is no place to put more information. At the end of the read of the second block at most two items have been used from the first block. The first block is not available until the fifth match of changes and masters; the earliest time this can occur is on the second record of the third master block (under our assumptions).

Since the third master block is delivered at 780 ms, the processing of a match with the last change of the first change block can occur at 880 ms, and at that time a new change block can be read. This block will be available at 1040 ms. During the 160 ms interval between the availability of buffer space and the delivery of the third block of changes, the end of processing of the three unchanged records of the third master block will have occurred at 910 ms, and the processor will idle until the delivery of the fourth block of master at 1040 ms.

It is necessary, in order for there to be a delay (due to the unavailability of change records from the third block), for all records in the second block of changes to be used in any interval between the delivery of the third block of changes and the fifth block of masters. The latest that the new changes would be read is at 910 ms; they would then be available at 1070 ms (match on last), and consequently all changes in block two of changes must be used in 30 ms for a delay to occur. This is impossible, since the processing of a match on the first record would take until past that time, and there would still be four change records left in the second block.

We can show in general, under the assumption of any activity level with any distribution of changes for this program, that no delay of a read or write is possible as a result of the input pattern of change records. At any occurrence of a depletion of a change block, there will be at least five change items available in storage. In order for a delay to occur, it is necessary that these five changes be processed. This will take 250 ms to accomplish.

The replenishment of the depleted block will require only 160 ms. Therefore, no completion of a change block will interfere with processing by causing it to suspend waiting for new changes, and necessary changes will always be in before the corresponding block of masters. Since writing will never be delayed, the hypothetical run time of 520 sec (time to write 2000 blocks of five records each at 260 ms per block) will be achieved even at 100 percent activity with the attendant increases in change record read time. Truly here the R/R capability is important to us.

13.5 WHAT IS INVOLVED IN UNDERSTANDING A PROGRAM

This minute analysis of this particular program (and not an especially distinguished one at that) is valuable for a number of reasons. Primarily it demonstrates to us again an underlying technique for achieving speed through parallelism, albeit in a preliminary way—a willingness to abjure optimization of a given component in order to achieve total minimum elapsed time. Further, it demonstrates that certain commonly accepted good techniques do not have general effectiveness at all, but must vie with other techniques equally accepted but which in the presence of a constrained resource represent tradeoffs and not enhancements.

In general, we have seen the considerations that are involved in mapping a program onto a machine in an efficient manner: the tradeoff for use of core, the channel allocation decisions, and the sensitivity to variations in activity levels.

A dramatic further example of tradeoff is in choosing the order of a disc merge and the size of a block from disc, given a certain amount of core. The higher the order of merge, the smaller the block size can be, since buffer space must be distributed among more files. With smaller block sizes the number of seeks of data will increase.

The lower the order of merge, the greater the block size may be;

however, with a lower order of merge the number of reads of data will increase, since it will be necessary to undertake more merge passes in order to accomplish the goal of a final ordered string.

It is theoretically true that a higher order of merge is always preferable to a lower order or merge. However, in both tape and disc machines real examples of poor performance due to too high a merge order can be shown. In this situation the increase in block reads and attendant seek times may dominate the efficiency achieved by higher merge orders and result in poorer performance.

In our case we found that optimizing tape by maximum blocking ruined our performance. Certainly, each program has its own balance, and individual design decisions must be made for each one. In some cases the attempt to achieve speed through parallel operation and the use of buffering fails absolutely, and best performance is achieved by abandoning read/write/compute overlap and using core space to hold the largest possible block. Consider our example if start time, instead of being 10 ms, were 1000 ms.

It is true that instances can be found in which a sequential machine will outperform an overlapped machine and that characteristically these are in situations where some overhead feature (like seek time) dominates all timing and the distribution of that overhead over the broadest possible basis is critical.

13.6. BOUNDEDNESS

We have been working with what we intuitively feel to be an I/O-bound program. The classical absolute form of I/O boundedness is the case of a single input unit and a single output unit at work with a processor. Each record requires an amount of time to process, and this amount of time is always less than the amount of time to read it or to write it (in an R/W/C machine). If the dominating time is read time, then we are read-bound; if write time is longer, we are write-bound.

The firmness of the concept falls off somewhat if we soften our definition to be that the sum of all time to read is less than the sum of all time to process. We are suggesting by this that there will be individual records whose processing time is greater than read time, and by implication from that there will be local periods of time when the run will be computebound—that is, where a delay will occur in writing or reading because of processing activity.

In I/O-bound programs we expect that CPU utilization will be low and that I/O utilization will be high. In the ideal we expect what we have shown for our example—a 100 percent utilization of the input equipment with periods of idle in the CPU awaiting the availability of another record to process.

With the softened definition we can expect what we have seen in the first analysis of our program—delays in all streams due to interference and bottlenecking.

In complex input and output environments in programs of any realistic complexity, we find that there are relationships between write, read, and processing times for individual records and local sequences of records such that it becomes grossly inadequate to characterize a program as being limited by I/O or processing—inadequate in the sense that it gives no real clue as to how the program will actually perform.

13.7   A MACHINE DESIGN PROBLEM—INTERACTION
OF ASYNCHRONOUS ELEMENTS

Consider the relationships illustrated by the following example. A manufacturer of a medium-size computer oriented toward the commercial market was at one point considering the impact of a doubling of the speed of memory on the performance of the machine. The machine had state-of-the-art peripheral equipment on line, operating off special-purpose channels, so that read card, punch card, print, and compute were all simultaneous operations. In addition, the system had a tape channel that allowed for simultaneous read and write of tape.

Some concern for the performance of the system with tapes had developed from some timings taken in the field. The tapes were rather high-performance tapes for a machine of this class, and the disappointing performance in some situations was ascribed to the inability of the processor to drive the tapes. Characteristically it was felt that with regard to tape operations this would be a computer-limited system. Since the system was intended largely as a satellite support system and tape-to-tape operations were not to dominate its program population, this was not considered too serious a drawback. A doubling of memory speed was being considered to improve performance, however, in order to enhance the machine and perhaps increase its market scope.

The machine had an interesting interrupt associated with its I/O operations. In order to avoid the possibility of overrun (a condition in which the demand for memory cycles in a given interval exceeds the capacity of the memory to grant them and which results, therefore, in a loss of data from one contending device), an inhibitory interrupt was generated which precluded the submission of an I/O order unless a system guarantee of no overrun could be made.

In the simple I/O environment of this system a hardware algorithm could be devised which, on the basis of a review of active devices, could recognize situations where a memory overload could occur. In more complex environments it is common to develop the interrupt when an overrun has actually occurred and then to attempt corrective action.

During the period of time that a card reader was active the reader would take large numbers of consecutive cycles and then (in interrow time) let large numbers of cycles go by. Tape operations would characteristically take one or two accesses during two cycles. Since it was impossible for the system to guarantee that a tape access could be granted when required if the reader was active, the system returned the inhibit interrupt for the reader whenever a tape was active and for the tape units whenever the reader was active. Further, the reader tended to stay active for longer periods of time, since it was unclutched and the logic of the reader software was to commit as many cards as were necessary to fill reader buffer areas. The effect of this was to delay tape operations quite seriously.

The central processor was not at fault in this delay, nor was the speed of memory. A faster memory cycle would have the effect of increasing CPU power but would not reduce the elapsed time for tape interlock due to the reader, since the problem was not memory's ability to grant cycles but the particular pattern that it was required to grant.

An analysis was undertaken of a program which read cards and which wrote cards to tape at a given blocking. The intent of the analysis was to measure the impact of a doubling of memory speed on the performance of a common utility operation. The program used a common output buffering technique, which set aside two write areas. The use intended for these areas was that one area would serve a tape unit while the other area received records from the processing activity.

The limiting device, of course, was the card reader, which was deeply buffered in order to keep it in operation. The processor delivered records to the output buffer and then requested the tape write to start. Because the reader was busy filling input buffers, the write of tape could not start, and the system queued the write. The process then delivered records to the alternate output buffer and requested a write. This write could not take place, because the reader was still active.

At this point there existed a wealth of input records to process but no place to put them, since the two output buffers were full. The process could not continue, and the reader stopped because it could find no place for another card since depletion of the input buffer had stopped. At this point the first write of tape started. The ability to compute and write tape could not be used at this point, because the processor still had nowhere to put output records until that write was complete. On the completion of that write processing could begin, and records were delivered to the output buffer while the second write was performed. After the number of records that filled the buffer were processed, the process again stopped. At this point the second write was still in operation, input buffers were partially depleted, and both the CPU and the reader were stalled.

At the completion of this write, the reader started to refill the input buffers and the process to deliver output records. This quickly stopped because of depletion of buffer space, and the system was reduced to running the reader alone until buffers were full, then again running the write alone until completion.

The impact of doubling memory speed was merely to double the number of nominal memory cycles where the CPU was inactive. Naturally, much of the problem resulted from the fact that an eccentricity in the way cycles were granted to the reader turned this into something less than a read/write/compute machine in the card/tape situation. But the impact on the system went beyond this to the point of disabling read/compute and compute/write capability that was truly there. In this situation a reasonable approximation to total time could be achieved by summing write and card read time, since there were no instances of compute alone. But the interrelationship between the functions due to the software buffering and I/O device/core relationships is clearly shown.

13.8. MORE ON BUFFERING

Consider a further example of this interaction between "asynchronous paths." A tape-to-tape program is running which requires 100 ms to read and 100 ms to write a block. Processing time for a block is expected to be 90 ms but has a maximum of 150 ms and a minimum of 40 ms. Double buffering is used for both input and output, so that there is provision for achieving R/W/C in the machine. Since the program averages 90 ms per block, we can say that it tends to be I/O dominated and determine that if 10,000 records are involved the running time will be 1000 sec.

Notice from Fig. 13.5 that the occurrence of two consecutive "long" blocks causes delays in writing block 3 and 4 and two delays in reading, for block 5 and block 6. Since it is usually not possible to predict precise distribution or occurrence of long blocks, it is impossible precisely to determine how long the program will run.

If we know the occurrence level, we could postulate a worse distribution and therefrom calculate the number of delays and from that true time on the input path and output path. We notice that the gapping effect occurred despite double buffering, demonstrating that a technique almost universally considered by young application programmers to be the solution to all problems of balance is not adequate to smooth performance gaps here.

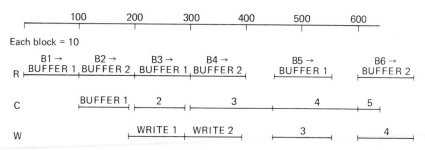

Fig. 13.5. Effect of nonpredictable "long" blocks.

In more complex environments, where there may also be a variation in the times for individual reads and writes, secondary effects may develop. Variations in block read and write time are, of course, common to disc environments.

A common secondary effect is to encounter a long write, which precludes the processor from continuing to deliver output items, because it cannot find an output buffer. This suspension, caused as a result of interaction between compute and read, later on causes a write delay because the processor cannot deplete an input buffer in time to submit a request for a new block, causing a delay in the read line. If the read line is the limiting path, then that gap will extend total running time for the program.

The intent of buffering in software is exactly the same as buffering in hardware: at the interface of two functional units of inherently different speeds to provide a mechanism to smooth the relationship between them so as to provide an increased probability that the depositing unit will have a place to put something and the withdrawing mechanism will find something there when it needs it.

When the interface is asynchronous, the size of the buffer is a limit on the independent operation of the units. The size of the buffer is equivalent to the number of buffers in common usage. Effectively a buffer is a queue dynamically representing the intersection of the input of one function and output of another.

Traditionally, among programmers the software buffer has been thought of as a way to support processor operation. In runs where a variation in record processing time occurred, a technique which would increase the probability that the processor would have work available to it was considered adequate. An alternate concept of buffering, particularly appropriate for a multichannel environment, is to reverse the intent of the technique and provide buffer space so that an I/O channel would always have a place to put something and to maintain activity on channels.

At a time when CPU power is coming to represent less and less of

total cost of a system, buffering techniques and levels that tend to optimize channel utilization and overlap should be seriously considered. There
are some serious system implications for program organization and even
for scheduling involved here, which we shall mention below. We might
look at some various buffering techniques before talking further about
buffering impact.

13.9. BUFFERING SCHEMES

There are two fundamental buffering schemes. One is to assign a set of
input buffers and a set of output buffers and physically to move records
as they are processed from an input to an output buffer. Variations involve the movement of an entire block from the input buffer into a
working buffer and then to an output buffer, or more commonly to
deplete an input block item by item, moving each item into working
storage and then building output buffers item by item.

Alternatively, buffers can be rotated by function. Given three buffers
initially, buffer one is an input buffer, buffer two a work buffer, and
buffer three an output buffer. At the completion of processing of the
block, pointers are exchanged so that the write buffer is switched to the
previous work buffer, the work buffer to be the previous read buffer, and
the input buffer to be the previous write buffer. At each completion of
processing the rotation occurs. The scheme can, of course, be employed
over more than three buffers. The important concept is the ability dynamically to hand buffers over to a function, and it is a first form of truly
dynamic buffering.

Beyond this, buffering schemes vary as to the depth of buffering, the
flexibility of buffer assignment, the amount of core that is allocated, and
when it is allocated.

A minimum core usage technique is sometimes used effectively when a
prediction can be made as to which device will provide the next input
record. In tape merging programs it is always possible to examine the last
entry from each input block to determine which block will deplete first. If
there are four input tapes, the technique requires only five buffers. The
"preselected" block is read into the back-up area. The block read in is the
next sequential block from the device that contributed the blocks that
will deplete first. When the depletion occurs, the successor block is available to join the merge. At that time another depletion test is made, and a
new block is read into the standby area.

The intent of the technique is to take advantage of the predictability
of successor blocks in order to approximate the effect of buffering for

each device with less core. The technique is effective when long sequences of contribution to output come from a single input device, that is, if output blocks tend to find their members from the same input block. Other input blocks will tend to deplete slowly, and the probability of depleting more than one block is low. Since the back-up buffer is made available to any device, the technique demonstrates a desirable feature of good buffering in a primitive way. That is, the availability of buffer space is dynamically adjusted to reflect the arrival and depletion rates for a particular device over a recent period of time.

The technique does not work well when there tends to be a balanced contribution to output blocks and input blocks tend to deplete at the same rate. Then the probability of simultaneous depletion is high, and there will be in core only the block that depleted first. A delay in processing is caused while waiting for the replenishment of additional depleted blocks. In core-constrained environments such core-preserving floating buffer techniques may be the only alternative to abandoning buffering. As the number of back-up blocks is increased so that they approach a limit of two for each input tape, the need for prediction becomes less and less severe, and performance approaches that of standard double buffering.

A technique that has been used involves the assignment of a single buffer area to each input file and a pool of unassigned areas behind. The unassigned areas may be used on a preselect basis or may be used dynamically to support the maintenance of channel activity on the system.

If buffer space is used in this way then the reading of additional input becomes more independent of the processing pattern against individual input blocks. A read is issued to a channel when it becomes free, if a buffer is available to hold a record.

Floating-block techniques, which exceed in space double the number of input devices used by double buffering, have also been used in support of the concept of sustaining channel activity. A Honeywell buffering system (on the H800) used a current block and standby block for each input (double buffering), plus a pool of floating unassigned blocks to hold records contributed by channels in order to sustain channel activity. The pool is again as large as the number of input tapes, so that the total of buffers is three times the number of input files.

The floating-buffer technique is a way of trying to approximate the enormous number of buffers that would be required completely to guarantee no delays in performance due to the interaction of input, processing, and output times that are not constant.

The implication of floating buffers for program organization is the necessity of structuring input files so that buffers may, indeed, float. This implies constraint on the variability of block sizes so that it is possible for one input file to use a buffer as easily as another.

The concept of the floating buffer, that buffer space should be dynamically transferable from one file to another based upon the need of that file for more space, is based on the observation that the rate of need or the potential rate of arrival of elements in a file should determine the length of queue it may have available to it.

There are some implications for multiprogramming systems. If it is possible to trade buffer space between input files in a single program, is it not also useful and possible to trade buffer space dynamically among programs? If system conventions about buffer pooling allow this tradeoff of buffers, then by a simple extension of the idea of floating buffering, system balance across multiprogram processes might also be achieved by exchanging buffers, and perhaps the allocation of buffer space to any program might be a scheduling decision based upon priority and current I/O compute-bound ratios.

By extension, perhaps processes should be scheduled on the basis of record arrivals and I/O should not be scheduled for the convenience of processes. In general, buffering should be thought of as a dynamic process and not a predetermined condition. Since a real program will experience dynamic variations in read time, processing times, and write times, the space at the interface between these functions should vary as well in order to maintain the balance that it is the goal of buffering to maintain.

13.10. BALANCING

Regardless of the perfection of balance that buffering may achieve for generally I/O-bound programs, it can do considerably less for chronically compute-bound programs, programs where every record requires longer to process than its input or output.

Notice Fig. 13.6. Using double buffering, we can achieve a continuity for the processor, a guarantee that a record will always be available for it. We can effectively overlap reading and computing, which would not be possible without the buffer. Without buffering there would be gaps in the compute leg waiting for the read of a record after computing.

For the particular example shown, we achieve intermittent read/write/compute alternating with periods of unoverlapped compute. Since we know that the period of unoverlapped compute will always exist, we are tempted to see what the impact of relieving this program of one of its channels might be. We have rather low channel utilization across the read and write channel, low enough so that it looks as if we could sustain performance if we combined reads and writes on a single channel. There is no particular benefit sustained from the parallel read and write; this chronic compute-bound program would perform as well on a compute and read or write machine.

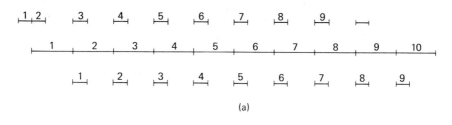

(a)

(b)

Single channel does as well.

Fig. 13.6. Double-buffering a compute bound run.

The serious implication of buffering as a smoothing technique at an interface is that an off-the-shelf technique like "double buffering" is not sufficient, that a great deal of statistical information must be acquired about a program before its buffering requirements can be determined, and that there are currently in the population of running programs a large number whose performance characteristics might be seriously changed by an alteration of buffer allocation. In this population, however, there are also programs that are so basically unbalanced either for the CPU or for the I/O system that it is impossible to achieve equilibrium. In a broadly parallel I/O machine with an interrupt system, an alternative to attempting fine honing of a single program is to attempt to "multiprogram." The goal here is to balance the utilization of components across a population of programs rather than across a single program.

Chapter 14

INTRODUCTORY CONCEPTS
OF
MULTIPROGRAMMING

14.1. INITIAL OBSERVATIONS

From our highly I/O bound program (the update) we notice that even when the program runs in its hypothetical minimum time, the CPU is still very lightly utilized. Nothing can be done to increase the utilization of the CPU over the period of time of the running of the update run. For this period of time, the user is paying for a machine that he does not, in truth, require, and he is losing potential productive power. Depending upon various installation requirements and possibilities, he may desire to correct this situation by multiprogramming his machine.

Multiprogramming is the phenomenon of having more than one process "active" at a point in time. The concept of "active" is a little imprecise, but it denotes a status in which a program has acquired a working set of resources in some manner and can, if given the control of the central processing unit, do some meaningful work. We shall further refine this status a little below.

The classical combination of "active" programs is a chronic I/O-bound program and a chronic compute-bound program. The goal is to balance the system and achieve greater utilization of each component by taking control away from the I/O-bound program when it is unable to continue because it is waiting for an I/O activity to complete and to give control to the compute-bound program during this interval. When the I/O activity for the I/O-bound program is complete, the I/O program is given back control and allowed to proceed until the next time an I/O service must occur. Notice that the ability of the I/O-bound program to continue at any time it desires I/O, that is, to overlap its own I/O with its own processing, is partly a function of its buffering capability. If it is entirely unbuffered, every I/O request will cause it to become "blocked" on the processor.

Let us postulate a program which we wish to run with the update in order to utilize the central processor during the time of the run. We need to assure ourselves of some initial conditions before we can undertake this, and we must also make some policy decisions about exactly what we wish to accomplish. We see that the I/O interrupt mechanism becomes very important to us. It provides a time at which the decision to continue with one program or to switch to the other can be reasonably made. Without the interrupt, this would be considerably more clumsy (although it could be done). From this point we shall assume the presence of an interrupt feature of some sort in our machine.

Beyond this, we must determine that there is a sufficient basis of resources to make multiprogramming feasible. We require a certain amount of core in which to hold both programs. If we have sufficient core for both programs, then the potential for multiprogramming them is available at no additional expense for the resource of primary storage. If we do not have sufficient storage, we may elect to acquire more storage or we may determine to use storage allocation techniques that permit a movement of storage dynamically from process to process.

As we expect, there are natural limits to both approaches. The increase in the work performed by the system in a given period of time due to the better utilization of a given component of the system must justify the addition of the equipment that is needed to achieve utilization. If it is necessary to acquire additional core to multiprogram, it must be shown that the value of the improvement in the performance of the system justifies the core. Additional storage space is a common requirement for multiprogramming systems.

Those techniques which undertake to reduce the need for additional space by increasing the effective utilization of current space with dynamic relocation, roll in/out, paging, etc. (as discussed in Part 7) are limited by the requirement that the system perform more meaningful work in a given time period despite the additional burden of those efforts required to support dynamic techniques. The dynamic sharing of core between programs causes delays because of increases in contention in just the same way that contention caused delays in Part 1.

In addition to core, we must determine whether or not there is sufficient resource in the I/O system to run programs together. If we had only three tape units, it would be manifestly impossible to run two programs together if the total of their requirements while running was more than three tapes. If we have sufficient tapes, we then face the problem of sufficient channels.

At this point we become concerned not with feasibility but with desirability, and we have a policy decision to make. We can decide that under no circumstances do we wish to delay any program's completion time because of multiprogramming. If this is true, then we may not allow any contention for any resource between the programs. We may not allow a delay to develop for a program due to the existence or presence of the other. In an absolute sense this is impossible, because at minimum the program running on the processor will experience a loss of memory cycles because of the I/O activity of the other. But beyond this unavoidable phenomenon, how much contention should be allowed?

Ideally in a situation of perfect balance and no contention, each program would run at exactly the same rate in the multiprogramming as in the uniprogramming environment. But this ideal must be achieved (if at all) by the acquisition of sufficient equipment to guarantee avoiding contention. In this, of course, lies an inherent paradox. The utilization of equipment that is acquired in order to increase the utilization of other equipment and yet guarantee no contention must necessarily be low. The only way to avoid its being low is to have a population of programs whose characteristics are so exactly known that each instance of its use of a resource can be precisely predicted. Since we cannot normally achieve this (since in fact even in our decisions about buffering a single program we are dependent upon statistical assumptions, averages, limits, probabilities), we cannot, without a danger of contention delays, expect high utilization from the tapes, control units, etc., which we acquire in order to multiprogram.

What we depend upon is that the additional resources that we acquire in order to support multiprogramming will be sufficiently inexpensive as compared to processors that the value of achieving additional processor activity will overbalance the expense of achieving it. The minimum multiprogrammable machine in any product line is characteristically more expensive than the minimum acquirable configuration.

There is another inherent paradox in multiprogramming. The concept of no delay is basically an unstable one. It is reasonable to expect that with the addition of only two tape drives on the existing three channels which we have for our update program we may run a highly compute-bound program with very low contention if the update program is unbuffered. This is because we have gaps in the use of the channels. If the compute-bound program does all its I/O while the update program is running, then that I/O will be performed in those gaps, and no delays due to the interference or contention on I/O will occur.

If we attempt to run update in its buffered form, however, no gaps develop in its use of I/O and any use of I/O whatsoever by the alternate program will cause delays to update. Therefore, the equipment required to run these programs together must include additional channels.

Later we discuss concepts of "feasibility" and "profitability" in more detail, but we already see here that the tradeoffs and judgments as to proper hardware and performance are complex indeed. Further than this, we must realize that in any realistic environment we are not searching to provide a multiprogramming basis for two given programs, but for a dynamically changing mix of programs which we desire the ability to run together. The precise behavior of each program is never known and is dependent upon input and subject to variations in the programs with which it is running.

The desire, therefore, to run with no delay is largely unrealistic. However, there are still acceptable benefits to be obtained. A secondary goal of multiprogramming (fall-back position, as it were) is to achieve the completion of a set of programs in a shorter elapsed time or in the same elapsed time on a lower cost basis than if they were run sequentially. Therefore, although particular programs may be delayed, the expected completion time of the set will be lower, or if that is not true, the cost of processing the set will be lower.

14.2. TIME AND COST TO A RESULT

Users are accustomed to this compromise cost-oriented approach and to the problems of balancing individual program completion and installation work loads. Consider the decision to introduce a satellite machine into a computer center. The reason for doing this is to disassociate a computer from those devices which are so slow that they represent a limiting resource for any process running on the system: readers, printers, and punches, for example. The direct reading or writing of these devices will necessarily force a low utilization of the rest of the system. Every process using an on-line printer will be write-bound. Because of the length of printing, the processor will find that it cannot use output buffers because of the length of time these are required by the printer. If the process is I/O-bound with respect to the CPU and other units, the unavailability of output buffers will (as we have seen) cause delays to CPU processing, which will cause delays to its reads which will cause input gaps. If the process is CPU-bound as opposed to other I/O, its CPU-boundedness is enhanced for the same reason.

Unless the process is such that the multiple of records required as input for each output is enormous and the complexity of processing is just right, the system is inherently unbalanced. An attempt to balance the system and achieve greater usage of other resources results in a decision to take the limiting device off-line to a support computer and run the "main-processor" tape to tape or disc.

The multicomputer systems are outgrowths of this decision. The effect of this decision is to produce more work on the main processor in any given period of time. The utilization of components of that system goes up, and with that its throughput. Consider, however, that what develops is a queue of tapes to be printed by the satellite and that the rate of printing is not faster on that machine than on the main processor.

There is no faster way to achieve human readable results from any program than by printing it on the on-line high-speed printer. The turnaround time for any particular job will be minimized by allowing the on-line operation. If job 1 requires three hours of print time and job 2 and job 3 the same and these are the only jobs in the shop, the minimum amount of time until human readable results for all jobs is nine hours. The average turnaround time is three hours. The turnaround for each job is, respectively, three, six, and nine hours.

Which job will receive which turnaround is a priority decision made by the management of the corporation or installation. If there is no contention between them, that is if job 2 is not ready until job 1 is already committed to the system (and no preempting is allowed), then, of course, job 2's turnaround time is less than six hours, because he submitted later.

When this system is altered so that jobs 1, 2, and 3 are tape and require one hour each, a tradeoff much like the multiprogramming tradeoff is being made. Job 1's human usable result will be ready at four hours after its submission, job 2's at seven hours, job 3's at nine hours. There has, indeed, been a delay in the availability of usable result. If, however, a larger population of jobs existed, jobs 4, 5, 6, etc., then the main processor would in eight hours have processed eight jobs. There will be a pool of printing of some 16 hours of printing to accomplish at the end of the day.

The throughput of the main processor has increased considerably, as one would expect. Notice, however, that the total time of each program spent in the system will actually go up, and an hour will be added to the time to meaningful result. The elapsed time of the set of eight is 24 hours in either case, but by application of the principle of functional specialization in parallel operation, the cost of these hours is considerably reduced, since it is less expensive to run 1401's or Model 30's than to run 7094's or Model 75's. The user, at the cost of some delay in individual program finish, will run his set of programs in the same period of time for less money. We shall show various considerations in running a set of programs in a shorter period of time in a multiprogramming system by abandoning consideration for when any program in particular is available so long as it is available in the reduced time of the entire set.

14.3. THROUGHPUT, TURNAROUND, AND WORK

There has been some recent discussion of what is called the throughput-turnaround dichotomy. This is an appreciation of the phenomenon that we have been discussing: that there is an apparent tradeoff between getting a given program run in a given amount of time and utilizing a system so as to finish as many jobs as possible in a given amount of time.

This dichotomy is a tendency and not an absolute. It is possible to find examples where an increase in throughput causes an improvement in turnaround time. Surely if a system is doing more productive work over a given period of time, and more jobs are completed as a result, the time to turn a job around must tend to be lower. It will go down when throughput goes up.

The essential similarity between concepts of throughput and turnaround is demonstrated by the fact that the concept of average turnaround time over a set of jobs is basically a throughput statement, as the set of jobs being considered approaches all the jobs that have to be performed. Systems under certain levels of loading do appear to demonstrate an opposite movement in turnaround and throughput, however.

But the concept of throughput is not truly adequate to demonstrate this dichotomy. The true dichotomy is work versus turnaround. Work is a job independent concept. In measuring the utilization of a system of components of a system, we describe the amount of work that an element has accomplished over a given period of its time, regardless of the distribution of that work across jobs or tasks that it has been asked to do.

The abstract goal of multiprogramming is to sustain work independent of the average completion time or the number of jobs that complete in a given time. As we shall later describe, for various reasons an installation may wish partially to disable the work-enhancing function of multiprogramming in order to protect a throughput rate, or, for a given class of users, a turnaround time. He will be willing to see jobs selected for execution or preferred for control of the CPU or other resources on a basis independent of whether this will accomplish more work. In a real way he is saying that there are things more important than proper full utilization of the system. The completion of a given number of jobs in a time period or of particular jobs has a higher value than the intense utilization of components of the compute resource, which might tend to complete more partial jobs but not produce human usable results.

14.4. MULTIPROCESSING

If it is true that additional CPU cycles can be brought into use by adding core and I/O equipment in order to provide work for the CPU in situations where there are available CPU cycles, it is also true that the opposite can be done. If the I/O subsystem is being underutilized because of an insufficient rate of I/O requests from a population of programs, it is possible to add an additional processor, or processors. If this is done we have created a "multiprocessor" environment. There are some special considerations and problems in hardware and programming techniques that are associated with the organization of a multiprocessor or with its effective usage.

Many of the problems of processor interaction in a multiprocessor, however, are truly problems of interaction between processes or processes and subprocesses. That is to say, given m active process in an environment of less than m processors, certain rules and conventions for interaction must be established. These rules are essentially the same whether n (the number of processors) is 1 or greater than 1.

In a multiprocessor environment, however, certain problems of interaction become more critical and require more efficient solution because of the problem of crossing processor boundaries and of synchronizing processes that are concurrent in a stronger sense than in the multiprogramming environment. Various solutions and approaches to this are discussed in Part 5.

An insight into the fundamental identity of multiprogramming with either one or more processors can be gained by remembering the attention we paid to the concept of time in the first chapter. Multiprogramming is a time-sharing technique for multiplexing the attention of a CPU across multiple processes. It is, in essence, a simulation technique, attempting to simulate the performance of a system with as many processors as processes. It is a scheme for creating "virtual machines," virtual CPU's which are brought into existence for intermittent periods of time to run the process that has been assigned to them for as long as they can, or for as long as some system control mechanism allows them to, and which are then quiesced, giving way to the performance of another virtual machine.

We can think of the active jobs as running on different processors which, as a phenomenon of architecture, have been implemented as a single piece of hardware. Since we are simulating processors, the behavior of the system should be essentially similar to systems in which the processors actually exist for each process or for systems in which the ratio of processors to processes is less than in the single-processor case. That is, we can simulate five machines across one processor or two processors, or more.

The profitability of doing this simulation is a function of the utilization of the other equipment which can be maintained and of the utilization of processors by available processes. We expect that we shall require more core to run more processors, and we indeed do, unless we choose to simulate core in a way that will provide the logical (if not the performance) effect of actually having the core. Dynamic relocation techniques are essentially simulations of additional core by using addresses in a core box for multiple processes over intervals of time. Depending on the loading of core and on the loading of channels required to support the alternation of processes in space, we shall experience different delays in performance.

The extent of simulation is a performance decision. We simulate channels by assigning devices belonging to different processes to the same channel; we simulate devices like drum and disc by assigning space to different processes on the same device. The balance between actual hardware and simulation of hardware is struck on the basis of the contention, the probability of interference, and the level of utilization of the hardware.

14.5. MULTIPROGRAMMING INHERENT
IN PROGRAM CREATION

Another conceptualization of multiprogramming is a simple extension of the idea relevant to balancing a single program, especially with regard to I/O and CPU operation. The only true distinction between attempting to balance a single program against a machine by splitting off a parallel activity for every asynchronous element and therefore attempting to assure continuous usage and attempting to balance a set of "independent" programs against a machine lies in the need for additional synchronization.

This synchronization is generally externally imposed, but it may be built into a program so that in effect a single program with unrelated elements may be written in order to balance the machine. A great deal of this kind of intertwining is actually done in the creation of code for the solution of problems. It is done informally, but it lies at the root of system design.

In any processing system consisting of a number of computer "runs," each resulting in a transformation of data, functions are informally clustered together on the basis of a "feeling" for the balance of a run. Transformations are defined, the number of runs is defined, and the sequence of runs is defined in part on a feeling for the efficiency of each run. Additional processing functions that could be split into a separate run, which are essentially separable from other functions of the program in which they reside, are put into that program because it is felt that they can be performed there without changing the program from an I/O-bound to a compute-bound run. The processing functions are interwoven closely with each other, from line to line on a coding sheet.

In a more directed way functional modules may be clustered together into a single complex process with subroutine linkages or transfer points providing an inherent synchronization. Beyond this it is possible to spawn subtasks and allow synchronization by the mother program.

The so-called "single program" is actually an aggregation of independent functions cohered into a single structure called the program and synchronized by that programs operation. There is no inherent need for the existence of an external synchronizer relating two "programs" in a multiprogram environment.

Characteristically, such an external synchronizing element does exist, however, and its use is to relieve a program from a need to know of the characteristics or existence of another program or programs coexisting in the system. Internal synchronization implies a degree of preknowledge and planning that restricts the flexibility of a multiprogramming system, limiting concurrent activation to sets of programs that have been planned together and interwoven, and prohibiting the spontaneous and unpreplanned running of programs in an arbitrary dynamic mix.

Since there is no synchronization between independent programs, there is a need for an external synchronizer which essentially replaces the natural asynchronous reflexivity between an I/O path and a CPU path for a given program. The essential descendence of multiprogramming from overlapped I/O can still be seen in the current state of the art, where it is usually true that the switching of attention from one program to another is a result of I/O activity.

The proces of task switching in some early systems was part of I/O interrupt handling. The closeness of the concept is reflected even in the word "dispatcher," which is generally used now to describe that part of an operating system which actually causes the load of the address of the program to be next executed into the control counter (location counter, program address counter, etc.) of the CPU. This word was initially used in some systems to refer to I/O queue service and to a routine that contains interrupt handlers.

14.6. UNDERLYING CONCEPT OF MULTIPROGRAMMING

The concept of multiprogramming is so intertwined with the concept of the modern operating system that we often forget the true nature and fundamental simplicity of the technique. The idea of a control stream language to remove limitation on flexibility that might result from a too early determination of such things as device assignment, etc., is not an inherent concept of multiprogramming.

Dynamic data description, priority definition, and device independence are not central concepts to multiprogramming. They exist in large measure in single-stream systems such as the 7094 IBSYS/IBJOB monitor, and there are multiprogramming systems where there is no control stream, particularly real-time operating systems where job population may be largely preplanned.

Whether a control stream does or does not exist is relevant to multiprogramming, of course, because it implies something about the way in which a system will acquire jobs for scheduling the system, the dynamic nature of such acquisition, and the ability of a system to treat various kinds of information in connection with decisions to run or not to run a program. But basically there is no reason why an operator cannot request the allocation of resources and the initiation of a program in a multiprogram system from the console.

What is important to multiprogramming is the related concept of a queue of jobs that are candidates for performance. Since balance and utilization are the goals of the technique, it follows that the greater the population of jobs that a multiprogram system has available for initiation, the greater the probability that it can find programs to run together properly.

It is not always a multiprogram system option as to which programs will run together. Often the subset of the population of jobs which is to be submitted to it is determined by installation management, often by accidents of machine room operation. To the extent that the system's knowledge of jobs is constrained, its capability for balance is constrained. If the subset of submitted jobs is carefully prebalanced, then the effect of the limitation is not felt. If this is not so, if jobs are submitted in a dynamic way without regard for balance, then the number of those jobs enqueued on service at any time can be important.

There is an idealization and abstraction of an infinite population of jobs whose characteristics are known and stable and which represent a heterogeneous enough group that any time the system can find a subset of jobs which can balance the machine. The greater the number of available jobs, the greater the probability of finding a good mix.

Other concepts of modern operating systems that are related to but not part of the basic cluster of ideas in the kernal of multiprogramming are various concepts of core and device independence. These preceded multiprogramming as features of operating systems, and they relate to the potential of the technique on any system, but they are not fundamental. Partitions may be defined which represent stable allocations of particular addresses and particular devices to particular processes.

14.7. OPERATING SYSTEMS AND MULTIPROGRAMMING

The reason for these observations is that in viewing multiprogramming systems many people in the field develop an inappropriate appreciation of the cost in system features and services to support the multiprogramming capability.

Operating systems features that are costing a user time and space in order to relieve his programmers of writing I/O support coding, to provide him with reliability by being able to replace a device address with another, to provide him flexibility in the organization of program structures, to provide aids for operational staff, aids for debugging, aids for system maintenance, etc., are seen as overheads which necessarily accrue to multiprogramming per se.

There have been, because of the burdens that modern operating systems impose in order to provide their services, some nasty surprises in the performance of current systems, and the estimation of the feasibility or profitability of multiprogramming has suffered by association, as it were.

Further, multiprogramming has been disappointing in some situations because of an inappropriate mix of jobs. It is conceivable that there are installations which cannot find among their populations of programs balanced mixes. But it is more usually the case that the disappointment comes from a poor understanding of the nature of the technique and from poor approaches to its usage.

Many decisions must be made in planning the installation of a modern operating system: decisions as to the residence of operating system functions on various devices or in core, decisions as to the blocking or buffering of given system files, decisions as to the sizes of queues, queue disciplines which could be applied to various queues, etc. These system configuration decisions are as much a part of defining a computational capability as the initial configuration decisions for the hardware. Good or bad decisions will have an impact upon the performance of individual programs, or of sequences of individual programs, as well as a multiprogramming mix.

For example, if an installation does not use multiprogramming but has a population of very short jobs, the activity of the scheduler, the mechanism which causes the allocation of resources to a process and its forwarding to the "in process" condition, is still critical to the system. If the scheduler in such an environment is allocated residence space on disc and is brought in from disc for each operation, the amount of time to schedule may represent a serious burden to systems performance.

If a drum exists in the system, or if it is possible to nominate some core residence for the most common scheduling activities, performance may be considerably improved. If, as in many operating systems, there exists a concept of SYSIN, the blocking and buffering of that file represents a critical system performance parameter. If system data sets like SYSIN are not properly allocated to separate channels, system performance will decay. These are system configuration considerations that are entirely independent of multiprogramming.

What will happen when multiprogramming is introduced is that any poorly planned system will have its problems more brightly illuminated because of the higher level of usage of system services and the greater general level of load put upon the system and its services.

There is a great deal of current interest in the timing of operating systems and in the timing of mixes. At the current time much activity is being undertaken to monitor the performance of possible mixes, to observe bottlenecks, to modify channel and device allocations, to relieve them when possible, and to change mixes when necessary.

This activity is all to be encouraged. So long as the concept of utilization is important to computer users (so long as the cost of systems is nontrivial), every effort which increases the success of multiprogramming by increasing understanding of its dynamics is much to be encouraged. Since the pattern of the industry has been (except for small machines at an ever lower and lower low end of the line) to increase capability for price rather than to reduce price, the technique that provides a possibility for using the increased capability will necessarily dominate our thinking for a number of years to come. In large measure the promise of multiprogramming has been fulfilled, even if only after, in some instances, a good deal of agonizing reappraisal.

Chapter 15

OPERATING SYSTEMS:
INTERFACES
AND INTENTIONS

15.1. OPERATING SYSTEM AS A SIMULATION

The concept of multiprogramming as a simulation of a machine population over time by multiplexing leads us to some further considerations in the nature of modern operating systems. The operating system is in itself a simulation. It postulates a "metamachine" or machine whose characteristics at the user interface are represented by the documents he receives describing the operating system.

The metamachine has certain characteristics. It has a set of programming languages; it has a set of conventions for data organization; it has a set of prepackaged functions; it has a set of operator characteristics. The image that the user has of the system with which he works comes from a concept of the system formed by the manufacturer in extending his basic hardware to form what he considers to be an attractive user-oriented system for some imagined class of users, hopefully large enough to make the system profitable for the manufacturer.

The characteristics of the operating system dominate the "user's" appreciation of his computational capability. From the edge of the hardware to the edge of the users needs and patience there exists a gap which is filled in by the metamachine, the simulation of the system the user would truly like to have (or some reasonable approximation thereof).

15.2. USER AND OPERATING SYSTEM

The "user" of a system has many essential personalities. He is a progmer who wishes to develop a problem solution that will produce potential for producing results on the machine in a minimum period of time and with minimum expense. In a sense, each new program developed for the system by a programmer further extends the machine and structures the image of the machine into that of the program he developed for users of that program. The programmer has two interfaces—the group or individuals who wish to solve a problem and the extended machine for which he will develop a program. If the problem solution is recurrent and used by a set of problem-oriented people, they may see only the forms of input data and job descriptions required by that program.

To the user of an application package, the system is that package, and the characteristics of the system are the characteristics of the package. [This is an ideal in a sense, since eccentricities of operating system design too often intrude themselves upon the consciousness of a "casual" user—a user who would be just as happy to have his problem solved by a roomful of slaves in the basement and who views any awareness of JCL characteristics (for example) imposed upon him as an annoying and unforgivable intrusion.]

This second aspect of the user, himself as a nonprofessional user of computational power, has received much attention. What he can be asked to know, what he can be protected from knowing, will surely affect the rate of growth of the industry over the next years. We have protected him from knowing clear and add; we are even in many instances protecting him from knowing $A = B * C$. Unhappily, we too often start him off with the job and data definition statements of the control stream. This user is simulating a special-purpose machine on a general-purpose computer. If there were no other problems to solve and the solution to his particular problems were valuable enough, he might well profit by having a special-purpose machine.

As much as a programmer paints an image of a system for the casual user, he has an image imposed on him by the operating system that is associated with the machine. He uses one of a member of a family of programming languages (usually FORTRAN or COBOL, but possibly ALGOL or some ALGOL dialect, PL/I, or in special cases SNOBOL, LISP, or some other list-oriented language). These languages represent concepts of programming style, programming needs, and programming efficiencies which may or may not map onto the basic hardware in a direct or efficient way.

Certain programming possibilities may be essentially disconnected by the language (bit manipulation in ALGOL 60); certain features of the language may be awkward to sustain for the machine (dynamic core acquisition and release for block structure languages). If the programmer chooses to approach the machine more directly, he discovers other layers of extension between him and the real machine. Input/out, for example, will be accessible only through software-provided conventions, a class of service functions having to do with allocation, synchronization and control will be provided by the operating system.

In another of his personalities, as a manager of a resource (or as a marketer of a resource), the user's image of his potential capability is entirely described by the operating system and its scheduling mechanisms and allocation policies.

In his last personality, a "user" is a planner of capability, and in this guise he begins to have insights into the true nature of the system he is using. In this role the user assumes some of the concerns and problems of a manufacturer, and "systems programming" function in a serious user environment has much of the flavor of manufacturer system development. If the manufacturer tests general system performance against classes of jobs to determine how to market and price a system and what options to provide, a user must performance-test against his own environment. If a manufacturer develops a language capability, a user may determine what particular subset of the capability will be used in his shop. If a manufacturer develops an operating schedule, the user must time and optimize that scheduler for his own work.

It currently requires a medium-sized donkey to carry the publications necessary to describe the complete metamachine to the aggregative user. It is almost impossible to find in an installation or a development laboratory a "man" in the Renaissance sense. The emergence of the metamachine has introduced new problems and has extended problems beyond points where they might reasonably have been through to be solved. It has done this by adding to an engineered hardware complex of components a large set of informally "engineered" software components. By consequence the complexity of a system is seriously increased.

Essentially no system is like any other system, even when basic hardware is fundamentally identical. At one time in the industry there were high hopes of ending machine incompatibility problems by providing essentially similar data interfaces. Then essentially similar processor architecture became a hope for achieving commonality. Hope for compatibility still exists in programming languages at higher levels. The phenomenon of being unable to run a program written for a given machine architecture and some subset of its configuration on another incarnation of the hardware is new to us in this generation. It comes from the variations in the choice of an operating system (a System 360/50 running under DOS, Disc Operating System, cannot run a program written for OS; an 1107 running EXEC I cannot run a program written for EXEC II interfaces) and even from variations in the specific generated form of the same operating system. It can appear at times that the specific architecture of a processor or system is almost irrelevant.

15.3. SUPPORTING THE SIMULATION

In classical system development the hardware arrives first, and programmers are asked to map a set of functional characteristics that marketing
thinks are required into the specified hardware. The designers of a hardware system are not entirely insensitive to user environmental requirements, but they can sometimes lack an appreciation for the difficulties
involved in developing basic software packages that will perform within
realistic time constraints. As with any simulation, the closer the hardware
is to the system being simulated, the more efficiently the simulation will
run.

Consider classical machine simulation on another machine. If it is
desired to save a population of 7094 programs for a 360 user so that his
investment in these programs is not suddenly rendered worthless by the
replacement of his machine, and if he desires to discontinue the operation
of an obsolescent machine, it is possible to run the 7094 program on the
360/65 by simulating the 7094.

In order, however, to produce a simulation that will run in a realistic
period of time and produce a tenable approximation of cost/performance,
it is necessary to provide a serious hardware assist called "emulation." The
emulator provides a hardware extension of the 360/65 which executes
7094 instructions. The total 94 simulation capability is a combined
hardware/software package that provides a meaningful 94 execution capability. Particularly interesting is the use of "programmed logic" in the
emulation package.

Programmed logic, "microprogramming," is also a feature of the low
order of the 360 product line, where many architectural features of the
360 are truly implemented in a fast read-only store by the execution of
very simple basic microinstructions. An interesting extension to microprogramming lies in the possibility of modifying machines and systems by
writable control stores.

The availability of programmed logic casts light on a problems of
particular interest at this time. In general, in the relationship between a
raw machine and a "system," how much should be done in hardware, how
much in software, and how much in "soft" logic, and who should make
these determinations?

Traditionally, programmers have followed engineers. There are dramatic exceptions, instances of programmer-designed machines. Best-known,
perhaps, is the Burroughs B5000 and its successors.

The desire to build an efficient processor for a high-level language and
to support the processing implication of the language led to the development of a radically different machine. A number of particular machine
characteristics developed out of an appreciation of what is involved in

ALGOL processing, primarily the hardware implementation of a LIFO stack mechanism, in which many of the characteristics that might be found in a software interpreter built to execute Polish strings are actually found in the hardware. In addition, a mechanism to support the dynamic acquisition and release of space is built into the hardware. Finally (and especially in the 6500/7500) mechanisms exist (in the form of bits in a data word which represent attributes of the data) both to simplify the compilation process and to reduce much of the supporting code that must be generated in compilers to support such things as mode conversion.

The engineers in this system would appear to have followed language designers and compiler implementers, providing instrumentation for functions felt to be needed for proper support of a high level language.

Over the years the concept of "architecture" has emerged. This word means different things at different times, but what it should mean is the design and conceptualization of a system. After the system is defined, determinations are made as to what to put into hardware, what in software, etc. These decisions are economic cost/performance decisions that must be made upon the study of the behavior of a total postulated processing image and capability which represents a goal to be achieved. The almost ad hoc mapping of major software upon frozen hardware must become a thing of the past. There is a developing pressure now for extended hardware support for language, operational functions, and applications.

The interrupt system, hardware dynamic memory management support, and instructions that perform task switches have come from a realization that there are just too many conventional machine instructions to be executed to make many necessary system features attractive.

Essentially, the simulation of the metamachine is requiring more hardware support. No user buys hardware; he buys a metamachine. This interface must have some satisfactory performance characteristics. If the proposed hardware cannot be a suitable host for the required metasystem, then the hardware proposal must be extended or abandoned. The metamachine interface specifications are inviolate.

15.4. INTERFACES IN AN OPERATING SYSTEM

It is no easy thing to define a user interface, nor is it easier to define the interface between a machine, an operating system, and a system of programming languages. An example of what goes wrong is the existence in current systems of separate languages for control of the system—the primary input stream language, "JCL" in OS 360. These languages have syntax and function entirely independent from the structure, function, and syntax of programming languages in the system.

In addition, much of the natural overlap between languages, i.e., the usability of common I/O packages, math libraries, and statistical libraries, are completely excluded because of the entirely independent implementation of compilers.

If the search for a universal language is abandoned, and if it is agreed that there may be multiple languages in a system, it is a requirement that each language represent in its own idiomatic form all of the functions and processes that must be described in the universe of actually getting a program "on the air."

Behind this capability the designers of the operating system transparently implement various features when they choose on the basis of an appreciation of the user environment and the class of machine they are dealing with, especially with regard to its ability to support various levels of dynamism without a disproportionate burden. For example, if a language has an OPEN file statement, that statement may cause various things to happen, based upon the system's design characteristics.

The OPEN for each file may be executed at the time the program is scheduled, at the time it is loaded, or for each file at the time of first reference. In any case the language form is insensitive to the system design. There should be no directives that are inherently compile time, load time, or reference time. The interface to the user should be uniform regardless of system implementation or level.

That considerable uncertainty exists concerning what is a "program" time or a control stream time determination is shown by the overlap in information between data descriptive features in programming and control stream languages. None of the major current languages takes the same approach to the interface with control stream. COBOL makes considerable venture into what might now be considered control stream domain, PL/I a little less, and FORTRAN and ALGOL almost none.

Equally variant from system to system is the relationship between programming languages, utilities, and the operating system. Some systems consider the languages to be "just another program" and require considerable control stream description to run the compiler (OS/360).

Other systems consider the processors to be very much the same as the scheduler or the allocation mechanism of the operating system: transient subprocesses enjoying a status as extensions of the system. In OS/360 utilities have their own private parameter languages and are run as separate job steps. In EXEC VIII utility functions are seen to be effective subprocessors of the compilers, and compiler options direct the performance of source precompile preparation and postcompile retention of source and output.

There is no best definition of the relationship among data management, utilities, language processors, job management, and supervision in a system, but the goal must be consistency and reasonableness within a conceptual framework so that the user sees a conceptual whole when he looks at his system.

15.5. DIFFERING GOALS

In addition to the relationships between an operating system and its machine host, its program guests and its component parts' operating systems differ in their fundamental ambitions and objectives. Part of this difference derives, of course, from an appreciation of the limits of the machine on which they are running. Multiprogramming simulation of a fifteen-processor machine on a machine with an upper bound of memory of 64,000 characters and two channels is not an appropriate ambition.

There are classes of operating systems, therefore, which are functionally differentiable. DOS/360 and OS/360 are examples of systems intended for different classes of machines. Just what the break point might be is much in argument, of course.

Whether systems should be functionally restricted or provide the same functions at various performance expenses is a point of debate. The point of view of identical function at different rates is idealistically to be preferred, but it flounders on the details of implementation in the real world. But other classifications cut across specific capability in specific functional areas. There are perhaps three major classes of operating systems: the generalized batch operating system, growing out of the job monitors of the late 1950's, of which OS/360 is an example; the real-time operating system, of which UNIVAC's REX for the UNIVAC 490 is an early generalized example; the time-sharing operating system, of which GE's BASIC service is an example, along with IBM's QUICKTRAN.

There are examples of composite systems for IBM (TSS 360), UNIVAC (1108 EXEC VIII), GE (MULTICS), and for the RCA SPECTRA 70/46, among others. The composite system is often a real-time support extension of the basic batch system (for GE GECOS-GERT) or a time-sharing extension of the basic batch system. Some question must arise as to the extent and cost of generality in a system. "Dedicated" single one-language systems like BASIC and QUICKTRAN have fallen into disfavor to multi-language systems, and composite systems are growing in attractiveness at this time.

The problem of the cost of generality has been particularly felt in real-time, where special-purpose operating systems have tended to proliferate either independently or as grafts onto the basic services of the basic supervisor.

Some fundamental differences in the problems faced by these classes of systems exist. The time-sharing system is oriented toward the quick preparation and debugging of programs. Users tend to be private in their data requirements, and the basic goal of the system is to provide reasonable response time for subtask elements to an active terminal. In real-time systems, however, applications tend to be prepackaged and systems are inherently data-base oriented, any pattern of messages over any period of time having a high probability of using the same collections of data. Multiprogramming in this environment serves the purpose of achieving the simulation of parallel paths to data.

15.6. VIRTUAL MACHINES

An interesting aspect of considerations about operating systems is the phenomenon, mentioned of the interface between programs and operating systems, of the fundamental inability of programs written for one system to run under another.

In a sense the metamachine can be simulated as easily as the hardware machine, and the concept of the virtual machine as used in a number of environments involves a simulation of the metamachine. A goal is to allow different operating systems to run with each other and thereby to make various applications written to run under different operating systems available to the installation without switching the operating system. The solution is "virtual metamachines," where an interface is provided behind the operating system that runs the operating system. On an IBM system DOS, OS, and other operating systems can be run together. Each system thinks that it is interfacing with the machine basic, but its I/O and service requests are fielded by the unifying underlying supervisor. The user has the impression of a "clean machine."

In the chapter on multiprocessor organization we discuss schemes whereby the effect of a processor in a system is achieved by the addition of equipment that is of considerably less magnitude than a full processor. The question is joined as to how much logical and arithmetic circuitry should be duplicated in a processor and how much should be time-shared in a pool of capability. When a machine is simulated, a similar question is raised. Consider a virtual machine system; how much of the system coding that will support operations should be duplicated in each simulated duplicate machine?

The IBM M44-44X system was the purest implementation of the virtual machine concept. There was a machine, the M44, which truly existed. It was a version of a product line machine, the IBM 7044, which had been modified to support dynamic memory relocation and to permit considerable extended address range for programs up to around two million words of the 36-bit words of the 7044 architecture.

In addition, the now common feature of reserving a set of instructions for use only by a "privileged" program (the executive) was added to the system. The "virtual machine"—the 44X—is mapped onto the M44. The 44X is a machine that pretends to have two million words of core and to be able to execute all instructions and systems. Multiprogramming on the M44 is achieved by the invocation of 44X's. In time-sharing usage the terminal is handled as the console of a 44X.

The operating system environment is imposed by two components: The first, which runs on the M44, which simulates on behalf of a 44X those functions that are executable only by the M44, is the modular operating system, representing the interface between the real and virtual machines. The second component is the operating system of each 44X, which includes a compiler, an assembler, control language, and loading and debugging aids. Conceptually, each 44X has a copy resident in its pseudo two million words of storage of each element of the programming system. Actually, they share a copy of these elements in real core.

Notice that in a standard contemporary multiprogramming system running (if it could) two simultaneous compilations the compilation would occur independently with independent copies in each process. The concept of sharing, although it exists in other systems, is limited to explicitly directed sharing or to very basic supervisory kernels. Among those things often not shared are the language processors.

In the M44/44X the real machine maintains a copy in real core, mapped to by the virtual machines. The question of how much to associate with the real machine in this manner, how much sharing the virtual processors should do, is equivalent to that of how large a "processor" should be. Problems of interference, contention, and cost performance are basically identical.

The virtual machine of M44/44X implementation represents explicitly what is happening in a multiprogramming environment. Its later elaboration into CP 67, allowing the running of different operating systems kernels, shows how flexible and variable the concepts of real and virtual truly are. The essential character of the M44/44X virtual machine comes out of the concept of virtual memory, and there are no fundamental concepts that differ broadly from a more usual image of multiprogramming, but the specific awareness of multiprogram as a machine simulation seems unique to the authors of that system.

Chapter 16

PROCESS RELATIONSHIPS
IN
MULTIPROCESS ENVIRONMENTS

16.1. ASYMMETRIC MULTIPROGRAMMING—
ON-LINE PERIPHERAL OPERATIONS

We have been describing various aspects of the simulation of a multi-processor in an operating system environment on a single CPU. The simulation that we have talked about is fundamentally "symmetric," a simulation of multiple identical processes. It is not necessary that this be true, and indeed it is an extension of an earlier concept of on-line peripheral operation called by IBM "spooling."

Spooling is a simulation of an asymmetric machine of the ASP or DCS type or of a main and satellite computer installation. In its basic form a "main chain" program alternates with a utility program or a family of utilities in gaining the CPU. The "main chain" represents the main or computational processor; the utility program, the support processor.

Since the goal of the system is to use the idle device and channel capacity of the system with minimum increase in CPU usage, the best effect is obtained with a compute-bound main chain. The systems generally run with a priority bias for the utility package in an attempt to keep the limiting low-speed resources being run by the utility in constant operation. The utility runs whenever it can run, that is, when it is not "blocked" awaiting the completion of an I/O function.

The operation of such a system affects program development in exactly the same way as the acquisition of a stand-alone satellite or the implementation of an ASP-like hookup. Programs avoid the use of slow-speed peripherals for input and output and depend upon interfaces with high-speed discs or tapes for I/O. The "dispatcher" in the simplest systems is an I/O component that fields interrupts and then forces control to the utility.

The interface between a main chain and the asynchronous utilities that read the card reader and run the printer and punch is a buffer file. The buffer file provides a communication mechanism of rather general power of which the use in spooling is a relatively trivial example. Even in spooling, the buffer may be used in different ways, depending upon how one views the level at which an interaction between processes may be defined.

If the buffer file is considered to be a nonsharable resource such that only one process may be using it at any time, then by definition the preexecution, execution, and postexecution stages of a program must be sequential. The "main chain," for example, must release the file before printing can get under way.

The pattern of usage of such a system is identical to what we have described for ASP or for the off-line satellite computer; the results of program 1 are being printed while program 2 is producing an output file. In ASP a file destined for print or punch service is not released by the generating program running on the back machine before it is entirely available for printing or punching.

In OS/360 a spooling capability is implemented as part of the general multiprogramming environment through the use of asynchronous readers and writers. The relationship between a program running in a region (under an initiator) and a writer is essentially a simulation of the ASP interface. It is necessary not only for the STEP producing the file to conclude, but for the entire JOB to conclude before the buffer files can be scheduled for the utility.

In this environment, since there may be multiple active "main chains" (multiple active initiators), the selection of a file for printing or punching involves some priority and class decisions. It is possible for multiple readers and writers to be active at the same time. In effect, an operating reader or writer and an operating initiator simulate an asymmetric system. The intent of such an organization, of course, is to simulate on a lower-cost basis the operation of a two-processor system. Closer approximations of ASP are available in extended OS/360 environments. Among them is HASP (Houston automatic spooling), an independently developed spooling system initially based upon a version of the fixed partition release of OS/360. ASP itself is available in a simulated system, called LASP (local ASP), where the support machine functions of the hookup are performed in one partition and the compute (execution phase) functions are performed in another partition.

16.2. EXTENDED USE OF BUFFERS AT AN INTERFACE

Another approach to the use of a buffer file has been taken by the developers of the UNIVAC 1107 EXEC II system. This system is also available for the 1108. In this development the buffer file is considered to be a resource shared by a main chain and an asynchronous utility called a symbiont.

The main chain and its attending symbiont run together, the main chain contributing records to the buffer, the symbiont printing records as they become available. The system provides a symbiont for each peripheral unit. These programs are not resident but are brought into core as required.

Resident with the EXEC is a family of routines called cooperatives, which provide an interface with the buffering system between the worker program and the symbiont. The main chain has a set of standard linkages to the resident cooperatives of the EXEC system. These linkages call upon a functionally specialized cooperative like CPNCH (card punch) or PRINT (printer) or CREAD (card read) to take or deliver an output or input record, specifying (for example) the core location to (from) which the image is to be put or taken and the size and any device control options which might be appropriate, such as printer skip or line advance.

The cooperative takes the image and places it into a buffer area on the system drum. A portion of the drum is set aside as a symbiont buffer area. The buffer space is organized into fixed-size blocks, and a file directory organizes collections of blocks by the files to which they are currently assigned—input card file, print file, output card file, etc.

In addition, a population of core buffers is organized into the same fixed-word-size blocks. These are dynamically acquirable and releasable by system-provided macro linkages. It is in these buffer areas that a symbiont called upon to operate finds residence space. (Space in drum buffer area is also dynamically acquirable and releasable.) A symbiont is called upon to operate by its corresponding cooperative by use of macros, which request activation of the symbiont loader and the activation of a symbiont or the placing of its name in the dispatcher queue (up to 43 symbionts may be active in the system). The activation of a symbiont is at the discretion of the cooperative, which does so on the basis of an output buffer being filled and ready for printing. A symbiont may also be initiated by an operator from the console.

The effect of the system is to provide simultaneous output while a worker program is still in operation without waiting for the output file to be closed and released. In normal operation in multifacility systems with more than one printer or reader or punch, it is normal to see one printer finishing the output for a completed main chain while the other printer is working on output for the current run.

Remote job entry with a remote UNIVAC 1004 was essentially handled in the same way. A beneficial side effect of this is the ability to hold a communications line after transmission of a remote job in the anticipation that output will be returned almost immediately, as soon as, for example, some initial compiler output has been produced. The necessity for local storage of output and then recontact and transmission is eliminated.

16.3. IMPLICATIONS FOR PARALLEL DESIGN

Some comments about the precedence and "parallelizability" of tasks or stages in a processor in a job come to mind here in association with ideas about the bases in which interactions around resources are perceived.

The fundamental difference between the IBM and the UNIVAC approach lies in the perception of the level of sharability of the interfacing buffer files. Since a requirement for parallel operation is that the intersection of the output set of one path and the input set of the other path be null, it is intuitive that a main chain and its output writer cannot work together. However, we see that an early and obvious parallelism has always been the computational path of a program and its I/O path, where by definition the input of one is the output of the other, and vice versa.

It is clear that we can relax our intersection rules when we know some additional things about the interface. Here we know quite a bit; we know the placement rule and the gather rule, and we know exactly the sequence in which each member of the I/O intersection will be put into the interfacing buffer. Further, we know that the processes are entirely independent except at this synchronization point. We perceive that true synchronization, true "lock-out" and sequentialization, is required not at the level of the entire collection of records but at some smaller unit, theoretically a record, but actually some collection of records organized for efficiency into a unit that is handled by the interface. The relationships are made meaningful because of the presence of the high-speed drum; in its absence the system would reduce to a traditional I/O overlap organization.

In searching for parallel opportunity in programs or algorithms, one finds many partially parallelizable processes. A purely parallel process is one in which a number of paths may be activated, each one running independently of the other (on real or on virtual processors) until some final point, where they all independently conclude. Beyond this a parallelism is partial, there being some point in time where a process must communicate with another process in order for either or both to continue.

Whether it is worthwhile to attempt parallelism depends upon ratios of time and cost for the interactions required by the paths as opposed to how much is being done truly in parallel or truly independently. In the 1107/8 environment it is highly profitable to undertake the parallelism for a number of reasons, primarily because it enhances the performance of the system as a turnaround machine by eliminating the additive time of the entire file-generating processes. Second, the amount of independence is seen to be highly significant. If the symbiont cannot continue its operation because of lack of items to print, it can simply be suspended, and another symbiont that does have a collection of material to work with can be given processor time in its place.

16.4. GENERALIZING PROCESS RELATIONSHIPS THROUGH BUFFERS

The use of buffers as interface communication mechanism in 1107/8 spooling is merely a trivial example of the use of buffers to enable relationships between processes and to simulate machine systems. In the most general usage any population of programs, processes, or tasks may be interfaced through the use of buffering, given the proper constraints about the flow of information. The buffer becomes a general intercommunication mechanism between independent or partially independent processes. Synchronization of programs is controlled at the point of placing data into or drawing data from a buffer file, and, as we have already observed with buffering performance considerations, the rate of progress of any program can be controlled by modification to the amount of input and output buffer space associated with it.

The level of synchronization, the degree to which "partial results" can be passed from one program to another, is inherent in the control of buffer size and pointing. Quite complex relationships between program elements, including spawning and subtasking, are realizable by the proper generation of synchronizing buffers at the interface and the development of a generalized buffer control mechanism that will recognize the availability of the buffer and the relative position of the buffer input pointer and buffer output pointer of cooperating processes. A quite good description of a buffer file-oriented system is given for the GE 645.

16.5. THE COMPILING MACHINE

Perhaps one of the very interesting examples of an approach of this type comes not from operating system considerations as such nor from data management, but from an article in the design and development of compilers by Conway. The use of buffers to simulate various machine configurations and the relationship between process elements is nicely seen.

Consider a machine system that has three special-purpose processors, a scanner, an analyzer, and a generator. Each machine was built to perform only one function in the process of reducing a symbolic language representation of the program into machine code for a general-purpose processor, a fourth processor in the system.

Consider that the scanner has an input device from which it reads the original source coding, and it has an output device on which it writes a transformation of the source code, let us say postfix Polish. The only input to the analyzer is the output of the scanner.

The analyzer replaces all symbols in the Polish string with machine addresses and reduces the Polish to a triad form. It has a single output device on which it records these triplets. The generator has only the input from the analyzer. It generates machine code and writes it on an output device, which is input to the general-purpose processor, which will execute the code and which has other inputs and outputs.

We have defined a multiprocessor system of some unusual characteristics. Particularly, we have mapped a machine population perfectly against a multistage process of compile and execute. If we visualize the system as going from left to right, we see that each processor causes a transformation in the data and that the processor to its immediate right sees only that transformation as its input. There is a series of transformations. The nature of the interfaces between these processors may vary considerably.

If we visualize this interface device as being a tape unit and each processor as having private memory, then the system is implementable as a series of alternating processors and tape units.

If the tape unit is a switchable device available either to a left processor for output or a right processor for input, then no processor in the system may run with any other processor, and we have a sequential process of staged compilation and execution. Parallelism across the continuum of a single source program is excluded; however, the geography of the system allows parallelism across source programs. Scanning of program 2 can continue in parallel with execution of program 1. The processing of program 2 can, indeed, continue up to the point that it requires the interface between the generating and executing processor.

If the ititial action of execution is to load program 1 so that the interface tape is released for written generated object code, then program 2 may reside at that tape while program 1 is executing; a program 3 may be advanced as far as readiness for generation, and a program 4 as far as readiness for analysis. Program 5 may be waiting for the input device of the scanner processor. At the end of execution of program 1, each program shifts right one machine, and the work of the system is seen to be the left to right movement of source programs across the family of processors. Utilization is a function of the relative amounts of time required for each stage limited by the constraint that a processor and its immediate successor may not be active simultaneously (except in the case of the generating and executing processor). This constraint comes solely as a result of our definition of the nature of the interface between processors.

If we allowed simple "device sharing" such that the interlock on the tape was at a block level, the fundamental behavior of the system could seriously change. All special-purpose processors could be active over the life of a single program in the system with interlocks only on tape read and write contention. Consequently, statements of the same program could be in different stages simultaneously with other statements, and parallelism over a single program could be achieved.

If we remove our series of processors and reduce to one processor, we find ourselves dealing with the traditional phases (approximately) of a compiler. The movement from left to right is replaced by a movement in time from one phase to another, and in a sense the concept of the movement of data across processors is modified into a fundamentally different concept, the concept of the movement of procedure against data.

If we visualize that each phase of the compiler occupies processor memory for an interval of time, produces its transformation on an external device for all of its input, and then releases control of the processor to its successor, the sequential behavior of the special-purpose multiprocessor is perfectly simulated.

Consider that the output/input tape is truly a buffering unit, a queue, and the interface of processors having certain attributes, among them that information flows in one direction. We may certainly consider that a process is independent of the specific character of the buffering units that surround it.

Particularly, we could change the tape to a disc or we could define a core buffer so that all output was placed in core and processes were passed against the internal files of transformations. If the process is truly independent of the nature of the buffer, the buffer emerges as an independent element of the system, something to which processes are hooked to form dynamic paired relationships for intervals of time and which itself is subject to allocation.

Consider a situation in which there is enough space in core just for the scanner phase of the compiler. The buffer associated with the process would be on tape, and the phase would operate sequentially over all source statements, the size of the buffer being that amount of tape necessary to hold all the transformation accomplished by the scanner. Consider an increase in core size of the system such that it is now possible to hold all the phases of the compiler in core at the same time.

If the buffering mechanism is now allocated to core, then it would be possible to transform the operation so that a single statement is taken across all phases. Phase one would process a statement and release control of the CPU after processing and placing the record in the buffer. Phase two would then operate, etc. The effect is the simulation of the shared-device multiprocessor system. Whether the compilation would proceed by running a phase against all statements or a statement against all phases becomes a function of policies affecting buffer management and allocation independent of the definition of the process.

By implication, the same compiler design would run on machines of various memory sizes with the support of a buffer allocation and processing manager. In general, program elements may be run together or sequentially in this way with the support of (perhaps) some additional editing and formatting capabilities to smooth the details of the interface.

16.6. DRAWING INFERENCES

There are two interesting things about observations relating to buffer interfacing. First, an insight into the nature of the parallizability of programs through buffer management is provided, and second, an insight into the transformability of real-time into batch or batch into real-time applications is provided.

Batch applications tend to be process driven, in that buffer management tends to reside in the process that has GET- and PUT-like functions that interface it with the I/O system whose vanguard elements are the buffers. I/O activity is scheduled as a result of process operation. Usually batch processing involves a series of programs run against transformations and reorganizations of data sets.

Real-time applications tend to be event driven, in that processor activity is scheduled as a result of an I/O event, such as the receipt of a message or an inquiry.

Usually, real-time processing involves the movement of individual data elements against series of sequences of processes. But the fundamental difference between batch and real time truly is the size and allocation of the process interfacing buffers and in the location of buffer management. In real time processes tend to be fed data by buffer (queue) management elements independent of the process; in batch buffering control tends to be internal to the process.

The fundamental analysis of a batch program, which could lead to its transformation into real time, is the determination of those points that really represent interfaces between subprocesses and at which point buffer-processes relationships are truly being found.

This is exactly the analysis required for the transformation of sequential algorithms into parallel algorithms. It is, indeed, the analysis inherent in the definition of a modular system of any sort. Operating systems built around generalized queue management are an obvious extension of the buffer interface concept. It is, of course, something of a simplification to indicate that process (and processor) intercommunication and linkage considerations find a total answer in the concept of an intervening buffer unit or file between every process or processor.

The one-way movement of information, the assumption implicit in Conway's "Co-routines," suggests that not all relationships can be mapped in this way. In addition, there are considerations related to the passing and control of resources between processes. But many of the complexities of program linkage come from the iniative taken by a process to call another process, and to drop out if the linkage is imposed through the external queue manager.

Chapter 17

STATEWORDS, STATUS, AND SWITCHING

17.1. STATEWORDS

The fundamental concept that supports the notion of a virtual machine in a single-processor environment, the switching of tasks in a single- or multi-processor environment, and in general the assignment of processes to processors is the idea of the "stateword" or state vector associated with a process at any point in time.

A process can be viewed as a series of transformations of its stateword components. The stateword consists of that information about a process which changes over time and upon which the continuation or resumption of the process depends. The basic element of the stateword is a pointer to the instruction that the process is next to execute.

Beyond this a fundamental stateword consists of representations of values in machine registers (arithmetic and index) and condition codes. In any system the fundamental stateword elements are partially implemented in hardware. The program state word (PSW) of IBM 360 and the program state register (PSR) of the 1108 are representations of stateword elements. Beyond location counter, condition codes, mask bits for the prohibition of interrupts, perhaps, stateword is still fundamentally a software concept, and the elements of the stateword are distributed through various software structures in the operating system.

For each process the software system defines a control block, which represents that task to the system and through which system and process interaction is performed. It provides a place in core where the contents of registers may be placed, and from which registers may be restored. It provides a place for the representation of any relationships between the process and processes that have invoked it, or which it has invoked; it provides a place for the description of events that must complete for the task to operate; it provides a place for pointers to other system control blocks, which represent the allocation of core and the allocation of devices to the process.

Operating systems differ broadly in how the stateword information of a process is distributed throughout the structures of the system and what paths must be followed to achieve associations between facilities and processes, processes and other processes. The control block concept, as such, is fairly universal. Each process admitted to the "mix" of jobs that may receive CPU service is represented by a control block for as long as it remains "active" in the system.

We promised, earlier, a more careful definition of the concept of "active." We have been using the word to this point to mean a process that has a control block in the system. The collection of control blocks linked together in some fashion represents a queue, where some members are actually in control of a CPU (as many as there are CPU's), some members are ready for the CPU—they could immediately execute if they could assume control—and some are blocked; that is, they are waiting for some event to happen in the system, and they cannot take control of the CPU until that event occurs.

17.2. STATUS OF A PROGRAM

Various systems use various words to describe the three fundamental states in which a program may be. READY, ACTIVE, BLOCKED, WAITING, and SLEEPING are commonly used. In some systems an additional status, SUSPENDED, may exist. A suspended program is a program that has been partially deallocated in order to provide a resource for a more critical program or which has been queued in order to reduce the load on the system when it has been observed (as it can be in some systems) that the load is too heavy.

Some systems, particularly time-sharing, distinguish between READY and ACTIVE so that the ACTIVE's represent a population of processes that are going to receive service over an immediate period of time and the READY's are those which can gain this population but which will not do so until service has been completed for the ACTIVE group. A program moves back and forth between ACTIVE and READY on the basis of some algorithm for the distribution of time among requesters.

Similarly, BLOCKED and WAITING or SLEEPING are generally synonymous, but local definitions vary. Some workers prefer BLOCKED to mean that a program is delayed pending the availability of a shared resource (its release by another process), whereas WAITING implies the completion of some system event, like a terminal action or the delivery of a "page" from memory.

17.3. SWITCHING

The multiplexing of CPU attention implies that there are going to be intervals of time when a process will not have control of the CPU. In a uniprogramming environment a program's "run" time and elapsed time are identical, since it receives continuous service from the system from the point at which it is activated until the point at which it is complete.

The essence of multiprogramming is that this should not be so, since there may be intervals of time in which components of the system are not actually in use if this is allowed to occur. A mechanism is required for the disassociation and association of processes with processors recurrently over time. What is required, when the decision is reached to switch a program, is that the stateword is saved in such a way that it is possible to restore the processor to exactly the condition (or a sufficient approximation of the condition, since disc arms may have moved, drums will be at different locations, etc.) in which it was at the time of switch so that the switched program may be restored to the processor.

The act of switching tasks, therefore, is the act of "saving" the stateword of a process and loading the processor registers with the stateword of another. One step in this is to reflect the hardware registers in the "save area" of the control block. In an active process actually running on the CPU, the stateword is represented in part in the hardware registers and in part in the control block. Since values in registers vary more quickly than other elements of the stateword, there is sometimes a tendency to consider the stateword in the restricted sense of these values. More properly, the entire control block is the stateword, as well as all information represented in such structures as file and device description block and memory allocation maps that relate to the process. It is a representation of the status of the process, and the definition of this status includes all of the capabilities (resources) associated with the process.

To understand the full extent of the state word, one may consider a system in which there are a number of processors and, instead of representing control blocks in shared, commonly assessible memory, control blocks are held in the private memory of each processor. In order to run a process on the processor, it would be necessary to load the registers, of course, but beyond this it would be necessary to provide addressability to all resources associated with the process and the time of its previous interruption. In order to do this, pointers to file and memory control blocks would be necessary, and the act of associating a process with a processor is the act of moving the entire control block to its private memory.

Consider machines that have dynamically changing segment populations in memory as they run. If the process blocks on a processor, the processor picking up the process would require that it have a map of the core assigned segments of the process at the time of interruption.

A task switch occurs as a result of a program's exercising a system macro informing the system that it wishes to block itself, or of a system service's deciding that a program must be blocked, or as a result of an interrupt (timer or I/O, or, on machines where they exist, internal program-generated interrupts). The interrupt situation is interesting because it implies a transient preemption of control, which may not actually result in the switching of processes.

17.4. TASK SWITCHING AND INTERRUPT HANDLING

Strictly speaking, a task is switched for every interrupt, since the interrupt handler replaces the running task as controller of the machine. However, this handler has a special status in the system (or may have) as a system task. The system task may not have a control block associated with it. If no mechanism is provided for the stateword retention of the system task, then that task must run as an uninterruptible task. Interrupt handlers often do run with interrupts disabled.

Since system task resources and capabilities are known to the system designers, it is possible to allow for interruptibility of system tasks with a mechanism less developed than the full control block used for user tasks. In any event, the function of the interrupt handler will involve usage of some of the registers of the processor. It may be necessary to store the registers of the interrupted process in order to run the interrupt handler. Many systems have conventions for which registers may be used by an interrupt handler, which registers it must save, etc. These register conventions are part of the general register usage conventions of the system. For example, in calling subroutine or system services users are told that certain values must always be in register X and that the result of service will be in register Y, etc. A commonly familiar convention is at the interface between a user task and I/O support, where a given register always has the pointer to the current record advanced by the dynamics of GET and PUT.

The time of interrupt handling is generally considered to be critical time for the system. The ability to keep the channel continuously active may depend on it, and certainly task switching time includes interrupt handling as an important component. Certain machine features exist specifically to provide efficiency in interrupt handling. The length of interrupt handling determines the delay in processing other interrupts, which may be locked out during interrupt handling. Even, therefore, where CPU cycles are available (in a system characteristically I/O-bound) to handle interrupts leisurely, quick interrupt handling is still a system requirement.

The requirement for storing the registers of the interrupted program represents a considerable burden for an interrupt handler. It is resolved by providing for very fast load and restore of multiple register populations or by providing for the use of the system a set of alternate registers available for system tasks running in supervisor mode. When the system enters supervisor mode, all references to registers made by a running process refer to the supervisor set. It is, therefore, possible for the interrupt handler to process an interrupt without disturbing the contents of the registers used by the interrupted task.

Consider that during the course of system activity interrupts will occur which will not lead to the switching of control from one user task to another but will result in the restoration of the interrupted task to the machine when interrupt handling is complete. To the extent that task switching is infrequent relative to the need for exercising supervisor tasks in supervisor mode, the payback from the additional registers may be high. It is a moot point as to whether this payback comes from the division of registers into a user or supervisor set or whether the doubling of register population as such might provide equal efficiency with proper conventions.

In another machine, the UNIVAC 494, a successor to the UNIVAC 490 in very much the same spirit as the 1108 is a successor to the 1107, the register population has been increased by a doubling of index registers (the machine has only two arithmetic registers, A and O, which also link as one register in a scheme reminiscent of IBM's 7094), but there is no private executive or user set.

A further problem in interrupt handling is I/O. In System/360 the PSW contains a set of bits which represent the enabling or disabling of interruptions on a channel basis. This "mask" may be preset so that the pattern of interrupts allowable need not be determined or set by the interrupt handler. On first glance, it appears that there is an inherent ability to permit multiple channel interrupts in the system. But because there is only one location that is available to receive information about the channel (the channel status word), interrupt from other I/O must be inhibited for at least as long as is required to free the channel status word. In the UNIVAC 1107, lockout of interrupts is automatic and is released by an "enable interrupts" instruction. In the 360, of course, a switch in the PSW automatically changes the lockout permit status of interrupts.

17.5. DISPATCHING

One of the determinations of an operating system design relates to the definition of when an interrupt handler should exit directly back to the program that was interrupted and when it should exit to the dispatcher. This determination involves concepts of scheduling and the nature of the interrupt system, particularly the use of the interrupt system to service relationships between programs and between user programs and the supervisor services.

In a minimum system with no "supervisor call" (SVC) facility which causes an interrupt when certain supervisor services are requested and no priority, it is possible to visualize a design where programs always release control only by voluntarily executing a WAIT. In such a system the interrupt handler would always return to the program that was interrupted. For normal I/O handling with other I/O inhibited, it would be reasonable to leave the location counter of the interrupted program in low core and transfer directly to it.

In a more elaborate system there will be times when the interrupt handler exits to the interrupted program and others when it exits to the dispatcher to select a new program for running. The dispatcher may or may not choose to reinitiate the interrupted program. This depends on the dispatching rules defined for the system and the status of ready, waiting, and active programs at that time.

Normally an I/O interruption is reason to go to the dispatcher. The determination of who should run involves accessibility to the list of control blocks currently in the system. These control blocks may be linked together so that movement through the list involves following chains from block to block. The chains may be rather complex, allowing a movement in a number of directions, depending upon the criterion that is being used for the switch decision.

Alternatively, a separate mix list may be maintained so that transversal across control blocks is eliminated or reduced. The mix list is a list of addresses which denote the origin of all system control blocks.

The exact design of control block linkages and system addressability is partially a function of the queue organization on which ready and blocked programs are represented. On the completion of the selection of a new task the system must load from the control block the registers and execute an instruction loading the PSW into the control counter of the machine.

17.6. OS SERVICE CALLS, SHARING, AND REENTRANCE

One of the more interesting aspects of task switching and interruptions involves almost the entire spectrum of operating systems consideration. This is the call by a user program upon a service of the operating system. In current systems this is supported by an executive service interrupt (on OS/360 an SVC).

A conceptual decision that must be made is whether these system services, running on the request of a user, run as system tasks or as the user. It is easy to see that this is partially a function of what the service is. If the service is to block the program, we are conceptually content that this is a system task. If the service is the execution of a sine calculation routine, we are not content that this is a system task.

Many system services have the characteristic that they run in the "privileged mode" because they must use instructions which are only available in that mode, or they must address into areas which are only addressable in that mode. This can be true on a bounds register machine, where addressing beyond bounds is an executive capability. Beyond running in a privileged mode, many system tasks have the characteristic that they run disabling all interrupts (or all interrupts of their type). Essentially this gives them the highest priority of any task in the system for the duration of their execution.

If one extends the concept of shared services to include a population of services which are represented in one body of coding and which are available for the use of programs but with no reason for increasing the priority of these programs during the use of these services, or granting them increased access to machine function, then alternate solutions to the user/system interface are required. In fact, the covering of all system service requests by SVC-like interrupts does not seem to be required.

What is required is only that these services be addressable by the program through its control block, that is, through its list of capabilities. Beyond this, however, since it will be possible to preempt a process while it is using these services, it is necessary to organize these services in such a way as to allow such preemption. The development of the concept of reentrant coding is a result of the desire to allow programs to share system code (or even arbitrarily designated islands of code among themselves).

A reentrant island of code is one in which no change occurs as the result of execution at any time. All parameters passed to it, all the intermediate values that it develops, all results, etc., are considered to be objects external to the coding itself.

If, for reasons of priority, for example, in a multiprogrammed uniprocessor a user is preempted from the use of the shared function before it is complete, it is necessary to have a mechanism that allows him to resume when the preempter passes through the function. A very similar mechanism is required for the simultaneous execution of the code on a multiprocessor. In the multiprocess case the mechanism is required in order to reduce the burden of lockout. As we shall describe, the "lockout" problem is particularly sensitive on multiprocessors, since the potential loss of efficiency is high. It exists, however, in any multiprocess situation.

The mechanism required to support reentrancy is the provision of work space for the values that are local to a given user of the system. In virtual memory machines sharing is accomplished by placing a copy of the function in the virtual memory (the apparent address space of a process). By then arranging auxiliary store maps the same physical code in the system library may be used, and by arranging memory maps (register/segment) tables copies of resident core code may be shared.

In nonvirtual memory systems without paging/segment hardware support, sharing may provide some interesting problems, and the solutions may demonstrate many of the problems involved in designing a hardware/software system. Consider a machine which has a base register that is always added to location counter to form an address, a system of privileged and nonprivileged operation, and a register which specifies the range of addresses that can be referenced by a task. A running task now wishes to use a common subroutine. The convention may well be that in order to do so it is necessary to execute an executive service request call. That is, all such requests are supported by the supervisor. The supervisor and the user must now exchange some information. Particularly, the input must be passed, and a work area for the function must be defined.

There are two approaches to this linkage. One is to consider the work space as the property of the caller and have the function address all of its data through a pointer provided by the caller. If the value pointer is a relative pointer (must be augmented by the base register to be effective), a problem may exist. An instruction must exist which either allows the caller to materialize an absolute pointer or allows the supervisor to materialize this pointer. The reason for the absolute pointer on a machine of this design is because at the time of flipping into "supervisor" mode the base register of the system becomes inoperative or must be set to addressability for the supervisor function. The function must properly stabilize instructions in its own area not offset by whatever base is used by the calling program. Without the materialization of an absolute pointer, the function would find itself able to address either user data or its own instructions, but not both.

An alternative is to consider all work space the property of the function, perhaps in a dynamic area contiguous with it. A further question arises as to whether it is desirable and necessary to run in supervisor mode. If the system has a convention that interrupts are disabled during supervisor state, it is highly undesirable to do this. If the system has no such convention, running this function in that state exacts less of a penalty but provides the possibility that pathologies may develop which will result from the function having available to it the full machine and all privileged instructions. In order to avoid these pathologies it is necessary to have developed another conceptual status for the machine. The status required for the machine must provide for a mode in which it is possible for a program to read beyond its address limits but not to have the ability to execute privileged instructions. It is, therefore, possible to run in this mode while common subroutines are executed.

17.7. FINAL REMARKS

The organization and design of an operating system is very complex indeed; logical and performance tradeoffs abound, and the interrelationships are subtle and often mysterious. Very often designers are tempted to settle for a first feasible design. This has been generally true in the development of general-purpose batch operating systems in the past. The atmosphere of the vendor's system programming shop is one of almost incredible pressure. In an interval of time from which the machine is relatively stable until the delivery of the first machines, a new machine must be defined—the metamachine represented by the operating system.

Much of this pressure may be relieved by a joint design of hardware and software, and by the continuation of the design programmers on the development project through the definition of the production product called the operating system. Up to and including the current generation of equipment, the systems programming activity has been the scapegoat caught in the middle between the needs defined by marketing groups and the functions supported by hardware groups. There has been little or no time to develop and objectively evaluate alternative designs, to measure the tradeoffs that must be made, and to impose a concept of service and function that is consistent, unified, and efficient across all components of the system. Even less time has been available to develop the tools that might be used to evaluate designs. Systems programming shops are either, apparently, too large or too small.

The problems of system programming are reflected by the high reorganization rate of the function. We see cycles in which system programming is a unified function organizationally coherent (on paper) and reporting at a level with marketing and engineering. We see other cycles in which it is distributed between engineering center managers, machine project managers, distributed between engineering and marketing groups. Within the shop itself we see both functional organizations, where programmers are organized into assembly development, compiler development, and schedule development, and machine-oriented organizations, where programmers are organized into groups representing each particular machine in the vendor's product line.

There are no measurements for what is a good system or what is a good organization, despite many attempts among manufacturers and in the universities (and other users) to study the ground rules of the process of evaluating a system design. It is not clear whether systems transcend or reflect the organizational structures that produce them. Much work is necessary in the development of design tools, as well as in the development of insights into the system development cycles. These considerations are beyond the scope of this effort. But what is very relevant are the underlying motives, concepts, techniques, alternatives, history where possible, and flavor of the problems of systems and the elimination of the rediscovery of fundamental principles by each new group for each new operating system.

PART FOUR:

SOURCES AND FURTHER READINGS

Betourné, et al., "Process Management and Resource Sharing In The Multiaccess System, 'ESOPE'," *ACM Second Symposium Operating System Principles* (October, 1969).

Clark, W. A., III, "Functional Structure of OS/360 Part III: Data Management," *IBM Systems Journal,* Vol. 5, No. 1 (1966).

Codd, E. F., Lowry, E.S., McDonough, E., and Scalzi, C.A., "Multiprogramming STRETCH: Feasibility Considerations," *Communications of ACM* (November, 1959).

Corbato, F. J., et al., "Some Considerations of Supervisor Program Design For Multiplexed Computer Systems," Clearinghouse for Federal Scientific and Technical Information, AD 687551 (February, 1968).

Critchlow, A. J., "Generalized Multiprocessing and Multiprogramming Systems," *AFIPS Proceedings,* Fall Joint Computer Conference (1963).

Dennis, J. B., and Van Horn, E. C., "Programming Semantics for Multiprogrammed Computations," *Communications of ACM* (March, 1966).

Heistand, R. E., "Executive System Implemented As A Finite State Automata," *Communications of ACM* (November, 1964).

Hutchinson, G. K., and Maguire, J. N., "Computer Systems Design and Analysis Through Simulation," *AFIPS Proceedings,* Fall Joint Computer Conference (1965).

Huxtable, D. H. R., and Warwick, M. T., "Dynamic Supervisors—Their Design and Construction," *ACM Symposium on Operating System Principles* (October, 1967).

Leonard, G. F., and Goodroe, J. R., "An Environment For An Operating System," *ACM Proceedings,* 19 (1964).

Leonard, G. F., and Goodroe, J. R., "More On Extensible Machines," *Communications of ACM* (March, 1966).

Mealy, G., "The Functional Structure of OS/360. Part I: Introductory Survey," *IBM Systems Journal,* Vol. 5, No. 1 (1966).

Morris, D., Sumner, F. H., and Wyld, M. T., "An Appraisal of the Atlas Supervisor," *Proceedings of ACM* (1967).

Ossanna, J. F., Mikus, L. E., and Dunten, S. D., "Communications and I/O Switching In A Multiplex Computing System," *AFIPS Proceedings,* Fall Joint Computer Conference (1965).

Patel, "Basic I/O Handling on B6500," *ACM Second Symposium Operating System Principles* (October, 1969).

Rosin, R. R., "Supervisory and Monitor System," *Computing Surveys,* Vol. 1, No. 1 (March, 1969).

Tonik, A. B., "Development of Executive Routines, Both Hardware and Software," *AFIPS Proceedings,* Fall Joint Computer Conference (1967).

Vyssotsky, V. A., Corbato, F. J., and Graham, R. M., "Structure of Multics Supervisor," *AFIPS Proceedings,* Fall Joint Computer Conference (1965).

Ward, B., "Throughput and Cost Effectiveness of Monoprogrammed, Multiprogrammed and Multi-Processing Digital Computers," Clearinghouse for Federal Scientific and Technical Information, AD 654384 (April, 1967).

5

Multiprocessing

Chapter 18

SYMMETRIC MULTIPROCESSOR:
CURRENT DESIGN
AND CONSIDERATIONS

18.1. SYMMETRIC SYSTEMS

We have discussed two extreme situations of parallel operations, the microcosmic "local" instruction level parallelism of a high-performance uniprocessor and the parallelism between large loosely connected computer systems. In each of these cases, the issues of symmetry, distribution of intelligence, contention, etc. that appeared to us in our description of the two television repairmen were essential elements of the description of the systems.

Each instance of multicomputer configurations that we described had the characteristic of asymmetry. We would like now to investigate a class of machine that is generically called the "general-purpose multiprocessor," which displays varying degrees of symmetry.

In discussing symmetric systems we must remember that the concept is an abstraction that derives from our point of interest and the level of task and system function that we wish to describe. No system is perfectly symmetric in the sense that any component can undertake any element of any task presented to the system. Functional specialization, localized asymmetry, is inherent in computer systems.

The symmetric quality of these systems derives basically from the availability of more than one CPU-like device available to undertake the execution of instructions that we commonly associate with direct execution on the CPU.

18.2. MULTI- VS. PARALLEL PROCESSORS

These systems are fundamentally organized to facilitate either the cooperative effort of a population of processors on a single task that has been organized so that elements can be run in parallel or the simultaneous execution of unrelated tasks.

The motive for the cooperative execution of elements of the same task is to finish that task more quickly. The elements may be identical, as in parallelizing iterations of a loop when that is possible, or nonidentical. The motive for parallel execution of unrelated tasks is the distribution of system cost over a larger base of programs as well as the increase in the throughput of the system.

In the literature of the field the designations "multiprocessor" and "parallel processor" are variously used. This discussion will not make an attempt to distinguish between these two designations, since they are inconsistently used. In customary usage they tend to refer, in the use of "multiprocessing," to machines that appear to be oriented toward independent job parallelism and, in the use of "parallel processing," to machines oriented toward the parallel execution of related job segments. The reader will experience neither insight nor comfort, however, by attempting to infer any particular qualities or characteristics from a given encounter with either term. We shall intentionally use the phrases interchangeably.

It is not clear that the general concept of a general-purpose machine need be sensitive to whether the user intends to run related or unrelated jobs in parallel. The fundamental mechanisms of initiation, termination, and control of processes executing in parallel are the same, and they bear a striking resemblance to the mechanisms (usually software) associated with the control of a multiprogramming environment where the activity of initiating a program in a mix by an operating system scheduler is conceptually identical to a request by a running program for execution of an associated subtask.

Performance considerations appear, however, in the fine design by virtue of the various anticipated levels of contention for the memory resources of the system, which will tend to be more severe when multiple processors are running the same code on related data. Since all systems dynamically relate "independent" programs through common usage of system service and utility routines, the issue of contention does not entirely fall away for the "independent" environment.

In a later part (Part 7) we discuss contention for space in a multiprogram environment. Space contention is only one aspect of contention; the cycle contention problem is pressingly present in multiprocessor situations. Further, since running job elements in parallel implies smaller "modules" of concurrent processes, the mechanisms for initiation, etc.

should tend to be more refined where one expects the execution of developed parallel algorithms in the same program.

An interesting aside concerning what is or is not critical in system design comes from the observation that multiprocessing organization has to this point seemed to have had little effect upon the architecture of the processors that exist in the system. In fact, a number of research efforts at IBM and at universities achieved advance points of development without giving serious consideration to the architectural properties of the processors involved. The researchers assumed the architecture of some given machine of the 7090 or 360 type and concentrated on the nature of the relationship between components of a postulated system.

18.3. CURRENT SYSTEMS

Most manufacturers in their current offerings of multiprocessing capability are presenting systems in which a standard model of their product line becomes part of a system where another such processor may be present, or more than one may be present. IBM's MP65 is a system containing two 360/65's; GE's multiprocessing for MULTICS (the time-sharing-oriented operating system developed cooperatively with Lincoln and Bell Labs) involves two 645's; UNIVAC's multiprocessing involves multiple (3) 1108's; CDC's 6700 involves two 6600's.

In general, these processing units have independent paths to memory; they share a common memory pool; they have varying capability to refer to each other's channel/device population; there are various mechanisms for direct interruption of one processor by another. In general each processor contains all the logic required to enable it to perform as a single member of a system. Register populations, whatever local stores might be involved in the architecture, and all functional capability are replicated in each machine, which is truly an 1108 or a 360/65, etc. These systems represent a final step of the line of development that begins at the coupling of computers in a loose way. An important, perhaps predominant, motive in this development is the desire for reliability, for fall-back capability.

Throughput capability for these systems is often less than proportionate to the increase in nominal processing power. That is, a two-processor system often does not do twice as much work as a single-processor system. A number of studies on the efficiency of multiprocessor systems have been undertaken with differing results; we shall discuss these in a later section. These studies variously attempt to measure the increase in computational power of a multiprocessor version vs. two stand-alone systems of the same processing power, the cost performance of such systems, multiprocessing vs. multiprogramming, effectiveness of degrees of symmetry and asymmetry, etc. The results are rather mixed, but they tend to indicate that much more must be understood about the nature of multiple-processor systems and their use before an "optimum multiprocessor" can be projected.

18.4. PROCESSOR INTERACTION

Clearly, the amount of sophistication and elaboration of processor interface mechanisms is a function of the appreciation one has of the degree of interaction one expects between processors, and this is determined in part by one's image of the system's goals.

The basic interactive mechanism available to any shared storage multiprocessor is the ability to pass information from one to the other through storage in an efficient, painless manner requiring no data transfer. This sharing of memory basically distinguishes the design from all other computer associations. This ability implies most of the problems and potentials of the generalized multiprocessor.

The memory interaction capability, however, is insufficient in that it is passive in nature. The processor receiving a message from another processor has complete discretion as to when it will investigate the possibility that a message has been received. We have seen a first example of the use of memory as the interactive mechanism in the interface between the control peripheral processor and the main frame of the 6600. Here the CPU lodges I/O requests which the peripheral processor investigates on a time cycle basis.

An analogous mechanism exists between the CPU and the devices of a B5500, where devices coming "on-line" to the system post ready bits, which are sampled by the CPU from time to time to determine if the device environment has changed.

The most significant use of general storage communications in a generalized multiprocessor is in the representation of system information in a single place in core accessible to any processor at its discretion. Changes in job status due to the asynchronous progress of jobs on various CPU's are posted in the tables in memory.

A second fundamental method of processor interface is through channel devices of the type we have seen used for ASP. This is a less passive mechanism in that within some time frame the receiving computer must respond to the receipt of a message. However, there is no concept of immediate preemption as exists with the third or "preemptive" interface mechanism which allows a processor to demand the immediate attention of another. In general, in a multiprocessor system one finds the preemptive (interrupt) and the passive (memory) communication capability used.

The exent of the elaboration of communication mechanisms is a function of our perception of the need and use for dynamic processor interaction.

18.5. MEMORY SHARING

The ability of multiple CPU's to access the same collection of primary storage introduces potentially severe memory contention problems into any multiple processor system. The delays caused by processor interference on memory reference have been estimated at as high as 20 percent in some studies.

An amusing form of the contention problem is introduced by the occasional execution of the same code by more than one processor. This can occur when both wish to use a common operating system or when they are actually attempting to execute the same problem task.

It is a problem with all multiprocessor organizations. If two "processors" are active on the execution of identical coding regions, there will be a high degree of interference between them on instruction fetch. The relationship between the processors and the fact that they are executing the same code will not be recognized by the system, and no advantage will be taken of the availability of an instruction needed by more than one processor when processors should intersect.

This contention is a serious problem for multiprocessing systems, yet the ability to execute identical code at various times is a true advantage for multiprocessors, and indeed an initial motive for sharing of memory. The ability of any processor to address all of primary memory space and consequently to execute any routine in memory provides for the flexibility of assignment and system control that is a goal of generalized multiprocessing.

A variety of schemes exist for countering the problem. The provision of local storage units between the main memory and the "processor" is a proposed solution often associated with automatic contents control of the IBM 360 model 85 type. If concurrent execution occurs, each processor uses its generated private copy from its own transient local store, reducing or possibly eliminating continuing contention to primary memory.

In the use of private local stores in general, care must be taken properly to control changed information, since there may be an intersection of data reference as well as an identity of executed procedure. If local storage is sufficiently large, quite stable populations of code representing parallel execution of loops may exist and may reduce contention to a trivial level.

In some discussions of the problem of parallel execution of identical code, software simulations of local store are proposed. Essentially, these are compiler techniques which, either on the explicit declaration of a potentially parallel region of code or upon the discovery of such a region, undertake to "unwind" the code, thereby producing separate images with directions to a loader to locate in different memory banks. The cost of this technique is replication of code and a loss of storage space, a critical resource in multiprocessors.

The cost of local stores of potentially the same speed ranges as primary store (since speed balancing is not the problem here) might not prohibit private storage, which could be easily put into the system by the use of "private tails" as in Fig. 18.1. In a basic system without automated content control, however, the transfer of code into private sections would accomplish no more than proper allocation and is dependent upon the same techniques as unwinding.

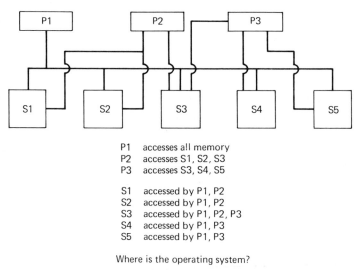

P1   accesses all memory
P2   accesses S1, S2, S3
P3   accesses S3, S4, S5

S1   accessed by P1, P2
S2   accessed by P1, P2
S3   accessed by P1, P2, P3
S4   accessed by P1, P3
S5   accessed by P1, P3

Where is the operating system?

Fig. 18.1. Private "tails."

As usual, dynamic monitoring is more effective, and the 85 type of mechanism should be considered a requirement for local storage management. This mechanism with the higher-speed memories (those that approach the speed of the processor) is, at this time, expensive equipment.

18.6. NATURE OF I/O INTERFACES

In addition to memory and whatever explicit interactions are designed into the system, processors relate to each other through the I/O subsystems. It is possible to visualize a multiprocessor in which each processor has an entirely private set of channels and control units and devices. In order for a task to operate on a processor, there must be sufficient availability of I/O equipment for that processor to satisfy the requirements of that task.

For completely independent tasks in a highly preplanned environment channel privacy might not be too great a constraint, especially if the processors were not also multiprogrammed. A difficulty would arise, however, in the joint use by the processors of system services and utilities. It would not be possible, for example, to have the operating system on a single device available to either processor. Those elements of the operating system that are dynamic and do not characteristically reside in core would have to be duplicated on devices private to each processor or one processor would depend entirely upon the other to acquire system elements. In fact, for reasons of reliability the entire operating system would have to be duplicated. Even though the processors may share memory storage and use common in-core copies of the operating system, the entirely private I/O arrangement causes an expensive duplication at another level in the storage hierarchy.

A common way around this is to develop the capability for a device to be accessible from more than one control unit. Each processor would have a private channel (with perhaps a unique address), which would lead to the control unit associated with the desired device. Alternatively, of course, a control unit could be dual-tailed to more than one processor unit. Both of these schemes are shown in Fig. 18.2.

In the UNIVAC 1108 multiprocessor this capability is supported by a device called the Shared Peripheral Interface. This unit provides an interface to 1108 processors or I/O controllers (IOC).

The IOC is a module of the 1108 system which enhances 1108 I/O capability by providing a residence for I/O commands that otherwise would be held in CPU core and would require a memory access for each adjustment of the fields representing word counts and target locations.

With the I/O controller additional channels may be added to the system, and the capability of scatter read/gather write—data chaining—is available. IOC is an optional unit interfacing to memory banks and to shared peripheral units (or directly to control units) for paths to I/O subsystem and to CPU's for receipt of I/O commands. It relates to the 1108 in a way roughly similar to the relationship of selector channels to IBM's 360/65, 75, and higher, except that the IOC has the capability of enlarging the path population by adding additional channel paths.

An IOC may interface with up to three CPU's, using an I/O channel position of each interfacing CPU. Any CPU associated with an IOC, therefore, has access to all I/O subsystem units interfaces at the up to 16 channels of an IOC. A CPU may interface with both the IOC's possible in

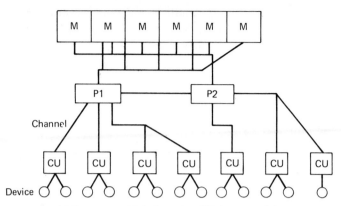

(a) Processors with independent I/O (asymmetric)

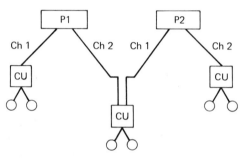

(b) Processors accessing a common control unit

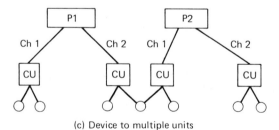

(c) Device to multiple units

Fig. 18.2. I/O Pathing.

the system. In the presence of an IOC, processors may share peripheral populations by connection with an IOC even in the absence of the SPI (shared peripheral unit).

An SPI may interface with up to four CPU's or I/O controllers at one end and with a control unit on the other. The path is, therefore, provided for a control unit to associate with an SPI and through it to be accessible to four processor modules of the system.

With the addition of a control unit, devices may be provided with dual paths to an SPI or to two SPI's. The multiple paths of access provided by this flexibility not only accomplish a highly elaborate sharing capability but provide for a high degree of reliability through the provision of alternative paths. From Fig. 18.3, a highly simplified representation of the system, the paths from three processors to a device are shown.

A similar flexibility is associated with the dual processor version of IBM's 360/67. This time-sharing-oriented system has a maximum of two processors. Each of these processors may be connected through a distributed crossbar to a device called the channel controller.

Like IOC's, a channel controller has an independent path to memory and interfaces with channels and CPU's. A channel controller interfaces with one or two IBM selector channels and an IBM multiplexor channel.

In System 360 the 2860 selector channel provides three paths to control units, each path capable of handling up to eight control units and

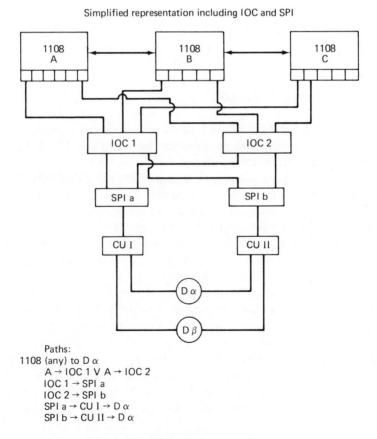

Simplified representation including IOC and SPI

Paths:
1108 (any) to D α
  A → IOC 1 V A → IOC 2
  IOC 1 → SPI a
  IOC 2 → SPI b
  SPI a → CU I → D α
  SPI b → CU II → D α

Fig. 18.3. UNIVAC 1108.

addressing 256 I/O devices. Connection of a channel controller to two selector channels actually provides six high-speed paths. The IBM multiplexor channel provides low-speed paths to slow devices of the reader, printer, punch, or terminal control unit type.

With two channel controllers on the system, either processor may reference any control unit and device. In addition, it is possible to bring a path from a control unit to more than one channel, providing alternate paths through the channel controllers associated with each control unit and also from a single device to multiple control units.

The multiprocessor (two-processor) version of the IBM 360/65 has a similar but somewhat less elaborated capability. The ability of the two processors to share I/O devices derives from the twin-tail capability of the control unit. A tail of each unit can be brought to the private channel of each processor, thus giving each processor availability of all devices on the system. In place of paths to IOC or controller units for the provision of alternate paths, it is possible (necessary) for a processor to request, over the direct line between them providing an interrupt path, for its colleague to submit an I/O request on its behalf if its own path is for some reason unavailable.

18.7.　I/O HANDLING—PREFIXING

It is not always possible to predict which processor will actually submit an I/O request in a flexible task management environment. This is because of enqueueing and task switching. Since the generation of an I/O request is typically handled by enqueueing a request with the I/O control function of the operating system, it is not necessary that a task associated with a processor at the time of enqueueing a request be associated with it at the time the I/O operation is submitted to the channel.

Since the I/O request queues are not necessarily (and characteristically are not) associated with each processor but are organized around channels, it is possible that at the time the I/O request is actually submitted it will be submitted by the "other" processor.

Suppose that task A on processor 1 requested an I/O that was enqueued and that the task was unable to continue so that it released control of the processor. In a general system the task would enter a wait list common to both processors. At some subsequent time, processor 2 was interrupted and as part of its response undertook to submit a new I/O activity. If the next activity was the task A request, then processor 2 would be the agent of submission and processor 2 would handle the ensuing interrupt. The reactivation of task A could occur on either processor.

The significant concept here is that activities and resources are associated with processes and not with processors. The 65 processor service request occurs at the time of attempted submission when a device is determined to be free, but if the processor executing the IO control coding does not find a path, it then "shoulder-taps" to request submission.

A problem associated with interrupt handling in general in a multiprocessor system is relevant here. In a shared storage system, where locations in memory are used to support the interrupt mechanism, it is necessary to isolate in some way the interrupts directed to one processor from those directed to another.

It is possible to avoid this problem entirely by using some local resource, private store, or a specialized set of registers in the processor or even to develop a specialized control processor to which all interrupts are directed.

In System 360 the problem of processor control areas is handled by defining a separate control area for each processor. In doing this, however, a system convention of the 360 common to many processors is violated. This convention is the establishment of specific address in lower core to handle the processor/IO communication through memory.

Specific interrupts affect specific locations. The definition of more than one interrupt area violates the addressing convention associated with machine hardware. Since all contemporary multiprocessors are built around the concept of a single operating system, it is desirable that the addresses generated by the system appear identical regardless of which processor is executing.

The problem is solved by providing a hidden upper bit extension to these addresses at the memory interface. In a two-processor system it is necessary only to identify which one of the processors is to be prefixed. When the unprefixed processor generates an address within the range of the control area, then the address is provided without alteration. When the prefixed processor generates a control area reference, the prefix is appended to the address and actual reference is made to the other control area. (This is the "multisystem" feature of system 360.) It is possible, of course, to prefix both processors or any number of processors. This relocation of the control area is a desirable feature even for a uniprocessor. If a memory module containing the control area should be "down," then the system could be sustained by relocation of the area to unaffected modules.

18.8. THE LOCK PROBLEM

The concept of "lock" is such a basic one to multiprocessor design that it is hardware supported in systems otherwise minimally designed to support parallelism. The concept is a resource-oriented concept not unrelated to the concept of hardware memory contention.

Contention for resources, as we have seen, causes delays. During a period of time that a shared resource is in private use, it cannot be used by another processor and the later processor must idle (or may be diverted from the task).

In the look-ahead uniprocessor the registers presented the problem of locking and synchronization; in a multiprogram machine, the problem occurs with collection of data sets internal and external. The general rules for effective simultaneous usage of such a resource we have seen to be that one may often allow simultaneous read, may allow read/write with caution and may deny simultaneous write. A method for locking registers is described in connection with our pseudo 6600.

Let us turn our attention to a system where any processor may execute the code of the operating system and to tables that form part of the operating system; that is, they may be referenced only by a processor in some "privileged" or "executive" mode. In a case where two processors have undertaken to enter the executive system to find work for themselves, the problem occurs in classical form.

Consider that processor A and processor B desire to inspect the task queue at the same time and remember that our early definition of "sametimeness" still holds—that is, any interval during which A or B desires access to a resource when A or B is still using it. It is inadmissible for A and B to access the task queue "simultaneously" even in read/read mode, because of the possibility that they will select the same job for execution. It is, therefore, necessary to develop a mechanism for the inhibition of access by a processor when the resource is in use. This mechanism may be purely a software mechanism, but as such it is complex, and all current multiprocessors avoid the software only solution.

The basic instruction added to processors in a multiprocessor system is called by both IBM (on the '67) and by UNIVAC (on the 1108) the TEST and SET. Basically the instruction tests a bit in memory and if it is OFF sets it to ON and goes on; if it is ON it leaves it be and goes on. The functional effect is to provide a means of locking and enforcing locks (if everyone cooperates).

On the 360 the instruction refers to a bit of a referenced byte. On reference this bit sets the condition code to either nonset (bit is zero) or set (bit is one). As the byte leaves memory, it is set to all ones and is placed in store with this value. During the time that the byte leaves store until it is reset into store, the byte under test cannot be accessed by any processor; therefore, it is not possible for a second processor to access the control byte until the restore operation is completed. It is, consequently, impossible for two or more processors to access the byte and set private condition codes to zero. With the convention that a zero condition code means that the associated core area is not in use and a one condition code means that it is in use, a mechanism for controlling sharing exists.*

A difficulty lies in the need for all programs that may conflict on a resource to know it and to use the test and set convention. If access to resource is gained by execution of the same code as in an operating system situation, then the enforcing of the convention is in no way cumbersome.

18.9. SOME LOGICAL LOCK CONSIDERATIONS

A truly nasty problem develops when it is possible for a resource to be locked by a process which then loses control of the processor for some reason without releasing the interlock. Merely releasing it on some observation of loss of control is not adequate, because no indication of the state of the resource is available and no guarantee that it is decent can be inferred. It is possible, therefore, for processes to wait for a resource indefinitely.

This is a strong form of a general-resource allocation problem common to multiprogramming and multiprocessing. If a process A takes a resource 1 and a process B takes a resource 2, and then if process B, without releasing 2, asks for 1 and process A, without releasing 1, asks for 2, we shall wait a very long time for progress in these processes.

If a process A is being multiprogrammed with a process B and A/B share a resource, the interlock problem is identical to the multiprocessor situation; the fact that A and B are running on "virtual processors" is irrelevant. A locks the resource and yeilds to B, who attempts to lock the resource and finds it locked. If A is dependent on B, the lock will never be reset.

In contemporary systems the problem is substantially avoided by restricting the sharing of resources and by interposing the operating system between lock-protected resources and users. In this environment one can avoid the interrupt problem by disabling (not permitting) interrupts when

---

*For those unfamiliar with 360 machine level coding, this is a machine that has a two-phase conditional transfer. The first phase compares values or generates conditions by arithmetic function such as register is zero, negative, etc. The second phase tests the results of the compare as set up in the condition codes and either does or does not branch. A typical compare sequence is therefore "COMPARE" followed by "BRANCH ON CONDITION."

in supervisor state, or by disabling those portions of the supervisor that will reference and lock resources. Hopefully, one might be able to leave certain system control interrupts enabled, such as the shoulder tap from the other processor notifying a serious malfunction.

The TEST and SET requirement exists for shared resource multiprogramming systems if the shared tasks can be interrupted before the resource is released.

It is somewhat undesirable to accord general interrupt protection to all processes that are using a lock-required resource. In effect, this raises the priority of such tasks beyond what may represent the best interests of the system. A compromise approach might be to allow interruptions of certain classes but to not allow the locking process to lose its processor. For example, one allows I/O interrupts to be fielded in order to maintain channel activity on the system but does not transfer control to processes enabled by I/O completion unless there is a processor free to take the process without interference with the execution of the locking process.

An interesting consideration in a system that allows processes which have locked a resource to be interrupted and to lose their processor is what to do when another process wishes the resource. Basically, as we have seen, one either task-switches or idles the process that cannot reach a locked resource. Consider a system in which task-switching time is characteristically longer than expected times for the release of a resource when the locking process does not lose control.

A proper strategy would be to determine if the locking process is active and, if it is active, to idle the locked-out processor. If the locking process is inactive (has no processor), it is proper to task-switch between the locked-out and the locking process. In order to do this effectively, it is necessary that the determination of what process has locked what resource can be quickly made, particularly in a small portion of task switch time.

A hardware mechanism associated with the TEST and SET might inspect an identification of the locking process associated with the lock word and then review a set of registers containing the identification of all active processes. If the locking process is found, the process attempting the lock would be idled; if the locking process is not in the active process registers, a task switch would be undertaken to force the process which locked the resource to be active.

Chapter 19

ALTERNATIVE DESIGNS:
THE PROCESSOR

19.1. MINIMUM MULTIPROCESSORS

We shall undertake in this chapter to describe some variations in the organization of multiprocessor systems which reflect approaches to actual development or to "paper" designs. Since we have seen a highly elaborated uniprocessor develop in Part 2, we shall begin our considerations at the point where the complex high-performance processor evolves into a parallel processor in the accepted sense.

At one point in the development of our machine we visualized an I box associated with two identical E boxes, the I box acting to pass off instructions to whatever E box was available and itself performing some instruction execution and/or doing some of the work associated with in-struction execution for each instruction. This is a highly parallel machine just a half step away from being a parallel processor. The constraint on the machine is that the I box pulls instructions from a single source and distributes these instructions. It is a single-stream machine, despite the elaboration of performance capability to execute the single stream.

To convert this machine into a multiprocessor, it is necessary either to build the logical capability into the I box to accommodate concurrent access of instructions from more than one source or to add an additional I box (or boxes). If we do either of these things with duplicate E's, we may also do them with specialized execution units. Whether we have identical execution boxes or specialized boxes depends upon our appreciation of the balance of the instruction streams that we expect, the speed of real-izable E boxes within cost constraints. We note that we might even have multiple or generalized I boxes fronting a single highly capable E box. Both the elaboration of I to stabilize multiple instruction sources and the addition of additional I's with both symmetric and asymmetric execution backing have been proposed or developed for parallel processor design. The effect of either approach is to simulate the logical operational characteristics of a multiprocessor of the type currently offered by major manufacturers on a reduced cost basis.

The analogy between this approach and multiprogramming and time-sharing systems on a single processor is evident. By providing multiple instruction sources to execution populations we are, in effect, "time-sharing" the computational capability of the system between (independent or related) users. The instruction streams contend for the use of execution power (attention by the E boxes), in effect "time-sharing" this capability.

In multiprocessor systems of current availability, massive resources are dedicated to each instruction stream. Each stream has the full power of a *large* processor. In our "simulation" computational capability is shared with each "processor" having access to a pool of capability.

In a multiprogramming system programs contend for use of the CPU, for periods of time in which they have use of the location counter and, consequently, computational capability. By adding additional CPU's the ferocity of this contention becomes reduced because of the availability of additional location counters. In a system involving many location counters, the contention for use is moved back one level into the relationship between the counters and the arithmetic/logical/I/O elements of the machine. Contention is for computational resource.

The question arises as to what an I box (a processor) should contain. A minimal solution is to have only a location counter (L.C.). Figure 19.1 shows a simple conceptualization of this. Each L.C. has access to storage and to decode capability. Each develops an instruction address and goes to memory, and the instruction is delivered to the decode capability. Since this function will characteristically expect a high rate of flow of instructions, it would probably have a buffer to hold instructions forwarded to it by the instruction fetchers (L.C.'s). It might have an output buffer as well at its interface to the E boxes.

A major characteristic of the E's must be their interinstruction independence. Since there is no way of predicting the sequence of instructions through the system, the assumption may be made that E's will commonly execute instructions randomly across the population of active L.C.'s. All dynamic information generated by execution must be external to the unit. But this was already true of look-ahead uniprocessors.

In fact, the assumption that a random mix of instructions artificially sequentialized by the decode mechanism will reach the execution units and that the execution units will switch their attention from one program

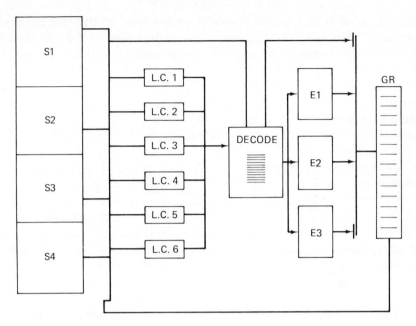

Fig. 19.1. A conceptual minimal multiprocessor.

to another as instructions are passed to them is fundamental to the concept of the machine.

A problem occurs with the use of registers in the pooled external population. In the single-stream environment the register usage was consistent with the logical intent of the program as interpreted (and perhaps optimized) by the compiler.

It is currently customary in program compilation to assign absolute register references at compile time. This differs from the assignment of memory addresses in current practice, which is discussed in a later chapter and which tends to be delayed until the program is actually going to run. This delay is to facilitate the assignment of addresses so that contention for memory locations between elements of a program or between independent programs is minimized.

In our situation the absolute assignment of register addresses would have the effect of forcing preplanning, since it would be necessary to avoid the referencing of the same registers by two active programs.

19.2. REGISTER REFERENCE IN MINIMAL MULTISTREAM SYSTEM

There are various solutions to the problem of controlling register references. One is to avoid it by abandoning the concept of specially referenced registers in a pool and implementing local scratch-pad memories associated with each L.C. This represents an extension of the L.C. to hold, in effect, the dynamic information of a process locally. We can see that this can also be done with a retention of register addressing.

If the common high-speed register pool is to be maintained, and it is potentially a highly flexible and efficient concept, a method of relative addressing of registers (or segments of scratch-pad) must be developed. A reasonably simple mechanism for accomplishing this is to add to the system a concept of relative register addressing.

An indexing mechanism would add a value to each register address to develop an effective true register address for the program. Compiled, a programmed register address would be fundamentally treated as relative addresses. The assignment of space in the register pool could be done simply by assigning a block corresponding to the basic register population of the apparent machine to each L.C. This would not be especially effective, since there would be a necessity of having at least as many register sets as L.C.'s (otherwise, why not just multiprogram L.C.'s?) at the cost of a more extended (and consequently slower) flow of data.

One would hope for the ability to assign register capability to L.C.'s in varying amounts as they require, therefore allowing more flexible balancing of register usage. Particularly, one would like to assign multiple register sets to an L.C. The duration of assignment might well be constrained to the length of a program.

In a time-sharing system all active terminals would be represented in registers. The system would be organized so that one would order as many register sets as there are terminals expected to be active. Hopefully, an optimizing compiler would produce both a statement of register need and a reasonably assured good usage of registers. The registers become a resource like any other resource, and very much like storage, for which programs contend. Various policies for overcommitment may be adopted, including task-switching on individual L.C.'s that cannot proceed because of register unavailability.

One interesting feature of a system where the "cost" of multiprocessing is so low is that the relative cost of quiescing a "processor" is very low in terms of system performance, whereas the relative cost of quiescing an 1108 in an 1108 multiprocessor is very high. Consequently, task-switching when a program cannot proceed for any reason (awaiting I/O or terminal interaction), so necessary to run current systems effectively, is not needed, since the penalty for letting a "processor" idle is minimum. The high contention for system services is compensated for in a "minimum processor" system by the very low cost of losing.

The placement of the register address index register presents a problem. If it is placed anywhere in the system but in the L.C. box, then a need to identify processors develops. (We are assuming that no explicit bits are taken in the instruction word to identify such an index register. There would seem little to justify doing this.)

Since we have so far not postulated any other reason for such knowledge in the decoder or the E boxes, it would seem that the logical place to do register indexing is in the L.C., where program identification is inherently shown. A small register index register could be placed in each L.C.

The instruction code of the machine might be organized so that a simple low-order bit test would identify instructions to which the indexing value should be applied so that the analysis of an instruction for modification might not be too burdensome for our simple, cheap L.C.

Analysis for indexing requires, however, that instructions pass through the L.C.'s. Our minimum processor would become a location counter, an instruction register, and a register index register (RIR). Additionally, the "processor" acquires a primitive instruction analysis function. The flow from the system is now L.C. instruction request to store, store to instruction register, instruction register to decode. This increase in traffic might make us consider separate paths from "processors" to store in place of the shared line shown in Fig. 19.1. Henceforth we call our L.C.'s "processors," since they have acquired a precode function.

The problem remains to define a mechanism for the dynamic acquisition and release of registers involving a setting of RIR. Parallel processors commonly require special instructions; defined within their architecture. The instructions to support the loading of a local index register might be among them.

19.3. ANONYMITY AND IDENTIFIABILITY

The problem of pooled register identification leads us to a consideration of a central concept of multiprocessing, the anonymity of elements in the system. In the Fig. 19.1 information flows basically in one direction in a circle from the L.C. to store via decode, E's, and registers. E's have explicit addresses to registers so that register reference is possible. E's may or may not be identified to decode.

The instruction passing procedure may involve a decision by decode to pass an instruction to a specific E after probing busy signals associated with an E, or decode may send an instruction on a common line to be picked up by a first encountered free E, which specific one being unknown to decode.

At the L.C.-decode interface, no information passes backward, and the flow is many to one. This interface may be handled in a number of ways. Decode may scan by some discipline the ready-to-send signals of the population of processors (multiplexing in a way not unlike the description in Part 1) and then specifically address a line to a given processor that wishes to send an instruction. It is not generally necessary for decode, however, really to know where an instruction is coming from, since its behavior for an instruction is invariant to the processor that forwarded the instruction.

It would be possible for processors to drop their instructions directly into a common buffer so that at the time decode begins to work with the instruction its source is lost. In this case processors are anonymous to decode, as we have seen that E's can be anonymous to decode. However, E's are not anonymous to registers, since the E requesting data from a register must be the E to receive it, and processors, for the very same reason, cannot be anonymous to core.

Whether this system is in a real sense anonymous is determined by whether or not the user of the system as represented by his code makes any use or assumption about which processor he is running on, whether he can determine which processor, or whether he can direct processors by reference to do one thing or another. This capability derives from the instruction set and its implementation.

In general, we shall call any system in which processor identification is not a program-usable concept an "anonymous system" and any system in which it is a usable concept an "identified system," regardless of the details of identification in the interfaces hidden from the program.

An introductory example of a usage in an "identified" system would be a command that tested to see if the processor executing it was processor 3 and branched to one point or another, depending on the result. This could give the programmer capability of specifically selecting the processors on which he might run his program.

19.4. ANONYMITY AND THE RIR

With regard to the register index register, it becomes necessary to identify the processor requesting a load of a base for register addressing because it is necessary to load the correct register. One design possibility is to have the processor itself execute the instructions referring to RIR. We have seen something very much like this in discussing address formulation in the I box, where we allowed I to execute index register instructions.

Let us postulate a LRIR (load register index register) for the system. We shall assume that there is a table representing a map of register availability in store accessible to all processors.

The LRIR instruction is part of a routine that acquires access to the table of registers' availability and develops an index to point to the first of the set of registers to be associated to this processor for this program. The process of acquiring access to the table is to determine if the table is being used and, if not, to set it to "in use" in some manner. All other processors fail the availability test and do something else. The something else is a matter of policy and design and may involve an interrupt or a looping on the instruction or an interlock.

Until the point where RIR must be loaded, it is not necessary to know who is executing the routine. The routine would certainly be a system utility available to all processors. Let us further postulate that the path of this index value is from storage to the RIR of processor (not through registers, since this would cost us a path from registers to processors, which we do not now need). What is needed is an identification of a processor at the storage interface. If the processor executes the LRIR directly, it can identify itself to the memory interface. If, however, the processor does not execute LRIR but passes it forward to shared decode/ execute elements, then some other mechanism for providing the identification is required.

One scheme is for an identifying codeword to be generated. In an identified system an instruction would send a built-in processor identifier forward to the shared decoder, which would then know that the instruction it was about to receive belonged to a given processor and that this identifier must be forwarded to execution units. These units would use the identifier to select the proper RIR. In an anonymous system the identifier would be generated by the processor if it recognized (in predecoded stage) an instruction of the nonanonymous class, or it could be forwarded in all cases, although this would be highly inefficient, since such an identification is rarely needed.

There are other approaches. One is to have a pool of RIR's placed between the processor and decode, that is, to externalize the register index register. If there is one RIR for each processor in fixed position, then the effect of the pool is logically identical to having an internal RIR.

With the pooling arrangement, however, certain possibilities exist not present with dedicated RIR's internal to each processor. If the entry in the RIR pool contained bits for processor identification, then it would be possible to associate multiple register sets with a processor, or any register set with any processor.

The advantage of dynamic register assignment is to provide greater flexibility in task assignment. The execution of an LRIR by a processor or on behalf of a processor would involve the formation of a bit pattern identifier of the relevant processor into the high-order bits of a word containing the base of the register set assigned to it.

Decode, in the process of preparing any arithmetic register referencing instruction for E execution, would scan the RIR pool to form a true register set address to be forwarded to E, or a true address could be formed on the way to the decode buffer.

This is an expensive process as far as time is concerned, which could be facilitated by an associative memory. If the RIR pool were an associative memory, then the search for the proper processor register could be accomplished very quickly.

An associative memory in its simplest form is one in which, given the partial contents of an entry, either the position in the memory or the rest of the contents of the memory is produced by the associative process.

It would be of further benefit to allow all processors parallel and perfectly concurrent access to the pool so that no processor is delayed awaiting availability of the store. This could be done for all instructions other than LRIR, since all references are read only. Whether it could also be done with LRIR depends on the logic of acquiring entry space in the pool.

19.5. USE OF THE POOL

Consider the possibility that some processors will be down or will be, for some reason, dedicated to a given job, so that in order to maintain a given number of active jobs in the system it is necessary to multiprogram some processors. This could be facilitated by allowing the multiprocessor to decay into the multiprogramming environment via the function of task-switching on a processor. One of the great expenses of task-switching, as we have seen, is the clearance and restitution of the dynamic portion of a program. At the point of "switching" the attention of the processor from one program to another and the variable information (contents of registers, condition codes, etc.) must be stored and replaced by the dynamic information of the new program. If multiple sets of registers could be associated with a single processor, then the continuing register status of programs could be represented, and the attention of a processor could be switched from one to the other without the overhead of reconstitution.

Further, the switching of a program from one processor to another is facilitated. To transfer a task from one processor to another, it is necessary only to change identifying codes in the associative pool. If a processor is to run multiple programs in this way, it is necessary to associate a program identification in the pool in addition to processor identification. Figure 19.2 shows an associative memory mapping register assignments for a multiprogramming multiprocessor system.

One last observation: If we return to our original postulation that "processors" are only location counters and that instructions do not go through them but directly to the decoder, then the problem of identifying

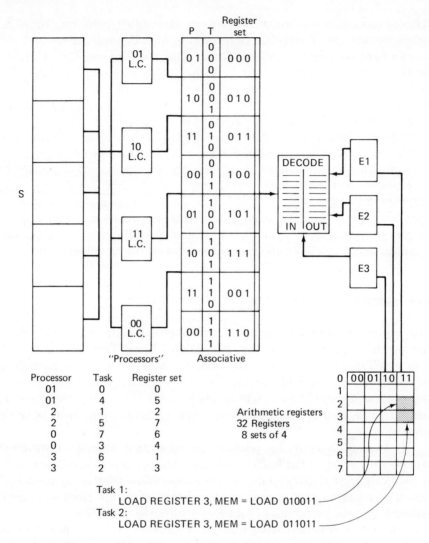

Fig. 19.2. Externalized relative index registers with associative memory.

a processor must be solved in another way. In such a system processor identification would not be necessary even at the L.C./memory interface, since the routing of instruction is directly to the decode buffers.

Consider that the associative store is used as a memory allocation mapping device in such a system. With the constraint that memory is allocated in continuous segments to processes, the associative memory would contain a pointer to the base of memory allocation, a process identifier, and a register set base pointer. In a high-order bit banking memory the memory could forward its identifying bank number. The associative memory could then find the corresponding register set.

We have described the hypothetical relative register problem at length, because it gives insight to the fundamental design considerations of a multiprocessor (or of any processor), which tend to concern themselves with the functional relationships of each element in a system and with the scope of responsibility that each element should have.

Consider that the problem of addressing main storage through base and index registers in a multiprocessor is an identical problem involving the same kinds of considerations about what should be local, what should be generally accessible, and what interface mechanisms are feasible. The fundamental design tradeoff is always based upon anticipation of interference and contention. To what extent will the reduction of private resources provide system economy and system flexibility enough to balance whatever delays might result from contention on the resource?

19.6. MINIMAL MULTIPROCESSOR AS A SINGLE I BOX

The minimum processor capability involving only multiple location counters can be implemented as a single expanded I box. This box would look very much like the I box in a look-ahead machine, except for the presence of multiple counters and the fact that the mechanism to which instructions are sent from store would represent a slot for each active program instead of a backup for a predicted flow of instructions from a single source.

It would certainly be possible to back each program instruction stream with a sequential buffer, but since the performance of the system is predicated upon maintaining the instruction flow from many programs, the need to have quick access to a successor instruction from any one is reduced, and the usefulness of sequential buffering falls off.

The I box function is to scan the collection of active instructions, fetching new instructions and selecting those for forwarding which have undergone whatever preparation is required prior to execution. Such preparation would include, as an example, applying an index register value to form an address.

The instruction fetch operation faces the location counters on one side and a queue of actual instructions on the other. When an instruction in the instruction queue is "cleared" for processing, the instruction fetch unit increments the location counter associated with that instruction and initiates a memory request for a new instruction to be sent to the active instruction list.

When an instruction is fetched from memory, it is set into the instruction list with a series of status bits set. An asynchronous processor for indexing operations scans the active instruction list cyclically, looking for instructions which require, but which have not yet undergone, an indexing cycle. When such an instruction is found, it is subjected to indexing; upon completion of the indexing operation, an indexing complete bit is set.

In addition to the asynchronous indexing function, another asynchronous unit scans the queue, looking for instructions whose addresses have been formed and which require an operand fetch. Instructions in this status are submitted to a queue of main storage requests, which is being constantly scanned by yet another processor, which has as its sole function the delivery of instructions and data. The target address of data is always a register, so operands are forwarded in effectively the same way that one sees in a look-ahead uniprocessor.

Through the operation asynchronously of an instruction fetch processor, an indexing processor, and a memory data control processor, a high degree of parallelism is achieved in the preliminary functions of an I box, preparing an instruction for execution. Instructions are simultaneously being fetched and indexed, and are waiting for data access. Because of the dynamics of operation, it is possible for a single program to have more than one instruction active in the system, as we shall see directly below.

When the memory reference associated with an instruction is completed, the instruction is "cleared"; ready status is reflected on the instruction list. The decode unit constantly scans the instruction list, looking for ready instructions. Some instructions, transfers, for example, are executed directly by decode, which resets the location counters, or by the I box itself.

Other instructions are handled by special-purpose interinstruction independent processors. The bulk of instructions may be handled by a population of special-purpose or generalized computational processors, which interface with the decode unit and with the population of registers.

The execution units (E boxes), when they are free, request from the decode box an instruction, which is then passed to them. In a symmetric system any instruction may be passed to any E box; in a nonsymmetric system E boxes may request the types of instructions in which they have interest and capability. When an instruction is requested, decode selects from the proper output queue and forwards to the requesting unit.

A location counter is incremented, and new instructions are brought into the initial instruction queue asynchronously with the request of an E box for an instruction from the decode queue. (Decode acts purely as a filter to determine in which set of queues to hold an instruction while waiting for an E box request.) It is possible for a program to have a number of instructions waiting for E box request. It is possible for sequence to be violated. It is necessary, therefore, to control source/sink references within the set of registers allocated to a given program.

We have here a system which differs little from the systems in Part 2 and which yet qualifies as a multiprocessor or parallel processor. It is not usual, however, to define a processor quite as minimally as we have seen here. The sharing of decode function is a touchy point, since decode is usually a bottleneck point in a system. This is countered here by the asynchronous performance of small I box/decode functions by independent units.

19.7. A LITTLE MORE PROCESSOR

The multistream organizations that we have been discussing to this point involve minimum extensions to an elaborate uniprocessor. We mentioned earlier that share of decode might provide a serious bottleneck for the system.

In general, the determination of what to include in a "processor" is a statistical exercise determined by where bottlenecks are going to occur.

A reasonable design point is to include in the "processor" all information that is unique to a process it is running. This would include the location counter, the instruction register, all arithmetic registers, index registers (if separated into a special group), condition codes, etc. If relocation registers or other address mapping devices exist, these also might form part of the "processor."

This approach is equivalent to duplicating I boxes and bringing registers into them. The processor would, like any I boxes, have sufficient logic to execute those instructions appropriate to it, to handle interrupts and interface with the I/O subsystem. Alternatively, a special control processor might provide interrupt handling. It might also forward operands to a pool of execution units in much the same way as in a high-performance uniprocessor.

Behind the "processor" is a pool of either differentiated or identical functional execution units. The decision as to whether the boxes should be specialized or generalized relate, as we have seen, to the economics of specialization within a given technology and the mix of instructions expected for the system.

The interface between the processors and the execution pool may be developed in a number of ways. There may be a path to each execution unit from each processor; there may be time-sharing of some form on a common bus; the execution unit may request instructions, or the processor may request execution units.

If the last approach is used, then the failure to find a free execution unit would have the same effect on a processor as the failure of the CDC 6600 to find a free functional unit. Within any "processor" local parallel capability for overlapping decode, instruction fetch, and address generation can be built in. Within any execution unit local parallelism can be built in to any extent.

The advantage of such a system lies in its open-endedness, its inherent ability to grow in small discrete components, and its potential to support very large populations of "processors" without the elaborate mechanisms of asynchronous instruction scanning we described for the multilocation counter I box.

With regard to a multiple full processor system, one would still expect a considerable cost/performance payback in the ability to share execution power. The problems of processor interaction, connection, and control are insensitive to whether execution power is shared or duplicated behind a "processor." It is these problems of element relationship to which we shall now turn our attention.

Chapter 20

A MEMORY INTERFACE
FOR MULTIPROCESSING

20.1. A NEW MEMORY INTERFACE

Let us consider a form of memory/processor interface called variously
"three-phase," "ring," and "conveyor belt."

In its basic form the system does not reduce contention as such, but is
a time-shared version of memory interface aimed at reducing the connection requirements involved in either "head or tail" (distributed crossbar)
or centralized switching designs.

Associated with each element in a system is a data transfer unit,
which contains space for data or address and an identification of the
element setting up the information transfer. At any time a given unit is
available to any element in the system; that is, processor 1 will have a
unit available to it, processor 2 a unit available to it, memory module 1
a unit available to it, etc.

On a fixed time cycle basis a unit is switched from one element to a
successor element. On the switch the new element inspects the unit to
determine if there is activity for it or if the unit is free (available for
establishment of a system request). If the element associated with the unit
is a memory bank, it will determine whether the address in the unit is an
address that it contains. If it contains the address, then the request is
removed from the unit, and the unit is marked as available, or, if memory
wishes to deliver a word, it is placed in the unit. If it does not contain the
address no action is taken.

At the time of the next "shift" a successor element will inspect the
relevance to itself. If the element associated with the unit is a processor, it

inspects to see if the unit is free (available to hold a request) or, if the element is awaiting data, whether these are the data it expects. This determination is made by comparing the element identifier returned from memory to see if it refers to this element. If it does, then the data in the unit are taken from the unit into the location counter (for an instruction) of the processing element. If at that time the element has a storage request to make, it may load it into the freed unit and send it down the line. If on a switch a processing element finds itself unreferenced by a data transfer unit that is not free, it must hold any requests until the next switch where it finds a free unit.

Memory contention is not precisely eliminated, since each processor element generates requests and retrieves data independently. A memory bank at a cycle switch may find a reference to itself but may well be busy and forced to enqueue the request, even if it is for the same location; therefore, the number of memory accesses is not reduced. A minimum delay time is inherent in the system, that time being the length of time required to execute enough shifts to effect a communication in both directions between a processing element and a memory element. This time is a function of the time to execute shifts and the number of elements participating in the switching before a complete circuit around the system is made. Switching is in one direction only. Representation is shown in Fig. 20.1. In addition to this delay, any contention in memory (plus memory access time, of course) adds to the time "around the ring."

The basic advantages of such a system as opposed to the more common memory interface lie in the relative simplicity of implementing decisions about routing and distribution and in reduction of hardware. In the more conventional system design now current, a processor is associated with memory by what is commonly known as "head and tail." Each memory unit has a port for as many units as are to have independent access. If there are three processors, there will be three ports in each memory bank. Each processor has a path to a line, which tails into all memory ports. If there are three processors, there are three such lines and three paths. A processor addresses a memory unit by sending a request up its path to the line associated with all ports. An interface at the memory unit accepts requests.

Contention is handled very similarly in the two systems, except that simultaneous requests cannot occur with the conveyer because of the delays in the data transfer unit organization. What is particularly interesting in the design is its potential in two areas, in one of which, the memory contention problem, we are interested now.

If an instruction once fetched could be made available to any processor requiring it, then only one access to store would be made for each such instruction, and contention potential would be reduced. The prob-

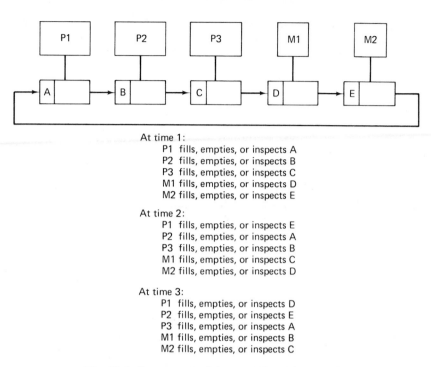

At time 1:
   P1 fills, empties, or inspects A
   P2 fills, empties, or inspects B
   P3 fills, empties, or inspects C
   M1 fills, empties, or inspects D
   M2 fills, empties, or inspects E

At time 2:
   P1 fills, empties, or inspects E
   P2 fills, empties, or inspects A
   P3 fills, empties, or inspects B
   M1 fills, empties, or inspects C
   M2 fills, empties, or inspects D

At time 3:
   P1 fills, empties, or inspects D
   P2 fills, empties, or inspects E
   P3 fills, empties, or inspects A
   M1 fills, empties, or inspects B
   M2 fills, empties, or inspects C

Fig. 20.1. Processors and data transfer units on a ring.

lem of direct simultaneous execution of an instruction can be handled. Since processors would receive their memory responses at a rate closer to ring switching than to memory access time, processor delays would also be reduced.

20.2. USED IN CLOSE FOLLOW

The fundamental approach is simple. Once an instruction is put into a data transfer unit and sent "down the line," the first processor to pick it up does not free the unit but allows it to pass down the line further to be inspected by another processor. The tradeoff here is that the reduction in availability of unit space, causing a delay in lodging a request until a free ring is found, will be overbalanced by the availability of an instruction to multiple processors. If many processors are executing the same instruction, this availability will be profitable, and less so as instruction duplication falls off.

The problems associated with this approach are manifest. If processors still generate their requests independently, then multiple data units will carry duplicate requests. The memory element will acquire and honor these requests as they arrive, unless there is some mechanism for recognizing duplicate requests in the memory interface. Such a mechanism could be built into a memory interface in a "head/tail" (distributed crossbar) system and is not a necessary characteristic of the conveyer structure.

Without the mechanism for recognizing duplicate requests a situation will develop such that processors will be picking up instructions from units and satisfying their needs while the memory unit is making unnecessary accesses. Delay time for a processor will be reduced, since it is no longer dependent upon the memory access, but other references to the same bank will experience no relief from contention delays.

Further if a processor always allows an instruction to stay in the unit after it has picked it up, then unless a "last processor" is identified, there exists no mechanism for ever clearing the unit, and the system will cease to operate. This might be countered by a fixed-time resetting of all data units.

Finally, a high degree of fruitless usage of data units occurs, because they are being used to carry instructions to processors that have already satisfied their requirements. The broadcast capability of the system, therefore, requires something somewhat beyond the retention in a data transfer unit of an instruction. What must be achieved is a recognition that multiple requests for identical data from memory have been generated and that only a single request goes down the line.

20.3.   GENERALIZING USE OF THE SHIFTING INTERFACE

A mechanism that might accomplish recognition of duplicate requests introduces us for the first time at this level to an issue of processor interface. This is the other aspect of the ring structure that interests us. Even in the simple form that we have been describing, the interposition of the data transfer units between a processing element and memory may be generalized to provide a communication between processing elements.

With only a slight generalization a processing element may load the data transfer unit with a message for another processor. This would involve the provision of a target address to identify the processing element that should pick up the message. We notice here that we are providing, at this stage, a simple unified interface between elements of a highly modular system, regardless of whether they are similar or dissimilar elements.

One way for a processing element to communicate its desire for I/O is to specify an I/O processor as a target for an I/O directive it sends down the line. On completion, the I/O processor sends back notification that I/O is complete. The most rudimentary form of this involves, as we have shown here, a specific knowledge of the processor that is to receive the message. Later we shall describe more sophisticated possibilities.

What is common to all processor interface mechanisms of any elaboration is the fundamental relief of the processor from participation in the interaction between processors. In this case the processor sends down a request or directive, and if its own logic tells it that it can continue, it does so without further involvement in processor interaction. The intent is to relieve processing units from burdens of synchronization, spawning, and control as much as is possible.

In general use the structure could support, as well as I/O, the spawning of any asynchronous task to be picked up by any free processor, or in very sophisticated development any processor of the proper class executing a lower-priority program.

Current multiprocessor systems have the rudiments of such interactions implemented largely in software supported by minor extensions to the interrupt system. Interactions largely involve execution of commands by which one processor can interrupt another to force it to execute instructions that undertake to find out why it was interrupted. One example is the DIRECT instructions (read and write) on IBM's 360.

With regard to memory fetch problems we may wish to support processor interaction and processor memory interaction with special mechanisms associated with additional logic, which establishes a functional difference between a processor and a memory at the conveyer interface.

We recognize a problem and a possibility in the cyclic nature of the ring structure. If each processor generates a request for an identical instruction simultaneously, then, as the data transfer units are switched, logic associated with the unit at each stage can recognize that the input is identical with its output request, and the unit can suppress the duplication. Therefore, the problem is to assure that no storage element inspects a data transfer unit before all processing elements have inspected it for duplication. Order by element type becomes a factor in the system.

If, as in Fig. 20.2, processors 1 and 3 simultaneously develop an identical instruction reference, processor 3 will allow processor 2's to go by unchallenged, but will suppress processor 1's. If processors 1 and 2 develop identity, 2 will suppress 1's; if processors 2 and 3 develop identity, 3 will suppress 2's, etc. If all generate a request, 2 will suppress 1, 3 will suppress 2, and $n + 1$ will suppress n's simultaneously. Therefore, it takes only one cycle to eliminate duplicates. No suppression will take place on a reference to the same bank, however, so contention will still be present in the system.

The memory elements will receive only one request for an element and will undertake one access to place it on the line. This mechanism implies a buffer area in which each processor remembers its last request

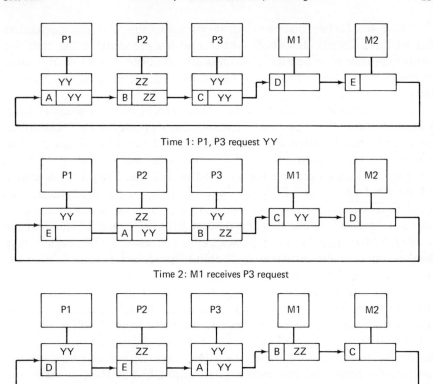

Time 1: P1, P3 request YY

Time 2: M1 receives P3 request

Time 3: P3 finds YY = YY, suppresses P1 request

Fig. 20.2. Request suppression.

and an interface between this buffer and the data transfer unit. The system identifies requests, not processors only.

As information cycles back from the memory units, it is sampled and allowed to remain until the end processor has seen it. The elimination of duplicates implies a large availability of empty data transfer units (DTU's) to memories, reducing the instances where a memory unit cannot put data on the line because of unavailable DTU's.

As we have said above, contention for core banks remains a problem, and in order for there to be any use of the system, it is apparently necessary for there to be absolutely simultaneous requests for an identical instruction. Characteristically, processors will not, unless they are specifically synchronized, follow each other that closely, even in executing the same code, because of later start times, contention delays for data, different local branches, etc. What would be desirable is to be able to relax the simultaneity requirement and detect identical requests within a broader time interval.

There is a certain degree of relaxation achievable with the mechanism that we have already described. Notice that if a processor to the "left" of another processor generates a request (a processor $n - k$) of the same address requested by processor n in the interval required for n's satisfaction, n can delete the request.

An example shows deletion on a three-processor, four-memory-bank system. At time $T = 1$ processor 1 is awaiting its response for location xx; processor 2 has initiated a request for yy, and processor 3 for zz. The response to the request for xx is passing by memory bank 4. At $T = 2$ processor 1 finds its request for xx satisfied and files a request for zz at $T = 3$. At $T = 5$ processor 3 finds the duplicate zz request and deletes it. This could have occurred at any time up until the zz response to processor 3 does not "follow" in the ring the request from processor 1. This is at the point $T = 9$ or just before z passes under processor 1. Processor 1 may then be up to eight instructions "behind" processor 3 in executing the same routine without causing a duplicate memory access.

On the other hand, if processor 3 generates a duplicate (it is behind one), not only will an unnecessary memory access take place, but a position in the ring will never be cleared, since processor 3 will receive the instruction from 1 and will have requested another instruction at the time the duplicate arrives. We must, therefore, have an end of processor line function for responses from memory left untouched by the processing set. To this point we have added relatively little logic to recognize the occasion of simultaneous request.

Whatever more should be done is constrained by the observation that on a generalized multiprocessor the execution of identical instructions by processor simultaneously is not a usual thing. In fact the characteristic leads to the consideration of another class of machine, a different elaboration of a uniprocessor, which is essentially a single-stream machine.

20.4. ELABORATION OF RING STRUCTURE

One notices a number of things from the foregoing example that should be touched upon before we leave the area that we have used to introduce processor interfaces and processor execution relationships.

If the number of processors is large and the number of memory banks is large, the time required to pass around the loop of data transfer units may become unacceptable. One solution is to partition the system into separate smaller loops with interconnections.

A module connecting the separate rings performs the function of switching from one ring to another. Elements on a ring may be mixed or identical. One effective use of a multiple ring structure that one can derive is the reduction of time a memory response must take in order to return to a processor.

In our example we see *zz* satisfied by memory bank 1 taking a useless amount of time passing by memories that are not productive. This costs processor wait time and occupies a data transfer unit for an unnecessarily long period of time. If a memory interface had the capability of breaking out of the ring, considerable efficiency would result.

In Fig. 20.3. there are two rings, a processor interaction ring and a memory ring connected by a device that has the general nature of any other interface. Any DTU now has two paths, one to its normal successor and one to the ring connecting unit.

Processor-to-processor messages are handled in the processor ring (perhaps instruction broadcast might be brought to refinement here). A processor reference to memory may be switched from the processor ring into the memory ring, if desired. More important, a memory response may be immediately switched into the processor ring. The rings must still be closed, since the interface unit will find any available DTU passing underneath it, and the search for proper memory will have to be fully conducted.

In a mixed system processors and memories might be on a single ring connected to other mixed rings. This introduces some interesting problems in the proper definition and usage of a given cluster. If a processor

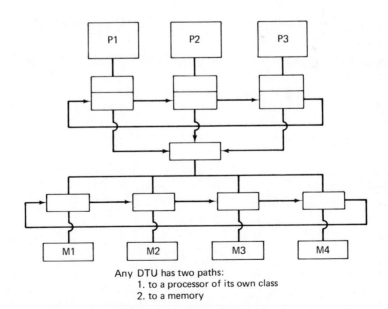

Any DTU has two paths:
1. to a processor of its own class
2. to a memory

Fig. 20.3. Two rings—homogeneous.

requires a memory reference to a memory in its own ring, it will profit from smaller ring size. If, however, it must be switched to another ring, the cost may be (worst) full traversal of its own ring (to find the memory not there) plus traversal time in the secondary ring plus partial traversal back in its own. This can be eased if ring placement is known in the system.

The implications for resource allocation and processor scheduling are that certain locations would be preferred for certain processors (those on its ring) and that the processors will cooperate more effectively with some processors (those on its ring) than with others.

Chapter 21

CLOSE FOLLOW
FORCED

21.1. HIGHLY PARALLEL SINGLE-STREAM MACHINES

We observed earlier that the execution of instructions in perfect simultaneity is a possible event in a multistream machine and one that might prove fruitful to do something about, but that it is not a generally expected event. An interesting class of machines which have developed and which require some comment is the class of machine which is designed to do this and only this. It is, in a sense, a variant form of the highly elaborated uniprocessor with multiple E boxes, in which instructions are passed off, not to a free E box or a specialized functional unit, but to all execution units at the same time. In our allusions to instructions broadcast above, we did not achieve perfect simultaneous distribution of instructions, there being at least conveyer delays with sequential access to the same instruction.

These machines are certainly oriented to the solution of large computational problems which can be more or less effectively partitioned so that subsets of data can be allocated to an execution unit which undertakes to perform transformations on the data identical to the transformations being performed by all other execution units synchronously.

As stand-alone systems they may not ever enter the "main stream" of system development, since by their nature they tend to be specialized machines limited in application to a set of problems having proper characteristics of symmetric partition ability.

As more flexibility is built into their design, however, they can offer some competition to machines of the CDC 7600 or IBM 360/195 class, which are also bidding for a large computational market. One promising development is the association of these machines with large-scale general-purpose machines in asymmetric systems. A development of this type is the association of the IBM 2938 array processor across a channel interface to a 360 and particularly the association of the ILLIAC IV with the Burroughs B6500.

21.2. SOLOMON

The machine most commonly identified as the classic machine of this type is the SOLOMON. In SOLOMON (Fig. 21.1) there is an array of program elements (execution boxes), each with a small private memory (4096 bits). The PE (program element) array is organized into a square matrix 32 X 32, giving 1024 individual PE's. Each PE can receive or pass data to its four immediate neighbors or use data from its local memory itself. The local memories are used exclusively for the retention of data to be manipulated by PE's.

The PE's at the edge of the matrix communicate with a private I/O interface, which provides for 32 data paths to a buffer unit, which has paths to the program memory. The program memory provides an interface to I/O devices and holds instructions to be executed. Since it is providing instructions to central control, receiving and forwarding data from the PE's as well as receiving I/O, the PM is interleaved into 16K modules to reduce contention. There is a high-bit interleave, which allows some modules to support I/O and others to support network and control access.

The central control unit fetches instructions one at a time and inspects to determine whether they are to be forwarded to the network, passed off the I/O control, or executed. Instructions passed to the network are decoded by a network sequencer that drives the PE's to execution of the instruction.

The PE's have a local possibility of nonperformance. Within each PE there is a mode register, which may be in one of four states. The instructions, distributed by the network sequencer, contain an indication of what state a PE should be in to execute the instruction. Any PE not in proper mode does not execute. Mode-changing capability derives from the instruction set whereby mode may be set by testing of internal values in the local PE registers. A use for this would be to avoid the division by zero. If a divisor in the arithmetic register were tested and found to be zero, the mode of the PE would be set to avoid execution of the subsequent divide. The system contains a mechanism for the broadcast of data, as well as instructions in order to eliminate the need for storing common constants in PE memory.

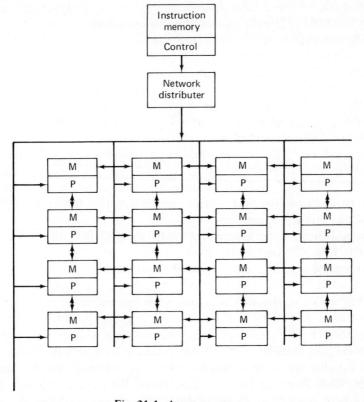

Fig. 21.1. Array processor.

The specific organization of SOLOMON has been criticized because of the relatively awkward mechanism for transferring data from one P.E. to another and because of the relative inflexibility accruing from fixed local allocation of storage.

Designs that allow for more flexible data access while maintaining the fundamental approach of SOLOMON have been suggested. The use of a commonly accessible high-bit interleaved banked store, which would allow a processer to reference an element in the store of any processer (perhaps with permission controls), might be combined with a central control which, in addition to setting up operations, would provide an address for each P.E.

21.3. ILLIAC IV

SOLOMON went through an evolutionary phase, which transformed SOLOMON I to SOLOMON II, the major differences between them being in the speed of the P.E.'s, in technology, and in the capability of inter-facing SOLOMON II with a general-purpose machine.

The current machine of most general interest is the ILLIAC IV, which is quite similar in concept to SOLOMON. Like SOLOMON, it contains a network of P.E.'s, but they are considerably more powerful than the minimal elements of SOLOMON. The total number of P.E.'s associated with ILLIAC IV is 256. In place of one control unit there are four central controls, each of which controls 64 of the P.E.'s. Unlike SOLOMON, ILLIAC IV I/O has access to each P.E.

A good deal of attention has been paid to sustaining the rate of instruction flow to the P.E.'s. Each control unit is itself a highly elabora-ted unit capable of parallel indexing, instruction fetch, and local arith-metic. In addition, the C.U. has an eight-instruction buffer. Each P.E. has a 2000 (64-bit) word memory.

In addition to the general access to the I/O bus, each P.E. has the SOLOMON-like communication with its neighbors. A possibility exists for the arrangement of the subarrays of P.E.'s into paired two-array operation or for uniting the entire set of processors into a single group.

This design is currently in a somewhat argumentative stage. Certainly, the potential power for jobs that can be properly partitioned into paral-lel-processable data segments is enormous. There is probably little doubt that good and intelligent study can produce a larger population of pro-grams suitable for such a machine than is currently known, and that greater flexibilities can be built into them. We have mentioned them in passing here because they are of interest as representatives of a parallel design; they are an interesting extension to our discussion of executing instructions in parallel.

Chapter 22

HIGHLY ELABORATE
INTERFACES AND
INTERACTION MECHANISMS

22.1. MORE DEMANDS

With more closely cooperating CPU's or in systems containing rather large numbers of CPU's the processor interface is generally very elaborate. An initial elaboration, of course, is inherent when a system goes from two to more than two processors. The inherent distinction between two and more than two is intuitive and is even reflected in some ancient languages. In Homeric Greek, for example, there were inflectional endings for one, two, and more than two (plural). In a two-processor case each processor may reference only the other processor. The destination of a communication is inherent.

In a more-than-two-processor case it is necessary in some fashion to identify the processor(s) that are to receive a message. In our description of the conveyer belt interface we described a simple way for a processor to send a message to another by direct identification. Without this mechanism it would be necessary for each processor to have a direct line to each other processor. This would of course, be faster, but in a large population of processors rather expensive.

Alternatively, some systems relate all processors through a centralized special switch. The difficulty with either the conveyer belt or other switching mechanisms is the inherent need for a processor to identify the processor with which it wishes to communicate. It is easy to visualize circumstances under which it is clumsy to do this.

Suppose that we have a dynamic unidentified system where it is possible for a processor operating on a task to spawn another task. This can occur at a number of levels. In a multiprocessing system of current design it is a formal event, in that it involves an interface with the operating system to register the new task and to enqueue it for execution. Resources allocation may or may not be involved; that is, the requested task may (as a system convention) be forced to work within the core, and device assignments may already be made to the requesting task.

Let us postulate that the task is placed upon a common work queue for all processors and that at a later time some processor will pick up the task for execution. This processor will mark the task as under execution. Let us further postulate that at some time after its request the process spawning the new task ceases to operate, or discovers that it no longer needs the spawned task to run. It is necessary to inform the processor executing the spawned task to cease its operation. An example of this is a situation in which a task has undertaken to search a file for a particular member. It requests another processor to help it with the search, so that it will search the upper half of the file, and the other processor, the lower half.

The first processor finds the item and wishes to abort the search. Initially it must determine whether or not the task has been picked up. This it can do by searching a list of system tasks in execution. It must next determine who picked it up. In order to do this, it is necessary for the processor picking up the task to identify itself and leave a trail in memory indicating its relationship with the task. In order to do this, an instruction enabling each processor to materialize its identity and drop it into core is necessary. In this way a record of every process assignment and the status of all processors may be maintained. Each process wishing to communicate with a related process goes to memory to find the assignment and then routes a message to the correct processor. A number of problems present themselves.

First, the inspection of the work table involves both processor time and memory cycles. Second, during the time the work table was being inspected, it would have to be made unavailable to other processors. Third, after identification the processor to receive the message would, if the mechanism were only an interrupt line, have to inspect to see what was required of it. This would not necessarily be true in the conveyer system, where the message could be "cease" operations.

Of course, it is necessary to undertake suppression only if it is expected that the temporary continuance of the now superfluous tasks will truly contribute to bad performance. It is probably bad policy to undertake less than the immediate release of processors from superfluous tasks but it is possible to consider allowing the task to run until it reaches a point that would in any case bring it to the attention of the operating system. For example, at the next request for I/O the process might ordinarily be "waited" on the I/O resource.

If the mother task had requested a bit to be set in the entry representing the subtask on the active task queue, indicating that this task should be terminated at its next wait point (forced or explicit), then the operating supervisor on handling the I/O request would recognize that bit and delete the task. The exposure of the system is limited to the interval from the recognition that the task is superfluous to its next interaction with the system. Notice that the deletion mechanism does not require CPU identification. It is dependent only upon the deletion of the task from the passively shared resource memory.

Perhaps a more elegant operation would be to enable a processor to send a message to "the processor executing task B" (whoever you are). If there were no other support but the interrupt mechanism, this could be implemented by the initiating processor's interrupting its neighbor, who would inspect a message in memory and either interrupt its neighbor or respond. Simultaneous broadcast of the interrupt might shorten the time only if all processors were not effectively interlocked on memory and resource (the code to handle the interrupt) contention. Since a large number of processor interruptions might occur before the processor was found, a great deal of useless processor interference would result.

The simplest alternative is to have the initiating processor inspect a table in common core to find its colleague. This is not the most general or fastest solution. If we took the basic ideas of the conveyer interface, IOC's, etc. and applied them here, we might find some profit. The fundamental concept is to unburden the processor from participation in attempts at interaction that do not concern him in order to allow an increase in the general level of interactions so as to allow closer cooperation between processors and to increase the number of situations in which parallelism is an attractive technique for increasing speed.

Essentially, we wish an interprocessor control subsystem. The first function of this subsystem is the control of messages between cooperating processors. A second function is the accumulation of system status data, and a third function is the assignment of processes to processors. A number of schemes have been presented for such subsystems; we shall discuss some variations in the next section.

22.2. INTERACTION AND THE RING

The communication of one processor with another involves two concepts: first, an extension of the instruction sets of processors to control interaction, and second, a mechanism to support the new instructions.

For the situation in which we wish a processor to be able to inform another already cooperating processor, we shall introduce the notion of an instruction called TELL. This instruction is an embryonic generalized form of a number of more specialized instructions, which carry meanings of STOP, WAIT, START, RESTART, and RELEASE.

Along with the TELL function, one associates "what" and "who." Who is the processor that is in a certain status or the processors that are in a certain status? Let us visualize, associated with each processor, a DTU initially like what we have seen at the conveyer belt interface. Let us expand the contents and function of a DTU to support more extensive interaction.

In each DTU is represented the type of processor (for example, I/O or general-purpose), the job or task on which the processor is currently working, and whatever else might be relevant to an interaction as conceived in a given system.

A given processor wishing to communicate with a processor sends to its DTU through the medium of a TELL instruction a message for "all processors working with me." The DTU sends a message down the line associating all DTU's. This message includes the task identification of the issuing processor. In minimal form all DTU's will inspect the message; if a DTU funds no correspondence with the task on which it is working, it will ignore it; if it finds a correspondence, it interrupts its processor. In a further elaboration it can interrupt its processor and force into its location counter a specific address provided by the initiating DTU. Only relevant processors are interrupted, and the interaction takes place without any knowledge on the part of the initiating processor of which other processors were associated with it.

In an effectively anonymous processing system interaction takes place without any effort on the part of the initiating processor to determine whom it wishes to attend to its message and without any delay accruing to irrelevant processors.

22.3. INTERACTION CONTROLLER

An interesting alternative to the conveyer belt for implementation of this mechanism is described in the literature by Dr. M. Lehman of IBM and his associates, including Driscoll, Lee, Mullery, Rosenfeld, Schlaeppi, and Wetzmann. In the described system the interaction box is called an interaction controller (IC). An interaction controller is like a conveyer register

(DTU) associated with each processor in the system, general-purpose, I/O, or any other specialized units. Instead of an arrangement on a fixed-time division circulating shift register, the IC's are associated with a single bus, the control of which they compete over.

The IC's are stored logic programmable ("microprogrammed," but this is not a general requirement for such a device) and represent true special-purpose interaction processors to the system. At any time one IC has control over the common bus. This is the controller that is in the process of issuing commands.

A processor requests an interaction by sending TELL to its associated IC. The IC then attempts to establish control over the common line. If the bus is not free, the IC loops in a continuing attempt to take over. When the line frees, the IC identifies itself to the line. If multiple IC's simultaneously attempt to seize the line, the tie is broken on the basis of identifying codes. Those IC's not in control of the line are in a receiving status. Processors that initiate interactions are free to continue with their work independent of the interaction subsystem activity. When an IC has control of the bus, it undertakes to attract the attention of potentially relevant processors by executing an IC directive, which will broadcast down the line a field of data identifying the processors with which communication is desired by status.

Initially it may be necessary to isolate processors of the same kind. For example, if a processor wishes to direct all processors working on the same task to execute a given instruction, there may be some I/O processors currently associated with the task which cannot execute that instruction.

The IC may initially send a directive down the line instructing all other IC's to determine if their processors are processors of the correct type. If they are not, the IC associated with them ignores all further action on the line. If they are of the correct type, they will accept the next directive to compare task number with the task ID being sent down the line.

At this point some further number of IC's will drop out, and only those executing the correct job which are of the correct type will receive the next directive. This is the directive that will instruct them to receive a forthcoming instruction and, at the end of the current instruction of the associated processor, to cause that processor to execute the instruction.

Here we see a possible means of instruction broadcast in which a single processor is generating fetches and distributing instructions to a family of processors cooperating with it on the task. Whether it would be feasible to execute strings of instructions in this way depends on the manner in which the associated processor is informed of the presence of the instruction.

If an interrupt must be forced on the processor, this would be an untenable means of broadcasting instructions. If, however, an IC could set its processor in a mode to suppress instruction fetch and accept instructions only from the IC until reset to fetch instructions from memory, this might be a workable approach to broadcast. Unlike the previous mechanism, instruction fetch is prohibited. We would have narrower application here, to cooperation in which it was anticipated that exact execution would be a requirement.

22.4. SPAWN AND CONTROL

With the IC providing a general way of any processor's transmitting a specific instruction to any other processor, we may gain an insight into the motive for such an interaction system.

We have previously described a number of levels of parallelism, starting with local possibilities for parallel execution of instructions in an instruction sequence and discussing new parallelism at the routine or task level. In the first chapter we discussed the cost of parallelism: the delays due to the amount of extra work that a system must undertake in order to achieve true parallel operation. Given any set of potentially parallel execution modules, the feasibility of executing them in parallel is partially a function of how much extra work is required. For example, if it requires 500 ms for a processor to request another processor to cooperate with it on a task, then the parallelism of the system is constrained to routines that are at least 500 ms long.

See Fig. 22.1. Here we have, in (a), a process A, which requires 500 ms to request another processor to start on task A'. If A and A' are both 100 ms long, then we are experiencing not only an increase in total processor cycles but an increase in total elapsed time to do the task. We notice from (a) (biased against parallelism, since no cost is shown for the transition A to A' on a single processor) that it will take a parallel system 600 elapsed microseconds to do the task set A', A as opposed to 200 ms on a single sequential machine. Further, a break-even point on elapsed time is not achieved at (b), where A and A' are 400 ms. Clearly, there is no profit in parallelization until the execution time is less than elapsed time for a single processor. This occurs only when the tasks run at least as long as the time required to associate more than one processor.

In all cases, except perhaps where specialized algorithms and coding techniques are developed for the creation of specifically parallel code, one may expect more total processor time across the system for the execution

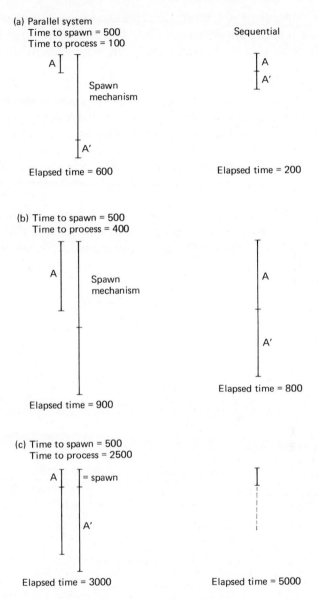

Fig. 22.1. Profitability of spawn.

of parallel tasks. We have shown here only "cost of parallelism" in terms of cost of initiation. Closely related and dropping directly out of our discussion of interaction to this point is the cost of synchronization.

Two tasks may be related in such a way that they may proceed in parallel for intervals of time but must at certain points communicate with each other in order to pass results or receive status or data. The impact of the required interaction of partially parallel tasks determines how long the sequences of parallel operation must be to be profitable on a given system.

Interaction mechanisms of the type we are describing attempt to reduce the amount of effort required on a processor to support parallelism in order to increase the amount of productive work that a processor may undertake in an interval and to enlarge the domain of profitable parallel algorithms. This could be done in part merely by providing an enlarged instruction repertoire oriented toward parallel processing, reducing task initiation and interaction time to instruction rather than to subroutine times.

Actually, such instructions would be appropriate for multiprogramming in order to reduce task-switching time and increase the amount of time in any interval that a CPU gives to user productive instructions. In more highly evolved multiprocessors the addition of CPU external supportive mechanisms would appear, from the literature, to be a generally accepted requirement.

Chapter 23

MORE ON
THE
LOCK PROBLEM

23.1. SOME PROBLEMS REMAINING WITH TEST AND SET

Some discussion is necessary to describe various TEST AND SET (lock) considerations. What should be locked, and when should it be locked? In order to prevent access to an area it is possible either to lock the area or, if it is a data area, to lock the code that manipulates the data. One notices in passing that the location that holds the lock bit need not be in any sense physically associated with the structure to be locked, but only symbolically associated through program structure.

A number of problems still occur in the presence of test and set. In current multiprocessing the primary function of the instruction is to protect operating system queues and tables, and it is expected that any processor wishing access to the area will express this wish by executing the code of the operating system.

A TEST AND SET imbedded in the code executed by any processor provides sufficient control over a locked area, if that area is in any case protected through the standard protection mechanisms of the machine imposed either by a protection key in core as in 360 or by protection keys explicit or inferred by the address-mapping mechanisms. The protection feature of the machine ensures that no other code can access the lock mechanism and that only one executor of the code can execute at a time.

Although TEST AND SET is not a privileged instruction on the 360 (it can be executed in the machine when in problem state), its use in the system effectively limits it to that class of instructions that are intended for use by the operating system. Access to the function still could be provided to any user program through a system service macro, of course.

In a more general usage involving sharing of a lockable logical resource by two different processes, each active on a processor, no way exists for a system to impose the lock if the TEST AND SET is not present in both processes. This problem has not been attacked by the industry; the various discussions and elaborations of TEST AND SET logic aimed at making it more general or more efficient now still depend upon the execution of a lock test or upon the protection mechanism to ensure logical use of a resource. These dual mechanisms are considered basically adequate due to contention of these resources.

Indeed, there seems no way around the use of TEST AND SET. This is a little annoying, because it defines some potential situations in which complex program structures that are running in a uniprocessor may not be carried unchanged to a multiprocessor environment, but must be adjusted to include TEST AND SET.

One searches for a mechanism that would lock and unlock resources in a shared environment without the TEST AND SET, but such a mechanism does not seem easily attainable. Much of it is simple, but it flounders in operation. For example, in a machine like the 360, where, associated with a block of locations, there are protection keys, one could consider adding a lock bit. This bit would be investigated just like the protection keys, and if a reference to a locked area were made, an interrupt would occur. Then any processor attempting to get into a locked area would be precluded without a test and set. However, there would be no mechanism for locking.

An implicit locking mechanism might be conceivable, but an implicit unlocking mechanism is difficult to see for the general case. One might contain the concept to a mother/daughter and place locks and unlocks only in the mother task, but this would not be generally usable. One could reset the lock after a reference or after a given time or on release of the processor, but these would also not be generally workable. The need for test and set lies in the nature of the time unpredictability of usage occurrence and duration and in the general fact that intersections of resource use cannot be generally predicted.

There is considerable interest in reducing the interlock time, the potential latency in the system due to interlock delays and consequently in defining those points at which a lock should be set. One approach is to lock the resource whenever the routine that is to manipulate the resource is entered. This allows for no reference to the resource until it is released, because any processor entering the routine will proceed only to the point of the lock and then be effectively disabled.

It is better to lock the resource at the time of specific reference to an element in it; then following processors can execute more of the code of the handling routine in common, idling just at the resource reference point.

When it is expected that all references to the resource will pass through the same body of shared code, one can lock the routine itself. Conceptually, what is shared is not the list or table, but the procedure. Again it is better to define the common procedure to be as small as possible, so that the maximum amount of work can be done.

The correct lock-out point is, of course, a function of the relationship between the sharers and the manipulation that the resource must undergo. The distinction is not made in current systems between various levels of lock, although the distinction might be made in the protection system. For example, the UNIVAC 1108 and the 360/67 distinguish between fetch and store protection but no fetch/store lockout. One possible extension of TEST AND SET might be to LOCK FOR READ, LOCK FOR WRITE, LOCK FOR READ/WRITE. An area in a lock for READ might be later (while still in read lock) locked also for WRITE. This would only apply, of course, when we are locking the resource directly, since all execution lock (locking the routine) would be covered by the LOCK FOR READ.

In a system (like the MP 65) where all lockable resources are under control of the supervisor, the grossest option is to set the lock whenever the operating system is entered, or whenever the "supervisor state" of the system is set. Considerable analysis of OS/360 has been undertaken to determine exactly what the minimum lock points in the system are. Locking at the entry to the supervisor is equivalent to transforming the system to one in which only one processor at a time is in control status.

Whenever the lock is set, the problem of what to do remains. The passive nature of the current scheme is, as we have said, potentially expensive. If a program has something else to do, it might enqueue on the resource in the same way that it requests I/O enqueueing and (perhaps nominating a return point when the response becomes available) go on about its business. If it had nothing to do, it might enqueue and then issue a WAIT, releasing control of the processor. Voluntary release of this sort would enable another task to receive the released processor and continue effective utilization. Along these lines one might provide an interrupt at the time of TEST and SET failure, which would cause an automatic release of the CPU. One might envision a TEST and PROCEED and a TEST and RELEASE instruction set. An IBM multiprocessor has an instruction DELAY, which idles a processor for a period of time before it reenters the test loop. Perhaps more important than the procedure of recognizing unavailability is the procedure when the resource becomes free (unlocked).

If tasks are to be switched on a TEST and SET lock-out, one can visualize that the task-switch procedure might take a longer period of time than the idle time and that greater true work-oriented utilization might occur if the switch had not taken place.

Although the ability to go on to other work should be available, what might really be desired if the processor has nothing else to do is merely to idle it entirely so that it does not interfere with other processors. If this is done, it is necessary to provide a signal that can allow it to continue, since there is no other way for it to be unlocked.

The idling, rather than the task switch, of a processor might be particularly attractive when the processors are small and each represents a small percentage of total system cost. One would pay a minimum penalty for idling and would receive a payback for reducing the amount of time a processor would be fishing about in the supervisor area looking for tasks, increasing contention in that area and perhaps running into further locks.

23.2. SUPPORT FROM INTERACTION SYSTEM

What can happen when two tasks are on two processors and one locks on a shared resource? An attempt is made to task-switch; i.e., the locked processor begins to execute operating system code. While doing this, it encounters a lock on the task queue. All it can do is spin on this lock, and the entire undertaking has been fruitless. If the processor is allowed merely to idle, then considerable unproductive task switching can be avoided.

In order to idle the processor and to reawaken it, use might be made of the interaction controller previously introduced. The idle occurred because of the execution of a TEST and RELEASE, which immediately put the processor into WAIT state (on a 360-like architecture the wait bit in the PSW might be immediately set).

When the processor having control of the resource unlocks it in current systems, it merely resets with a store of zero into the lock control byte. Another processor waiting will then pass its TEST and SET. If the other processors idle, however, it is necessary to do something more. The generation of a signal from the releasing computer is required. If the resource that idled a processor were identified in the interaction controller, and if the releasing computer executed an instruction that sent to its interaction controller the identification of the resource released, it would be possible for the device to broadcast the availability to all IC's. An IC, seeing the match on interlock and on released identities, would then force its associated processor to continue, that is, to execute the TEST and SET again, this time passing it.

If we have a situation in which more than one processor is waiting for the resource, it is necessary to select the single processor that is to continue. It is not necessarily true that one can predict which IC will provide the interrupt to its processor first and, therefore, that all but one will return to interlocked state. It is possible to introduce priority, implicit by position or explicit by task, but in general it is not possible to guarantee that any given processor will not be locked out indefinitely.

A similar mechanism might be implemented without an interaction controller. A queue of processes waiting for the resource may be formed in core. The processor releasing the resource executes an instruction reference, which begins execution of the supervisor, which inspects the queue and activates a selected member.

Since the process that was interlocked on the resource is still associated with the processor on which it was running at the time of the interlock (no task switch occurred), it is necessary for the supervisor specifically to address that processor to jolt it out of wait state. A combined approach would provide for enqueueing on the resource, execution of process selection, and then use of an interaction mechanism to get the processor moving again.

If the processor had other work to do and lodged a request for the resource, it is possible to handle availability passively by the normal event posting of an operating system. It is also possible to interrupt the process to inform it that the resource is free. Perhaps a much simpler approach to TEST and PROCEED is valid. Ignore it and wait for the request to come round again.

23.3. TEST AND SET AND LOCAL STORES

A final problem with TEST AND SET (or the various extensions of it) relates to its usage in systems having nontrivial amounts of local store. If, for example, a multiprocessor model 85 of the IBM 360 line were to be implemented with a private cache on each processor, it would be necessary to assure that the TEST AND SET truly worked.

If processor A executes a TEST AND SET from its cache and processor B executes a TEST AND SET from its cache, it is necessary that the lock control be passed back to memory and cache contents be guaranteed not to lead to a paradox.

If A and B were both executing the task assignment phase of the supervisor, it would be very reasonable to expect that both machines would have private copies of that supervisor section, as well as private copies of the assignment tables and the lock byte.

The execution of a TEST AND SET would cause a write through to memory, but it could not necessarily be predicted that the write through would occur in time to prohibit the other processor from passing a TEST AND SET from the local status of the referenced lock bit.

We are assuming that the processors with their caches would relate to each other and to a main store in the same way that a channel and CPU on 85 or 195 relate. On recognition by the storage control unit that the block of memory being read into by a channel also exists in the cache, the 85 system refreshes the image; the 195 sets the cache invalid and provides new data on reference.

This latter method would seem to have an advantage in a multiprogrammed system, where there is a reasonable probability that the process requesting the I/O would not have control of the processor and the effect of force feeding might to be refresh a sector that is about to be displaced by the reference pattern of the process that has current control of the processor.

The exposure would appear to exist because the inseparability of the TEST AND SET function available from main store is missing. With main store it is impossible for any processor to access the lock byte from the time another processor has referenced it until the time it is reset.

If the lock byte is duplicated in the cache, it is possible for both processors simultaneously to access the byte and set condition codes based upon its contents. By the time the new lock byte setting was sent to the cache of the other processor, the condition code would already be set, and both processors could potentially pass the TEST AND SET.

One way around this might involve an interaction controller. At the time of initiating a TEST AND SET processors would be delayed one cycle, while the IC would broadcast the intention of the processor and cause other IC's to force their processors into a wait state. The initiating IC would then permit its processor to continue; on notification of completion, the IC would tell all other IC's to place their processors back into running state. An alternative would be to construct a TEST AND SET processor on which all general-purpose processors must queue.

The problem could easily dissolve for the 85 if processors were interfaced on a common cache. This would retain many of the properties of the cache system—the faster processor/storage relationship and the isolation of the processor from cycle stealing by the channels. It might provide more efficient use of the cache to share a 32K unit rather than to isolate 16K to each processor. This would cost potential contention in the cache, which might or might not be tolerable.

We have to this point made passing reference to a number of things that deserve some fuller attention. Among them are the nomination of a processor to field interrupts and the various ways in which control can be distributed in a multiprocessor system. We have been discussing a system in which any processor can execute the operating system partially in parallel with any other. Two or more processors may be in "control" or supervisor state at the same time, and we have generally remarked that although we cannot necessarily predict who will finally submit an I/O to the channel interface, we can predict that the processor that gives the instruction will field the interrupt. These topics relate to our continuing discussion of processor communication and the possible usage of special processor interface mechanisms.

Chapter 24

CONTROL OF
PROCESSES

24.1. VARIATIONS IN CONTROL

There are basically three ways to introduce control into a multiprocessing system. One way is to allow each processor to execute its own control as it requires it and to execute whatever general control function might be required on behalf of the system at whatever time it is in a control status. This implies that whatever hardware is involved in the control function is replicated in each processor and that the system software is available to all. We shall consider a system in which each processor privately executes all control at all times a variant of this. This is a distributed control system.

A second way is to nominate a given processor as a control processor without in any significant way changing the hardware of the machine. The nomination of any particular processor as the "control" processor is fundamentally arbitrary, and the period of nomination may vary. A system in which, when a processor enters a control state, all control functions are executed by it is a variant of this approach. We may call this floating control.

The final way is to build a special control processor, different in design and architecture from all other processors in the system, which is responsible only and always for the control functions. A generally symmetric multiprocessor may have such a centralized controller without losing its essential functional symmetry. We may think of this as specialized control.

24.2. PROCESSES AND PROCESSORS

The essential acts of a control involve those activities which associate a process to a processor, which monitor the interactions required for a process with other processes, and which monitor the progress of a task in the system. We describe in the introductory multiprogramming chapter a number of various states in which a process may be. In multiprocessing the state concept that is significant is whether a process is or is not at any point associated with a given processor.

The separation of the concept of process from processor is a reasonably simple one, which we have been intuitively assuming. It is a multiprocessor extension of the initial concepts of device independence and memory location independence.

Similarly, a process need no longer be associated with a particular processor. This is obvious when we postulate that at some time a processor will operate to select from a common work queue a task to perform. At the initial activation of that task we see that in a dynamic self-scheduling symmetric system we do not necessarily care which processor selects the task. During the period of time after the selection of a task by a processor (or the assignment of a processor to a task) when the process occupies the control counter of the processor and is actually "running," we say that the task is associated with a given processor.

At any time the process becomes blocked, it is unable to continue; then a decision may be made as to whether the process should continue its association and wait until it can continue or if it should release the processor (or be released by the processor).

A process is associated with a processor, even if it is blocked, if the dynamic information of that process remains internal to the processor or if it remains (in a system, for example, where dynamic information is centralized and processors point to the stateword in a pool to reflect current process) associated with the processor in any way so that it is unavailable for execution by another processor and the processor is unavailable for another job.

The process becomes disassociated from the processor if it is placed by the system in a pending queue, where, upon the release of the block, it becomes available to any processor and the processor on which it was running is free to acquire or accept more work.

In current multiprocessor offering it is important to disassociate processors from locked processes because of the expense of idling a large processor (representing one-half or one-third, perhaps, of computational capability of the system). This implies that some degree of multiprogramming (time-sharing of CPU) continues to occur in a multiprocessor system of this type and that the extension of the resource to include additional processors is essentially equivalent to transforming a single-server system into a multiserver system servicing a common work queue in shared core.

The striking similarity between the multiprocessor 65 or 1108 and the single-processor versions of the operating system reflects the multiprogram—multiprocess similarity. The modifications to OS/360 MVT to support MP 65 were not severe and involved primarily the support of the lock/unlock requirement and elaborated I/O subsystem tables reflecting the dual paths of control units.

Therefore, given N processors in a system of this type, there is no implication that only N processes may be "active" in the system. One interesting observation is useful at this point. The resources, such as core and devices, that exist in a system belong not to a processor but to a process. If peripheral unit asymmetry exists such that some family of devices is private to a processor, then a constraint exists in the assignability of processes to processors, since in order to be a candidate for execution of a process, a processor must have a path to all the resources granted to a process.

24.3. DISTRIBUTED CONTROL AND PROCESS INTERACTION

In a distributed control system it is common that a single copy of the operating system be shared by all general-purpose processors. More than one processor may be in "control" state at any time, executing task selection and assignment, handling interrupts, placing requests on channel queues, etc. The only constraint imposed is the lock on "critical code" sections, the simultaneous usage of which may represent a potential for illogical results.

The presence of multiple processors is essentially invisible to a user except when he wishes to use resource-sharing techniques, at which time he must insert test and sets. Differences in programming style due to a desire to take maximum advantage of potential parallelism might, of course, lead to more intensive use of whatever system macros are available in the operating system for the spawning of subtasks.

Even in such systems, however, there is a usage for interprocessor communication in the area of control. One such use on the 65 multiprocessor is in the service of the concept of maximizing the economic utility of the system by assuring that at any time each processor is running the highest available priority job.

In an environment where priority exists, one objective for a multiprocessor control strategy is to ensure that the sum of priorities of tasks active in the system is minimum [or, in systems like OS/360, where priority designation is reversed, running from high number (greatest) to low number, that it is a maximum].

The source or basis for a given job's having a certain priority, how priority is acquired, and what it means are discussed in the following chapter. We do not mean to restrict our concept here to a number punched into a job identification card.

Let us postulate that at time 1 a process I with priority 2 is running on processor B and that this priority program becomes blocked on an I/O service request. The process is disassociated from the processor and marked as awaiting I/O event in a table in common store. At this time processor B executes the operating system and selects an available task. The dynamic portion of the program is transferred into the processor registers, and the control counter receives the first instruction of the process to execute. The new process on B has a priority of 5.

We have before seen how the I/O request of the queued process might be submitted by processor A. It will, therefore, be A that fields the interrupt. At the time A does this it has temporarily interrupted work on a task of, let us say, priority 1. Under the policy of keeping higher-priority tasks active, process I will not receive control of processor A. However, if B is still running the process with priority 5, it is not true that the sum of priorities is minimum, since a priority 2 task is available but has no processor. It is true that at some point in the near future the priority 5 process will request some service that will block it, and at that time the priority 2 task will become a significant candidate for the control of B.

The strong enforcement of the priority strategy, however, requires that there be preemption of the task on B and replacement with the now unblocked priority task as soon as it is unblocked. Processor B is unaware of the availability, since processor A handled the event that made it available. It is necessary for processor A to tell processor B to switch its task. It does this across the "shoulder-tap" read-write line connecting the machine.

In a broader system use of an interaction mechanism of the interaction controller type might be useful in optimizing priority activity. Whenever a process block was removed by a processor, the priority of the task unblocked could be sent out to the IC subsystem. The IC would then transmit down the line to find an IC that had a processor running a lower-priority process. This IC would then interrupt its processor and cause it to exercise task selection. By extension a selection could be made of the lowest-priority processor, "the most interruptible" processor in the system.

24.4. VIGOROUS ENFORCEMENT—
MOST INTERRUPTIBLE PROCESSOR

The concept of "most interruptible" can be significantly advanced. Consider the circumstance where since the time a processor submitted an I/O request, it has become the one associated with the highest-priority task. The process for which this processor will be interrupted will have a lower priority at the time of the interrupt signal than the process active on that processor, and will, therefore, cause delay in work on the highest-priority task.

It might be possible, in a priority interrupt system of the kind associated with process control in the industrial sense (where, in addition to "classes" of interrupts like processor error, I/O, etc., a member of a class might have a priority level associated with it), to lock out all interrupts associated with lower-priority tasks and to enable only those of higher priority.

Priority of this type is of itself interesting to investigate in the hardware. In current systems there are truly two priority mechanisms that conflict with each other. One is the software enqueueing mechanisms with their associated queue disciplines; the other is the hardware priority mechanism, which resolves ties between a CPU and channels for memory access, between individual channels for access to memory, between channels for access to CPU interrupt handling. Similarly, control units and devices have relative priorities. These priorities are characteristically fixed by position in the hardware. A channel with a low address has higher priority than a channel with a higher address, etc., and all channels have access priority over the processor.

More flexible schemes might relax the CPU/channel rule and associate dynamic priorities related to operation. For example, a drum unit might have an initial memory access priority lower than a processor, but as it was refused memory accesses and the danger of losing a revolution increased, it would be granted a higher priority. Ideally, one would like to associate access and interrupt priority with the priority of the task using the resource. For the request on behalf of this task, the channel has a priority of X.

Even, however, if it were possible to lock out the interrupt from a processor having an active higher-priority task, it would be undesirable to lock out the interrupt from the entire system, since there may well be processors running tasks of lower priority than the one that is attempting to force interruption. We would like the handling of the interruption to occur on the most interruptible processor. This could be accomplished either by a special interaction unit at the interface of the processors and the I/O subsystems or by use of the general interaction mechanism.

Some implication exists for the intelligence of the I/O subsystem. Many multiprocessor designs visualize highly evolved, very capable I/O computing elements similar in power to the general processors themselves. The more intelligent the I/O subsystem, of course, the more it can, in general, unburden the CPU's from the expense of interrupt handling.

The fundamental difference made by intelligence is how far the I/O subsystem will go. A highly intelligent system might handle the interrupt itself and even go so far as to execute any task-switch coding that might be needed as a result of interrupt. A less intelligent unit would initially recognize the interrupt condition (end of data transfer, end of channel activity, malfunction, etc.) and then tend to the distribution of the interrupt.

A suggestion is to put at the interface between processors and I/O controllers a mechanism for judging the relative utility and necessity of fielding the interrupt. All information relative to the task assignment and priority of tasks could be kept in a centralized small buffer accessible by the IOC.

In the IOC would be the task priority of the I/O request and in addition whatever dynamic device-oriented priority might be used to decide how critical it is for a processor to take this interrupt at this time. A comparison by the IOC selects the most interruptible processor and passes off the interrupt to it. It then enters control state to execute consequent system control actions.

24.5. FLOATING CONTROL

One can note where a design tradeoff exists between specialized, floating, and distributed systems. In the fully distributed system, the truly symmetric anonymous system, money is spent to implement and/or improve the performance of critical functions by elaborating the interfaces between generally capable devices. The design alternative is to define these functions into a specialized processor. If the system is not specialized so as to allow I/O access only through a single processor but to implement connections to all processors, a system being between distributed and specialized comes into view.

In this system only one processor at a time may be in the control state, but any processor may enter the state and, when it does so, execute control for the entire system. The duration of the control state may be months or microseconds.

If it is months, the transference of the control function from one processor to another is entirely a residual reliability feature; if it is microseconds, then the floating system resembles a distributed system with the broadest possible interlock on resources.

While in a control state a processor handles all interrupts that develop
in the system (except the processor local interrupts, which may occur in
many architectures—floating divide overflow, etc.) Noncontrol processors
enqueue all work requests, I/O requests, waits, and all supervisor functions
on the control processor. In a microsecond float other processors may idle
on the availability of the control status. The processor that has entered
control handles all I/O interrupts; when it leaves control, the processor
gaining control assumes that responsibility.

From the interrupt-handling strategy we may look at the "floating
control" system. It is possible to route all interrupts to a single processing
unit. Regardless of the computer that submitted the request, a single
processor would receive interrupts. It might be possible for any processor
to execute the coding to submit a request to the channel, and if this were
true then the possibility of finding the trail back to the submitting processor would exist only through the process. In other words, the processor
finding an interrupt must identify the submitting process and through it,
if it is associated with a processor, a processor using processor identification techniques touched upon before.

However, the most natural occurrence in a system is for the submission of an I/O request to be a function of interrupt handling—determining
that the path or device is not free and that a request has been enqueued.
Since this is so, some economies might be effected in the system by only
enabling the interrupt handler to submit requests. If this is done, the
system is truly transformed into a specialized control system.

24.6. SPECIALIZED CONTROL

The specialized system contains hardware elements that are unique to it.
There is a system control processor, which might characteristically contain
hardware or microprogram implementation of critical control functions
too expensive to duplicate in all machines in the system and yet with a
requirement for very fast execution. In those multiprocessor systems that
are only apparent multiprocessors, as we discussed, such a control processor would be almost an inherent requirement. Whether they should also
exist in more capable processor systems depends very much on an appreciation of the extent and nature of control in the system. Currently we
support multiprocessor functions with macros; the next step is to field
machine instructions. Whether beyond this it is necessary to build special
execution units for those instructions must fall out of the design performance goals and dollars available for future systems.

Currently there is still much to be said economically for specialization, but new technologies may turn the argument around. The interest in

the specialized control processor derives in part from the possibility that we are beginning to understand enough about scheduling to represent algorithms in hardware and that the task control and selection mechanisms associated with a dynamic system represent sufficient overhead to warrant particular attention to their efficient performance.

If there were in the system a processor organized for these functions, the task of balancing system load could be put "out of line." We have, in the generalized distributed control system, been putting things out of line with specialization of elements at interfaces. One constraint that must be kept in mind in the design of highly parallel systems is that the elements developed for the control of a system must not cost more than the savings achieved by pooling resources.

This constraint exists more or less strongly. In systems where the reliability motive is dominant, there is an awareness of a need for extra cost and a willingness to undertake it. Where multiprocessor capability is acquired for performance and efficiency, then the cost constraint is very real. The goal of design is the economic redistribution and enhancement of existing functions, and not the addition of functions.

Chapter 25

PARALLEL
TASK SPAWNING

25.1. FUNDAMENTAL FUNCTION

A fundamental control function that we have not described in detail is the
creation of parallel tasks in a system. There are two sources of instruction
streams in contemporary systems—streams admitted by the operation of a
operating system scheduler and streams initiated by the activity of a task.

The classical instance of the second activity is a request for I/O. A
working task requests initiation of a subtask whose function is to read or
write data for the requesting task. This task is performed in part by a
general-purpose processor and in part by a specialized I/O processor, the
split of work between them depending on the intelligence of the I/O
processor. After making a request, a task may or may not continue its
own operations. It may have other records to process or it may block
itself until the I/O is complete (or be blocked by the interfacing mechanism of I/O support in the operating system).

Conceptually, what occurs is that a task asks for the services of
another task, which by its nature is run on a specialized subprocessor in
the system. There are a number of ways to implement this activity beyond
current designs. One approach would be to take I/O queues out of core
and into a specialized buffer at the interface of a processor and the I/O
subsystem. A READ instruction would now cause an entry into a common I/O request queue. An I/O processor would pick up tasks from this
queue as it became "spontaneously available," that is, as it completed
work already under way.

In effect, the READ order is a request for space in a work queue associated with a particular processor or family of processors. Upon the completion of activity the I/O processor would post a completion notice and (perhaps) interrupt a processor. If specialized I/O processors existed, they might have private work queues, or they might share a queue. If I/O processors were sufficiently intelligent, then a selection algorithm could be built into them (software, hardware, or microprogramming).

Two observations come to mind. First is that the ability of the requesting task to continue after its I/O request is a measure of the true parallel capability of the task vis-à-vis its requested task. The issue is one of "effective" parallelism or "true parallel effect." Along this line, a task technique for increasing the true parallel effect is a synchronization mechanism, basically passive, which we call an I/O buffer. Second, the request for an asynchronous I/O subtask is not essentially different from the request for the operation of any subtask, although they are supported in systems that allow subtasking by independent mechanisms. One can visualize a system where a macro-like ATTACH is used to support I/O as well as subtasking. The current separation is due to differences in the detailed requirements in a multiprogramming organization where I/O support is expected to be resident, where synchronization is supported by interrupt.

As a further observation, no distinction need necessarily be made as to whether a spawned task is to execute private procedure on private or shared data or common procedure. With this generalization, we find that the question of true parallel effect raises itself in a more complete form.

25.2. TRUE PARALLEL EFFECT

Basically, true parallel effect is reduced by any resource contention that might arise as a result of the existence of multiple streams. It is further impacted by the time cost of initiating a new process. We have seen a mechanism for pushing initiation "out of line," but we have no way of limiting the amount of time until, on a system where there are more processes than processors, a processor becomes available to undertake the spawned process.

If, for example, the requesting process finishes before the new process finds a processor, there is no parallel effect whatsoever. Beyond this is the degree of parallelism that actually exists. In effect, is there sufficient independent activity to justify parallelism, or are the synchronization requirements and interdependency between the parallel paths so high that intercommunication time will be large enough to make parallel operation absolutely or relatively unprofitable?

The nature of initiation and intercommunication hardware will make different levels of true independence profitable on different machines. But there would appear to be many processes which are inherently sequential in nature and which cannot be profitably partitioned in parallel subprocesses on any reasonably conceivable machine.

The spawning of a task may arise out of two conditions: (1) The process has reached a point where it wishes to execute two paths in parallel, both of which are to begin from an identical point, that is, the status of the process at the time of initiation represents the initial status of both legs of the parallel operation. (2) The process has reached a point where it desires to initiate a number of subtasks, each with a private set of data to work on.

The difference between these situations is small, but reflects some interesting design problems in the area of task spawning. Under condition (1) all that is required is for the spawning processor to take a picture of the process at that time and duplicate it somewhere. This may be in the central memory of the system, or it may be in a memory associated with a control processor, or it may be sent to an interaction controller.

In the second instance it is necessary for the spawning process to generate some information for the other processors, most particularly setting index registers to define a sector of an array or matrix. This raises the point of what is truly included in the dynamic information of a process. Process starting point, initial values, and all capabilities in terms of memory and devices are the private data of a process.

If we visualize for a moment a centralized representation in core of the maps associated with a process—devices, core assignment, all available resources—then what must be transferred to an initiating processor are pointers to the locations in core that represent its capability. It may be that the contents of the registers of the processor doing the spawn are not relevant at all. Essentially, the linkage conventions between calling and called subroutines that apply in single-stream machines apply to multiprocessor spawning linkages, with the exception that if the spawning processor wishes its registers to represent a portion of linkage, this information must be externalized.

One of the classic descriptions of an externalization mechanism is Conway's multiprocessor. This system does not represent an exhaustive solution to spawning, but it is an interesting vehicle to use to touch upon problems and variations.

25.3.   FORK AND JOIN

In Conway's processor it is assumed that all that is needed to spawn a task is a snapshot of the dynamic information of the spawning processor. A machine instruction is provided, which duplicates the dynamic information and sends it to the control memory, which represents a queue of waiting tasks. This is an informal mechanism supporting informal spawns.

The limit of the mechanism is its implication that the spawned process requires no resource distinction from the spawning process, and the presence in core of the parallel path or a transparent paging mechanism is implied by FORK.

The control processor memory is associated with a control processor that has the capability to order the queue by any discipline. Any meaningful ordering, of course, other than first in first out would require that the priority to be given the requested task be part of the dynamic information. Each processor has an independent path to the control memory, so that simultaneous transfer of statements occurs in the system.

The FORK instruction is actually a transfer instruction that has the characteristic of initiating both legs of a branch. The concept of FORK, under various names and with various forms, is basic to multiprocessing. Conway's FORK as a true machine instruction allows three fields of specification: a location at which the spawned process is to start, a counter to determine conditions of coordination and a value to be placed in the counter.

The process of acquiring a task is implemented by a corresponding instruction JOIN. JOIN is a composite that actually combines two concepts, ACQUIRE and WAIT. The variations in JOIN provide a hardware implementation of the options to acquire a new task, to idle on completion of an event, or to transfer to another portion of the current task. JOIN tests the value of the counter set up by FORK, decrementing it as part of the test. If the counter is zero, JOIN advances to the location just after the counter, if it is not zero, JOIN either branches to an idle or alternate routine or forces the process to release the processor.

The method by which JOIN releases the processor is to signal the control processor that a new task definition is to be transferred into the executing processor. The interaction mechanism involves communication through shared memory (instructions and counters) and the active interaction through control memory. The two sources of a control memory entry are the FORK instruction specification and the dynamic registers of the spawning processor.

To watch FORK and JOIN, let us postulate that a processor A has a process 1 and at location 1000 desires to acquire additional processors to execute routines beginning at 1100, 2000, and 4000 (See Fig. 25.1). It sets a counter at 1500 with a value of 4, indicating that four routines are to joint at location 1501, the spawning routine on processor A and the three daughters.

The mother task in this instance is to continue on its processor and

```
1000 FORK 1100, 1500, 4
1001 FORK 2000
1002 FORK 4000
1003      —
          —
          —
```

Three forks sustain
four processors

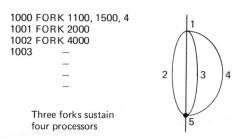

Fig. 25.1. Four-way branch.

three processors, for processes 2, 3, and 4, are requested. The three
FORKS cause three queue entries to form on control memory. It is impossible to predict how long they will stay on the queue and consequently
what degree of effective parallelism there will be. It is also impossible to
predict which processors are going to perform the spawned routines.

At the completion of each routine a private JOIN is executed, which
will decrement the counter at 1500. The processor that zeroes the counter
will be the processor that continues execution, undertaking 5 at 1501.
The processors finishing earlier will release by requesting a new task. If
there are more than three available processors (processors interlocked on a
JOIN at the time of the FORK enqueueing), then the processes at 1100,
2000, and 4000 will be picked up in that sequence and the four paths will
proceed in parallel.

JOIN must have a fetch interlock of the TEST and SET type while it
accesses the counter. This counter is a limited equivalent of an event control block in a multiprogramming system—the initial value represents the
number of times an event must occur (i.e., completions of parallel paths
in this case) before some successor process may continue. Since there is no
way of determining which process will finish first or last, it is impossible
to predict the processor that will execute 1.

Figure 25.2 shows a variation in which the only function of the
spawning task is to spawn, and after spawning it makes its processor
available. In this instance this is less efficient, since an extra initiation is
performed to gather four processors, and there is a possibility that on the
JOIN the spawning processor will pick up one of the spawned queue
entries. Notice that there is no reason why one of the spawned tasks might
not be an I/O routine, and we see that the distinction between I/O and
subtasking may disappear.

It is interesting to watch the dynamics of this situation in a one-
processor case. The process at location 1000 will complete first and reduce
the count to 3; finding the counter nonzero, it will release, and the
process at 1100 will take control, finish, and release to 2000, which

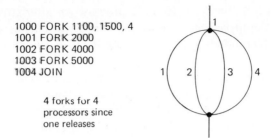

```
1000 FORK 1100, 1500, 4
1001 FORK 2000
1002 FORK 4000
1003 FORK 5000
1004 JOIN
```

4 forks for 4
processors since
one releases

Fig. 25.2. Spawn and quit.

releases to 4000, which will go on to 1501. The total effect of this will be to increase the running time of the collection of processes by FORK and JOIN time. Of course, if by reason of program misconception a process does not release on the join, it will interlock the system indefinitely.

This word of caution: Note that the process is not insensitive to the number of potentially available processors. If, for some reason, the concept of mother process was strongly enforced so that the spawning processor was programmed to idle (branch loop on join in this case, although a drop into idle status as earlier described would have the same effect) until other completion occurred, the one-processor case would not work, since the spawning processor would never become free to pick up the spawned tasks.

25.4. CONTROLLING REQUESTS

Another problem is that of available queue space. Conceivably, queue space could be made extensible by a software intervention at the time that an entry tried to enter a filled control memory. Such techniques are seen in the implementation of stack machines, where the top of the stack is implemented in "hard registers" and a pointer points to locations in core that hold remaining stack entries. In the B5500 two registers are the top of stack, and a pointer points to a program's stack area.

An interrupt on control memory full could be generated, and over-flow entries could be placed in main core. If this is not done, it is possible that the rate of task requests generated by active processes could force the system to quiesce a process at the time it requested a process.

A brutally unfair approach to this is to force off the process that has just caused overflow. This forceoff could mean that the processor was compelled to task switch or that the processor was idled. More fair, perhaps, would be to idle the lowest-priority task, and, even fairer, to idle the task that had spawned the most requests. Further, it would be possible to delete all requests for the forced-off process. An alternative and perhaps more effective approach is to limit the number of entries in the queue by using coding techniques that limit requests to a function of the actual number of processors already acquired.

The fundamental technique is to generate a request for a single process and have that process generate a request when it acquires a processor. See Fig. 25.3. The initial FORK spawns a routine at 1100 and sets the process finish counter to 4. The process then continues with its regular work. A single queue request is made for the process at 1100. If this should acquire a processor, it will then spawn the routine at 2000, which

Process 1:
1000 FORK 1100, 1500 4

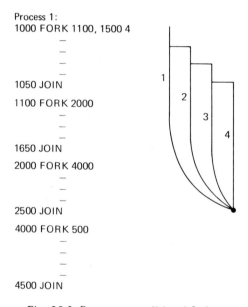

Fig. 25.3. Processor conditional forks.

will wait for a processor and then spawn the routine at 4000, etc. At any time only one queue entry is taken for this computation. Notice that in the one-processor case performance is conceptually the same, and the effect of a single mother with many daughters is the same as having each process mother its own.

25.5. PREEMPTION IN FORK ENVIRONMENT

Throughout the discussion of enqueueing requests we have been using a concept that transcends the implementation of FORK and JOIN on the system. If we wished to preempt processors from a task, either to maximize priority as higher-priority tasks were requested or to enforce a service rule for a time-sharing system, we would require additional mechanisms.

Using the control memory and control processor, we could force a processor to interrupt on recognition of a higher-priority task arriving in the queue or on a time basis. Alternatively, each processor could have its own clock settable by the control processor. The running time to interrupt could be set differently for each processor. In an IC-type system interrupts could also be forced down the IC line.

The time-sharing environment is unique, in that we have a situation where we wish cyclically to preempt jobs even though they are not blocked nor do they need to defer to a higher priority. A problem with the FORK/JOIN as we have described them is that there is no mechanism for preemption. Any given process wishing to preempt itself can issue a FORK/JOIN pair, naming its own entry point in FORK. The FORK effectively requeues the task and at the "resume" point the JOIN acquires another task.

It is not usual, however, for processes to do this, since in most systems a process is written in such a way as to be independent of the actual arrangement of the slices of time it will receive during its life from a processor, and it certainly cannot be expected to determine at what point it is necessary to relinquish a CPU for load-balancing service rate maintenance reasons.

If the control processor cycles through control memory on a timer basis, it must have some way to preempt a processor. It might execute an instruction that would interrupt processors and force them to execute FORK/JOIN pairs. If control were centralized to this point, it might also be the job of the control processor to allocate all resources and indeed to initiate whatever transfers to core were required to operate the process being taken on.

In whatever way it is organized, the point is that, depending upon the spontaneous self-directed availability of processors, completing a process and requesting another may not be sufficient in some environments, and interactive mechanisms of the type earlier described may be required.

An interesting problem is the determination of when processors should be preempted and whom to preempt. We can answer the last part of the question by remembering our most interruptible processor.

The first part is determined by environmental conditions having to do with the urgency of a process requiring a processor (in real time, perhaps) or the nature of the schedule and the loading of the system (in time-sharing). A property that is desirable in multiprocessors is dynamically to adjust the number of processors active in the service of a given computation or subsystem at any time, depending upon the urgency or load of the computation or subsystem.

The determinations of urgency and priority we leave to later, but the general capability of ' ganging'' or distributing the attention of processors is an inherent feature of a multiprocessor system. This capability rests upon the avoidance of processor-identified computations, the avoidance of dependence upon a given number of available processors in the system in order to execute any given process, and an ability to preempt processes as well as to provide mechanisms for the spontaneous availability and acquisition of processes by a processor. These considerations are both performance- and reliability-oriented.

In almost all multiprocessor systems special provision is made for reliability. This usually takes the form of an ability to isolate malfunctioning units combined with some level of ability to reconfigure. The state of this art in available systems concerns us next.

Chapter 26

RELIABILITY

26.1. FAIL-SAFE AND FAIL-SOFT

The fundamental concept of duplication of elements in a system is that of "fail-safe." The organization of modules is such that in the event of a failure of any component, the system can continue to operate at required levels. An extension of this concept is that of "fail-soft," which implies the definition of fall-back levels of operation appropriate to different patterns of failure.

The condition of fail-soft may be obtained by abandoning functions in the collection of processes or by sustaining all function but operating in a slower way. Achievement of fail-soft is at the edge of the art at this time, since it implies not only a profound understanding of hardware replacement possibilities across different devices, but quite extensive understanding and definitions of what fail-soft levels are reasonable to define.

For example, in attempting to achieve fail-safe, we may replicate disc units in a system and paths to disc units in order to achieve an acceptably low probability that a disc unit will not be available to a system function requiring it. We define a minimum system configuration that is needed and attempt to insure that a pool of resources contain sufficient members of all types, reasonably easily associated with other resources, to guarantee to a certain degree of confidence (usually a systems design parameter) a usable configuration.

Fail-soft attempts to define alternatives beyond this. In a system with an on-line real-time primary mission with peripheral support functions going on more or less concurrently, fail-soft presents a number of varying alternatives. One is to achieve the operational needs of the primary mission. Another is to reduce the scope of the operational mission (perhaps to servicing only a few critical lines). A third is to reduce the rate of service by relying on slower but functionally equivalent hardware.

Suppose that two processors were connected by a shoulder tap, and this shoulder tap failed. One would elect to partition the system so that these processors would effectively operate as stand-alone machines. The partitioning might include the closing of paths to shared memory and the closing of paths to shared devices.

An alternative would be to maintain the processor operation, but in a degraded hardware mode, using a channel-to-channel interface and reducing the load on the system to one that could be expected reasonably to run, using the slower interaction mechanism. If this mechanism failed, complete reliance could be placed on shared memory communication.

Each level of operation represents a degredation of performance and a special fail-soft mode. In many very large systems one feels that an insufficient definition of partitionability has been applied, perhaps due to packaging. For example, if a specialized execution box of a highly elaborated uniprocessor were to fail a floating-point multiply unit, it is now commonly necessary to retire the processor for maintenance. This floating-point multiply unit, however, is not critical for continuing system operation in a degraded way. If a fault signal could be generated while the unit was out, then every floating-point multiply instruction could cause an interrupt, and the multiply could be executed as a software subroutine using a floating-point add execution unit.

The first step, then, in achieving fail-soft or fail-safe is to define exactly those interfaces that are to be identified as isolatable. In the design of highly modular systems this process goes hand in hand with system design. When this is accomplished, critical points can be identified —that is, those points which are functional or where hardware duplication is necessary because of the high vulnerability of the entire system to a failure at this point. It is at these points that additional paths or components are placed in a system, despite a very low probability that the duplication will lead to improved throughput or performance.

The level of duplication is important. It should be possible to use alternate components to improve performance, as we mentioned in Part 3. The UNIVAC I contained a duplication of each arithmetic register and constantly compared one to the other, stopping the machine if a disagreement occurred. Reliability was achieved, but no advantage could be taken of the presence in the machine of additional registers. Further, the availability of the machine was low, since no support was given to correction once an error had been discovered.

Availability is a measure of the probability that the system or any component of the system will be available at any particular time. Reliability is a component of availability, being a measure of the probability of failure. The time to recover from a failure, however, also contributes to the availability of a unit, and a system design not only must be reliable but must provide features that speed the recovery of a malfunctioning unit by contributing to analysis and diagnosis of malfunctions.

26.2. RECONFIGURATION

The partitioning of a system is a basic form of the reconfigurability required for reliable operation. For example, IBM's, UNIVAC's and other multiprocessors can be run as two independent systems, and switching of I/O components can be accomplished.

In current systems the techniques of configuration depend upon the control tables of the operating system, the command language available to a console operator, and physical switches that open and close paths. Each resource in a system is known to the operating system. It basically maintains lists of processors, control units, channels, devices, and core spaces to support the activity of allocating resources to tasks. An essential step in partitioning is to assure that system tables are adjusted to reflect those resources that are not to be used by the system.

When a system is brought "on the air," it determines the population of elements with which it is associated and constructs proper control tables. In the Burroughs B5500 dynamic changes in element status are observed by the control system as devices "coming on line" set presence bits that are periodically investigated by MCP (master control program).

If it is desired to make a resource unavailable to an initializing configuration, physical switches (on a configuration control panel) can be set so that the initializing procedure recognizes these devices as unavailable. A system operator has available to him in the OS/360 environment a console command VARY, which he can use to inform the system that he wishes to take some resource "off-line." On the two-processor 65 a processor can be taken out of the system in this way with the precaution that it must be brought to an "orderly halt." Orderly halt merely means that I/O should be allowed to terminate, and all other pending functions should be completed or enqueued for the remaining CPU.

An important concept on the 65 is that of the malfunction alert sent across the connecting line between processors so that a processor that has detected a processor malfunction notifies the other processor, which is interrupted to undertake an attempt to help the malfunctioning processor stay "up."

The process will determine whether the malfunctioning CPU has experienced only a transient error or if the error is permanent, requiring maintenance, and the element must be taken off line. The system informs the operator of what reconfiguration activities are required on his part, identifying what elements are to be dropped or moved. It may be desirable for the operator to partition the system so that a defined independent system can be isolated for independent maintenance. Attendant on these procedures are various options for determining the extent of the salvage operation required by the malfunction and the ensuing reconfiguration decisions.

A similar capability is present in the UNIVAC 1108 through a unit called the Availability Control Unit, which interfaces with each CPU, each IOC, and memory banks. Associated with one of the CPU's is an Availability Control Panel containing switches for necessary physical reconfigurations. The ACU provides the capability of physically isolating system elements from each other.

Associated with error detection on the 1108 is a unit called the Automatic Recovery System. This unit basically assumes that the system is malfunctioning and will cause an error-connecting reinitialization of the system unless it is reset by the action of the executive system within a certain interval.

If a malfunction is actually detected, then an attempt at a system reinitialization is made. This attempt is based upon a predefined partitioning of the system, which defines two basic subpartitions, each with a capability for reinitializing the system. First one partition would undertake to "start up"; if it passed this, the cause of the failure would be known to be in the other partition. If it failed, then the other partition would undertake and would be successful (in the presence of one malfunction). This concept is called partitioned system recovery. Each predefined partition must, of course, contain sufficient I/O, core, and processors to undertake system recovery.

26.3. IBM 9020

A further elaboration of configurability and recovery is found in the IBM FAA real-time system, the IBM 9020. This is a multiprocessing system that contains three computing elements, each roughly of the capability of the IBM 360/50 and with the 360 architecture. These computing elements are associated by a distributed crossbar to memory banks of 128K bytes (data width of four bytes).

An interesting feature of the system is the presence of highly intelligent I/O control elements (IOCE) which interface with the I/O subsystem.

Each IOCE has a private storage of 32K bytes, which it uses to hold control words for the I/O subsystem. Each I/O element interfaces with general storage. There are multiples of these units in a configuration. (Configurations will vary, based upon air traffic intensity at the various airports where the system will be installed. The application here is air traffic control.)

In order to accommodate the need of an on-line real-time system for very fast reconfiguration a hardware feature of the 9020 is a Configuration Control Register (CCR). This register represents for each element in the system its current status and its relationship to computing elements and other elements in the system. Each memory bank, IOCE control unit, and computing element contains such a register. An associated instruction SET CONFIGURATION, executable by a processor only when it is in control status (a "privileged" instruction), is used to control the contents of the configuration registers.

A system configuration mask represents the state, configuration control, and communication fields of the CCR. Under control of a unit selection mask a SET CONFIGURATION places appropriate data into the CCR's of designated units. For example, the selector mask may include memory module 3 as a unit whose CCR is to be referenced. In this register in a memory module there is a representation of the state of the module and the identification of the computing elements that may change the CCR of the module, as well as identification of all computing elements or I/O elements that may communicate with this module.

Similarly, in an I/O element there is a representation of status, a designation of all memory banks with which it may communicate, and all computing elements, as well as a designation of those computing elements that may change the CCR.

The ability to change a CCR is equivalent to reconfiguration, and it occurs without any of the operator assistance or intervention at a console typewriter or control panel that is present in other systems. The state of a computing element determines whether or not it is allowed to execute a SET CONFIGURATION instruction. Partitions are differentiated from each other by use of the STATE code, each unit reflecting whether it is on-line, on-line back-up, off-line active (on a support function), or unavailable.

The malfunction alert is associated with this system and can be generated by a CE, an IOCE, a memory bank, or a control unit. An operational error analysis program provides the link between malfunction detection and whatever reconfiguration might be needed. Extensive diagnostic monitoring is undertaken by the system in an attempt to keep the system operational, to localize and salvage for recovery, and to collect error statistics.

The effect of extended hardware attention to interconnection and the effective combination of "logical" and physical reconfiguration is to allow for the definition of fail-safe or fail-soft partitions in a time frame realistic for on-line operations. In such environments reconfiguration time may be quite critical and may not allow for operator cooperation. An interesting additional capability of the 9020 is the ability dynamically to assign addresses to various storage elements (this is beyond the prefixing of lower logical core) so that address space may remain the same if a memory bank goes down.

Involved in the design of any high-availability system are four underlying concepts, which we present here in summary: (1) modularity—the definition of functional elements that are assignable, connectable, and isolatable, (2) multiplicity—the replication of selected units to achieve fail-safe redundancy, (3) configurability—the ability quickly to reconnect modules in the face of detected error, (4) analysis and recoverability—the ability to detect errors and to minimize the impact of errors in the system.

Chapter 27

PERFORMANCE

27.1. COMPARANDS

A generally accepted rule of thumb in the computer industry for some number of years has been an observation called Grosch's law. This "law" states that the power of a processor increases as the square of the increase in its cost. This would imply that the best approach to additional computing power is the development of faster single processors.

A multiprocessor concept, especially one that envisions multiple grouping of moderate-speed processors, would seem to run counter to this widely accepted canon of art. Actually it need not, since we have shown that within a given technology there are natural speed limits and that the spending of more money to develop a faster processor actually may mean the development of parallel techniques, which might well involve the presence of additional inexpensive instruction processors sharing high-performance functional units. In considering the performance of multiprocessor systems one must be careful to select what they are being compared with and on what basis.

There are a number of possible measures. One is the throughput of an N-processor system in terms of the total amount of work that can be achieved over a given period of time. Another is the cost of work performed by the system. We introduced some conceptual considerations in the first chapter to show how one would go about developing a value index for a multiprocessor system.

With either work capability or cost of work as a measure, a comparative system must be selected. There are a number of different comparative systems. One popular one is the single-processor configuration of the mul-

tiprocessor system. To make comparison on this basis more realistic, the
single processor might be expanded to include that amount of additional
core and devices required to support the multiprocessor. The additional
core and devices are then used to support a level of multiprogramming
equivalent to the breadth of task concurrency afforded by the multiprocessor.

An alternative measure is to compare the multiprocessor versions of a
system of processors of a given level of processor capability with a larger
and more capable uniprocessor in the same line.

Specifically, the first measure might compare a two-processor 65 with
1 million bytes of storage with a multiprogrammed single 65 with either
512,000 or a million bytes of store. The second measure might compare a
multiprocessor model 40 system (if such existed) with a uniprocessor 50.

It is generally true that an effective multiprocessor organization implies more resources in the system than a uniprocessor, although it is not
obvious that the increase in core or devices to support multiprocessing is
characteristically greater than what is required to support broad multiprogramming.

The increase in resources is a result of the increase in the number of
tasks that will hold various subsets of resources simultaneously and not of
the increase in processors. It is usually true that the total resources acquired to support a shared memory symmetric processor will be less than
that required to support stand-alone systems, but it is not obvious that
this will be true in all cases, as it is possible that the increase in communi-
cative and connection mechanisms will, in any given configuration, over-
shadow the economic advantages of pooled resources.

27.2. MIGRATION AND GROWTH

The great question, not at all resolved at this time, is how one will grow in
the future. Separate traditional patterns for growth have developed in the
industry because of the differing user environments. One pattern is to
replace a machine with a larger machine. This is called "migration" and
basically provides manufacturers with the motives for compatible (upward)
product lines of the IBM 360, RCA SPECTRA, or UNIVAC 9000 series
type. An alternative has been to get another machine of the same type.

In the future one hopes to see the third pattern—a real and possible
choice to add another processor to a given configuration in the way one
currently adds control units or devices. A current problem with this approach is the grossness of the intervals one must take in going to a multiprocessor system. To go from an 1108 to an 1108II multiprocessor adds
at least an 1108II, a device considerably more expensive than a channel or

a tape unit. Users who find themselves with a need for 10 percent more processor capacity in a system must increase the cost of the system disproportionately in order to achieve it. This provides a strong motive for the development along the lines of very small "processors" sharing high-performance execution capability.

Changes in computer use pattern over the next decade will provide strong additional motives for developing multiprocessor capacities. One would expect this to be particularly true as the population of large data base users increases. Additional stand-alone systems will not provide more effective parallel access to data bases unless the base is replicated for each system, an unacceptable alternative. Increases in computational power will not be particularly relevant in the area, since what is required is the proliferation of paths and accesses to the data base.

A problem in some multiprocessor configurations (particularly IBM MP65) is that it is often channel capacity that the user wishes to increase, not CPU capacity. The only way to increase the channel capacity of some systems is to acquire a processor that has ports for the additional channels that are desired. The development of boxes like UNIVAC's IOC eases this problem for systems where an element of this type can be added independently of another processor.

27.3. DETERMINING PERFORMANCE

The fundamental problem in determining the performance of multiprocessors is the isolation of the factors that contribute to performance. Memory interference, channel contention, device contention, and logical resource locks will all tend to degrade a multiprocessor from its performance goal. The raw goal of a multiprocessor is to reduce the elapsed time of a complex of work, jobs, subtasks, and path from a given elapsed time T to that time divided by the number of processors, T/N. That is to say that the amount of throughput is proportional to the increase in processing power. This number is correctable on a cost performance basis to a proportion of increase in system cost.

In a multiprogram environment there is already a considerable level of contention for resource, and it is difficult to isolate precisely what interferences and consequent delays are uniquely contributed by multiprocessing. Certainly, storage interference increases to a critical level and logical resource contention is introduced in a stronger form, but channel and device contention (for example, additional "surprise" arm movement on a disc device caused by use by processor A after use by processor B) as a multiprocessor problem is not isolatable. A study of IBM 360/65 multiprocessor performance is reported by B. Witt at the ACM National Conference in 1968. The results indicate that the system is a high throughput system as opposed to a single processor with double store, or a partitioned (stand-alone) two-processor system. Especially to be noted in this study is the sensitivity to job mix. Naturally one would conjecture that relatively little performance improvement would result from applying more computational power to a mix of work whose bottleneck is in I/O.

Significant improvements in throughput were observed in a CPU-limited job stream on a two-processor system over a single processor with increased store and over a single processor without it. This increase was significantly greater than the increase of the increased storage processor over the uniprocessor with smaller storage. Since the operating system is essentially identical (an extension of OS MVT), the difference in performance is ascribable to the presence of the additional processor.

The results for I/O-limited streams reflect the fact that the additional processor power is marginal and may not even be usable in the system. In fact, the increase in storage for the single processor results in considerably greater performance impact than the increase of the multiprocessor over the increase storage uniprocessor. The increase in throughput of the multiprocessor over the basic uniprocessor, as opposed to the improvement achieved from increased store, would not appear to justify the acquisition of the second processor.

Even the raw increase in throughput seems to have pleasantly surprised some M65 people. They had earlier made some conjectures on the nature of throughput on a system containing a processor twice as fast as the system containing two identical processors. On the basis of the most favorable contention assumptions for the symmetric multiprocessor, they speculated that the more powerful processor would always outperform the multiprocessor because of the lower probability that sufficient jobs not waiting for I/O completions (ready to go) would exist fully to utilize multiple CPU'S.

Their speculation leads us inferentially to a conclusion that more resources must be available to a multiprocessor if it is to run reasonably well. In an environment of limited resources the problems of the multiprocessor will be severely aggravated, and a single processor will experience less delay. It is difficult to counter objections of this type, since they are basically sound.

We may speculate that there is apparently some tradeoff or growth point where movement to a larger processor is required, especially in machine lines where the addressability of store is limited by processor model, as on the IBM 360. The counter argument must be based upon the assumption that many users will not need the increase in power available from a larger uniprocessor, that indeed many installations face the problem that their I/O populations cannot keep up with their present processor, and that they are seriously I/O-bound. The ability to add more devices and paths to devices is what they require, and they can achieve this by adding a specialized I/O processor or a small additional CPU to increase I/O flow.

The general problem of high throughput on a multiprocessor relates to the availability of resources and the availability of sufficient work.

Very early in the text we made the point that the level of definition of work tasks determines the extent to which parallel activity can be undertaken in a system. There must be a sufficiently large number of independent or significantly independent elements of work to keep parallel components active. In order to use the computational power of 25 processors, there must be at least 25 processes which can run at the same time. These should be large users of CPU power (i.e., computer-oriented). To utilize computational power of a single processor containing the power of 25 smaller units in a 25 processor system does not require independence of elements, since the work can be run sequentially.

Similarly an expansion of resource is inherent in a multiprocessor since it must be possible to allocate sufficient core and devices for the multiplicity of tasks active at the same time. If resources are not broadly available and contention becomes serious, then more and more of the processors will experience delays (periods of nonutilization) and the throughput of the system naturally falls off. The increase in core is critical since it is necessary to have room for the number of work elements required to keep processors busy so that processes will be "ready to run."

A general observation for the effectiveness of a multiprocessor in a throughput-oriented situation vs. a faster uniprocessor seems to be that if a uniprocessor with the required speed and breadth of I/O is available at proper cost, then it is probably at this time preferable to a multiprocessor. The constraints on this are the speed/price relationship in a product line and the flexibility of configuring the uniprocessor with proper amounts of core and channels.

There appear to be some real questions raised about the inherent performance capability of a multiprocessor as opposed to a large uniprocessor, but reasonable confidence that a multiprocessor will match or outperform or out-cost/perform a processor of its own type or two processors of its own type. What implication does this have for the future of multiprocessor development? Will high-performance machines of the future be descendants of IBM's 195 or UNIVAC's 1108, or variations of the minimal processor machines envisioned by Schwartz, Aschenbrenner, and others? The answer to this lies in the rates of development in technology and the direction of growth of user usage pattern. In any case, the use of parallelism as a means of achieving greater speeds seems fundamental.

Chapter 28

PROGRAM PARALLELISM
FOR
MULTISTREAM MACHINES

28.1. REVIEW OF LEVELS OF PARALLELISM

The preceding chapters have described the hardware and software concepts of parallel operations: the phenomenon of having more than one activity in progress over some period of time agreed upon by custom or convention to represent a span in which concurrency may be observed. Parallelism is a phenomenon that is independent of the level of definition of the activity. It may occur within the bounds of a single arithmetic assignment statement in a compiler language, at the level of a formal loop, or at task or at job definition levels.

Initially we have seen the operation of a highly evolved high-performance processor upon multiple instructions or the multiple stages of an instruction in a single stream of instructions. Programs are written as sequential streams, and parallelism is basically introduced into the processing by the hardware mechanisms. In effect, "task" scheduling (an instruction is a task) is performed by the hardware, because enough is known about the variables of task performance to make excellent local predictions about concurrency and duration.

The hardware mapping may be aided by a programmer, who orders his instructions in order to achieve parallel effect, or by a compiler, which reorders local operations in order to maximize parallel operations. In either case, the parallelism is essentially "implicit" in that no specific directives are provided for the programmer or compiler which state that concurrent operation is desired or which describe constraints and synchronizations between concurrent activities.

A further characteristic of parallelism at this level is that the logical effect of operation of the code is the same whether or not the concurrency actually occurs. The function is insensitive to the level of parallel operation actually achieved. This is an important characteristic which should be a goal of parallelism at all strata of task definition and which should be reflected in the hardware or software implementations of all explicit directives for parallelism.

Another form of parallelism at this level, which we have discussed in connection with machine organization and processor intercommunication, is the parallel application of a single instruction to different local data. This is accomplished in SOLOMON and ILLIAC IV class machines. Here also much of the parallelism is implicit; however, there is a need for some capability for a programmer explicitly to define relationship and constraints between the local incarnations of his process.

28.2. A BASIC PARALLEL ELEMENT—DO-LOOPS

One of the difficulties in discussing parallelism between higher-level process structures is due to the variation in the meanings and conceptions of what these structures are. One has a little difficulty in defining task, job, path, and segment so as to gain easy universal acceptance.

The example of a "path" that most readily comes to mind is a single iteration of a "DO" loop in FORTRAN or PL/I and a FOR loop in ALGOL. The final result of parallelism in a loop, where each iteration is considered to be a path, must, or course, be the result of the normal sequential operation of the loop. During operation, however, intermediate values may indeed differ; sequences of manipulation of the data may also differ due to the dynamic interleave of paths running on the hardware.

The parallel loop, of course, implies a multistream machine, since there is no profit, and there may be severe penalty in organizing for parallel operations when there is going to be none.

This penalty will develop partly as a result of the organization of data and data references in order to maximize expected parallel effect and partly as a result of whatever coordinating mechanisms are required if the paths are interactive in a nontrivial way.

The DO-loop, of course, is a conventionalization of a fundamental principle of computer programming, the iterative execution of local regions of code. The conventions of structure in a loop are inherent in all high-level languages, and their existence provides a possibility for the discovery of parallel opportunity in this essentially sequential form. Because of the formalism that surrounds them, DO-loops become susceptible to analysis by a compiler attempting to discover whether or not it is truly necessary to execute each iteration sequentially. This analysis is a natural extension of the highly sophisticated loop-optimization techniques in highly optimizing compilers, which determine on the basis of path analysis whether certain statements can be moved out of loops to reduce the number of times they are executed.

There exist suggested explicit forms for the direction of parallel execution of iterations of loops in parallel processors or highly parallel uniprocessors, but there is an increasing interest in eliminating the need for such directives (even as they are developed) on the contention that the use of parallel techniques for machines will become common only when the programmer is relieved of such directives.

The significant element in the form of a DO-loop is the availability of the loop control statement, which allows for the recognition of the extent of the loop in space (DO 10) and time ($I = m1, m2, m3$). If the m's are integers, the number of executions is known at compiler time; if the m's are variables, the number of executions is still known before the beginning of execution. There are, in the various languages, constraints on redefinition for I or m while execution is progressing within the range of the loop. The availability of the index allows for an analysis of the amount of overlap between the inputs to an iteration and the inputs and outputs of each iteration.

In order to run iterations in parallel, it is necessary that no iteration of the loop be dependent upon a previous iteration and that no successor be dependent upon its completion. It is, therefore, necessary to determine which elements (of the indexed data array referenced by subscripted variables in the loop) are both input and output elements and if they are unique for each iteration, and it is necessary to know the order in which things are referenced.

The result of the analysis is a matter of compiler policy. The compiler may insist upon perfect parallelism and abandon all analysis when it discovers this is not possible, or it may indicate possibilities of parallelism with an indication of what clarifications might be necessary if it is to be achieved. That is, the compiler may or may not undertake attempts to determine limited parallelism.

Since it is the index of the loop which provides for the capability of observing or predicting reference to elements in arrays, the analysis is restricted to those variables whose subscripts are the index. As one would suspect, the language most promising for this kind of analysis is the one that is most restrictive in its loop iteration definition capability, i.e., FORTRAN. In ALGOL, PL/I (and even BASIC) the complexity of the loop control definition considerably extends the problem. Consider Fig. 28.1, the loop that transfers one vector V1 to another vector V2, where both are of equal size. Considering the n iterations of the loop, we see that each iteration will transfer the Ith element of V1 to the Ith element of

(a) FORTRAN
        DO 1, I = 1, N, 1
      1 V2 (I) = V1 (I)

(b) PL/1
        DO I = 1 TO L;
        V2 (I) = V1 (I);
        END;

(c) PL/1
        V2 = V1   (DECLARE V2 (N), V1 (N);)

Fig. 28.1. A loop to transfer a vector.

V2. This loop may be run parallel, because no element's transfer is dependent upon a previous transfer and the order of transfer is irrelevant. Any iteration is commutative with any other, and the loop is completely parallelizable. The set of inputs (V1 (1), V1 (2), V1 (3) . . . V1 (n)) for n iterations contains no duplication, and the set of outputs (V2 (1), V2 (2) . . . V2 (n)) contains no intersection with the set of inputs.

For a compiler to have discovered this, it was necessary for it to form for each iteration a list of its inputs and a list of its outputs and to inspect intersection between the outputs of an iteration and the inputs of a successor.

Notice that the process is dependent upon a finer concept of what represents intersection than the normal intersection of all iterations on the array V1 and V2. We have another example of the scheduling process' being dependent upon a relevant definition of the appropriate level of sharability of a resource, just as we find more concurrency between tasks when we look for record and not data set dependency.

An interesting case is presented when there is an intersection. It may be possible to rearrange so that the intersection disappears. An analysis must be made of the reason for the apparent dependency. Consider the very common situation in which some unsubscripted variable is input to a loop. It is input to each iteration, but it is not modified by any iteration; its value is invariable throughout execution of the loop. Its value is set externally to the loop.

The condition of external definition can be determined by inspecting all the output lists for an occurrence of the variable; when none are found, it can be concluded that it is safe to allow parallel execution of iterations in the same spirit that read only access to an input file may be allowed for programs. The common existence of the variable in the input list is transcended by the fact that no intersection of input and output sets occurs.

It may be, however, that the variable exists in both input and output lists and consequently appears to be a temporary value set by one iteration for a successor iteration and, therefore, to preclude parallelism. This is reminiscent of the implied dependency between instructions we saw in compiling for a 6600-type machine. There were times when an apparent dependency existed, because the programmer chose to use a given arithmetic register as a "sink." The use of the register name implied a relationship of the data in that register with source references to the same register. Sometimes this relationship did not truly exist, and the parallel execution of instructions could be continued by simply using another register, the only apparent dependency being eliminated.

What we wish here is a way of renaming a variable so that it can be privately used by an iteration when the conflict on it is only a location and not a data conflict. How can a location only conflict be determined?

A variable is known to be an output if it appears on the left side of an assignment statement, since this is the only way of setting a variable to a value in an algebraic language (excluding I/O).

A variable is known to be an input if it appears on the right side of an assignment statement, since its reference implies that there is a useful value associated with it.

If it does not appear on the left side, then it is necessary that the value come from outside and that it not have been set during execution of the loop. If it appears on the left and right sides, then it is a value set during the loop and used during the loop.

For those variables that are set before they are used, it is impossible for any iteration to use them unless they are local iteration-dependent values, and consequently any conflict on the name must be a location-only conflict, since it is impossible for any iteration to pass values to another through that variable.

On recognizing location-only conflict the compiler may define the variable to be an array, indexed through the index variable, so that each iteration with input index i will set and use an independent local variable for its internal operations.

If, however, the variable is used before it is set, it appears on the right before it appears on the left; then by direct implication it must be assumed that its initial value comes from another iteration, and its value at the end of this iteration will be passed on to a successor. There is no other way of receiving the value (it could not be set from outside). Renaming is impossible, and a true data dependency exists.

It would be possible at this point for the compiler to generate a "lock" instruction to synchronize references, but this currently appears to be beyond the point at which investigation in this area, fundamentally a result of models based upon directed graphs, has gone.

Another complexity exists, also, with regard to the output of iterations. Consider the case where each iteration is contributing to a common accumulating result. At the conclusion of each iteration, a computed value is added to a common sum. The assignment statement accomplishing this would contain the variable on the left and right sides; this would constitute use prior to set and would, by the above rule, preclude parallel execution of the iterations.

Much useful work, however, might be done in parallel. The elimination of parallelism might not be desired here. What one might visualize in future compilers as one of the optimization options is the possibility of a programmer's indicating whether he desires the abandonment of parallel execution or the generation of synchronizing instructions for situations where the parallelism is not "pure," for example, in this case the generation of a TEST AND SET.

What might be profitable, in the automatic analysis of programs involving the use of graphs, is the addition of some information about the performance of each node so that the compiler might determine the profitability of allowing parallel operation at the cost of synchronizing instructions or procedures.

Two other considerations are involved in the analysis of loops for parallel operations. One is the use of conditional instructions, which might be based upon references to values of a subscripted variable dependent upon a previous iteration. The other is the occurrence of nested loops. The nesting of loops complicates the problem of determining when index values may coincide.

If a nest of Do's is purely parallelizable, then the number of parallel paths is the product of the number of iterations of the outer loop times the number of iterations of the inner loop. It is not necessary, in a nest, that inner Do's be parallel for outer Do's to be parallel.

Within a nest a single index may take on various characteristics. In an outer loop an index may take on any value in its range; in an inner loop it may be constant, or it may take on only successive values. Commonly, an inner loop will accept a constant value for a subscript for its computation and vary the subscript of its index. The subscript using the index variable may never have the same value over two iterations (in FORTRAN; the statement is not necessarily true when the subscript value may be a dynamic value of a complicated expression). The relationships between the input and output sets of nested loops become considerably involved, since the action of inner loops will affect the appreciation of the status of variables for the outer loops.

The classical example of a nonparallelizable loop is that in Fig. 28.2, of course, where each iteration is particularly dependent upon the execution of an earlier iteration for its input. If the iterations were allowed to operate in parallel without synchronization, it would be impossible to predict exactly what value any incarnation of the loop would pick up.

If the iterations were operated with some synchronization, the cost of

```
DO      DO I = 1 TO 10;      (DECLARE V2 (0:10), V1 (0:10);)
        V2 (I) = V2 (I-1) * B;
        END;
```

Fig. 28.2. A fundamentally nonparallel loop.

such synchronization would surely outweigh the benefit of parallel operation and would, in effect, only result in a longer-running sequential process.

There are certain algorithms and processes that are profoundly sequential in nature and which cannot be turned into parallel process without a basic reconception of the process and its problem.

The observation is often made that the development of generally
parallel systems awaits the development of parallel algorithms. A vast body
of solutions and techniques depends upon convergences and approximations derived from applying a function to a value developed as a result of a
previous application of the function.

28.3. PATHS/TASKS—EXPRESSING PROCESS RELATIONSHIPS

Other than loops, the concept of a path as distinct from a subtask often
derives from the formal eccentricities of a specific machine and system
and from the concepts of binding and modularity. Conway's FORK is
basically a spawner of paths, since it assumes that the executing programmer knows where the routine to be executed is. Much of the work in
evoking parallelism while developing a single program concerns itself with
pathing.

It is often entirely permissible to use path and task interchangeably;
we distinguish them in order to associate a host of considerations relevant
to the cooperative use and sharing of resources with parallelism at the task
level. We mean to imply that there is some possibility of differentiating
the resources granted to a subtask, that it may work with only a partial set
of the capabilities of the task or even perhaps with an extended set of
capabilities. The task is a named and structured process, which may itself
be registered with the operating system. It is at this level that the characteristics and extent of the language supports for expressing parallelism are
most complete.

The language of multiprocessing includes any expressive feature that
explicitly describes a relationship between processes. In earlier chapters
we have mentioned ATTACH, FORK, LOCK, WAIT, and JOIN. These are
expressions of an event having to do with the development and synchronization of processes and are relevant to multiprogramming or multiprocessing.

The essential identity between true and apparent simultaneity is especially obvious when one discusses what language features are potentially needed. Many of the macro services provided in or suggested for multiprogram operating systems are representations of parallel process functions. For example, Dennis suggests a series of metainstructions for resource coordination in a multiprogram environment which cover resource relationships between mother and daughter tasks and which should form an element of any multiprocess-oriented expressive technique.

They basically provide capabilities for the manufacture of a proper stateword for a spawned task. This stateword is a stateword in the expanded sense; it represents the status and capability of the task. For example, the GRANT macro specifies that the spawned task may have certain rights of access or ownership to certain portions of the mother task's capabilities and, depending on the liberality of the grant, may execute, execute and write, or may or may not destroy without permission of the mother task.

The remainder of this chapter will discuss some of the suggestions that have been made for the extension of current language to support a parallel capability and some of the programming considerations that may be involved in explicitly programming a multiprocessor.

The use of explicit language means that the programmer is in control. If he wants a DO-loop executed in parallel, he will say so, and not rely upon the compiler to accomplish an automatic analysis.

It is necessary to point out, however, that in many cases the possibilities for discovery of implicit parallelism are considerably more powerful than we have so far suggested. The parallelism may be implied by the nature of the definition of data or operation in the context of the program. For example, in PL/I the expression A=B*C, where A, B, and C are declared to be arrays, implies the do loop of $A(I) = B(I) * C(I)$. Once this loop is formed, the compiler may or may not go on to inspect for parallelism, depending upon the resource environment.

In a language like APL (based upon Iverson's published notations in *A Programming Language*) the matrix operators imply opportunities for parallel execution. The combination of the attributes of data and the operations relating to them almost approaches explicit definitions of parallel operation.

28.4. LANGUAGE EXTENSIONS AND PROGRAMMING CONSIDERATIONS

There are a number of suggested extensions to the ALGOL language to support the explicit expression of parallelism. These include ALGOL versions of FORK and JOIN and various forms of LOCK and UNLOCK, parallel FOR and the simple parallel connective AND.

Two specific discussions, those by Anderson and Dennis, will be used here, because the examples that are given with them serve to illustrate a number of programming approaches to a parallel algorithm.

In Anderson's FORK all labels of code to follow the FORK are explicitly represented in the FORK statement. There is no assumption of fall through to a statement sequentially following the FORK (as in Conway's hardware version). The FORK is restricted to two named entry points (a label pair defined metalinguistically in BNF as a label followed by a label). The corresponding JOIN may name more members on a list of labels, evidently anticipating that there will be a need for joining more than two paths.

There is a TERMINATE statement that acts like the execution of JOIN when not all paths collected at the join point are complete. The lock and unlock functions are represented by OBTAIN and RELEASE, each of which use as "operands" a new ALGOL structure defined to be a variable list and which represents the variable names that are being locked and unlocked by the statement.

To illustrate the formation of ALGOL program using FORK and JOIN as presented, a vector multiplication is shown. The coding is reproduced in Fig. 28.3. The code reveals an approach to the development of a parallel algorithm and the placement of certain functions. By the syntactic nature of FORK, the path that executes the spawn suspends during the operation of its daughter paths and is resumed at "continue." The necessity for the "go to continue" in First and Last is due to the necessity for the JOIN to be at the same block level as the FORK. The activity of the

```
FORK        first, last
First:      begin S1: = 0;
            for i: = 1 STEP 1 UNTIL N/2 DO
            S1 = S1 + (A(i) x B(i));
            GO TO CONTINUE;
            end;

Last:       begin
            S2: = 0;
            for j: = N/2 + 1 STEP 1 UNTIL N/2 DO
            S2: = S2 + (A(j) x B(j));
            GO TO CONTINUE;
            end;

Continue:   JOIN first, last
            S = S1 + S2
```

Fig. 28.3. Anderson's Vector product. (Reprinted by permission of J.P. Anderson, "Program Structures for Parallel Processing," *CACM*, December, 1965.)

mother path is, in addition to spawning, to set a variable to hold the final result to initial zero.

Notice that each path determines what partition of the vectors A(1) and B(1) will be used by it and essentially controls its own extent of operation. Each path will do one-half of the multiplication and will form the sum of products in a local variable. After reaching the JOIN point, each local variable will be added to form the final sum.

If we remember the compiler technique of renaming on order to avoid conflict, we see an example of that here. The use of the transient variables S1 and S2 eliminates a need to lock on a reference to a common S used by both paths. These two paths have no interaction and will run without interference (logical interference) until completion.

The efficient execution of the algorithm requires two processors; it cannot use more; it can run on one. There is no suggestion of any dependency or preferability for one path to run on any given processor or any awareness of processor identification. The cost to the processing system if only one processor is available is the cost of execution of the fork and join, the final S addition, the duplicate development of indices, and whatever cost is associated with the system for the invocation of an ALGOL block.

An alternative form for coding a product of this kind in ALGOL extension is presented by Dennis. Although not a specific recommended extension to ALGOL, the article introduces a number of concepts for controlling multiple processes in an ALGOL notation.

The coding for the product in part is shown as Fig. 28.4. The capabilities that differ from Anderson are the instances of PRIVATE and QUIT. Notice that, like the previous multiply, the mother task exists solely for the purpose of spawning paths. The PRIVATE declaration is in support of the passing of stateword from the mother to the daughters. The declaration of private for a variable indicates that that variable should exist in the stateword of the declaring block and that on execution of a FORK is passed on. QUIT is basically TERMINATE, except that it has the additional function of releasing the stateword.

The development of the definition of what each path should do is placed in the mother task. The FOR statement develops a stateword for each path on each iteration, and in this case it passes on consecutively higher indices to the arrays so that there will be as many paths spawned as there are elements in A and B. Each path will multiply one element by another, placing the value in a private X. Since each e will "execute" its own declaration, each will, of course, have a private copy of X.

Some number of multiplications may get under way before the mother task QUITS, and some may even complete before she is finished generating FORK requests. Each path places its product into a common "pseudo-algol"

```
begin       Real Array A [1:n], B[1:n];
            Boolean w Real s Integer t;
            Private Integer i;
            t: = n;
            FOR i: = 1 Step 1 Until n DO
            FORK e
            quit
            e: begin PRIVATE Real x
            x: = A(i) * B(i);
            Lock w;
            S: = S + x;
            Unlock w
            JOIN T,r;
            quit
            end
      r:
end
```

Fig. 28.4. Dennis's example of FORK, JOIN, QUIT. (Reprinted by permission of J.B. Dennis and E.C. Van Horn, "Programming Semantics for Multiprogrammed Computations," *CACM,*|March, 1966.)

sum S. Because this sum (declared in the outer block) is the input and output of each path, it is necessary to LOCK it while it is being updated and un-LOCK it afterwards.

28.5. NUMBER OF PROCESSORS

We might look at some general principles of parallel programming. Not all these principles are universally accepted, but in collection they appear to represent a consensus in the field.

The first of these is the question of the dependency upon the number of processors. In general, a parallel algorithm should be independent of the number of processors in the system. It should be capable of running on a single processor, if necessary, or on any number of processors. It may well be that the programmer anticipates what an optimal number of processors may be in terms of his representation of the algorithm, and he may also feel that there is a minimum number and a maximum number.

In the language we have seen so far no means for expressing these preferences exists. Such indication might be included in the job control language in much the same spirit that other allocation requests are made. In OS/360 each STEP might have on its EXEC card an indication of the desired and/or required CPU's. It is also desirable to have such a capability locally in the programming language. A verb for FORTRAN called RE-QUEST, where a request can be made for a number of CPU's for a sequence beginning at a given statement number, has been suggested.

It is interesting to consider why a programmer might indicate a maximum. There is a possibility that the addition of more processors in the execution of an algorithm that is programmed to run on an indefinite number may actually extend the running time of the algorithm. There are a number of reasons for this. The cost of starting up the path may become relatively large, compared to the amount of work an additional processor will perform. The physical interference between processors for access to input data or output locations may slow the execution of processors to a point where fewer processors doing more but operating more efficiently might finish the job in less time.

The phenomena of redundance and indeterminancy may contribute to a slowing of solution. In some algorithms it is possible for a programmer to be careless about the distribution of matrices of input data and not to predict what processor will process what specific partition of the data. This is particularly true if the program may run on a range of processor allocations.

Redundancy is a legitimate technique of parallelism. In the manipulation of matrices it might involve an overlap in the area of a matrix where any processor may pick up its input. This redundancy might be avoided by LOCKING, but the programmer may decide that the certainty of delay due to locking is less desirable than allowing processors to duplicate some operations.

The probability of redundancy may be low up to a certain number of processors, but beyond that number the amount of time a processor is doing useless work (effectively idling) becomes high enough to slow the total computation.

A simple example of this phenomenon is when there is a test in the algorithm that determines whether or not a value has fallen within a specified range. If the number of processors is large, the amount of work each is doing becomes small, and the probability that they will be working in the same locality of procedure becomes high. This increases interference in core but also adds to the possibility that at consecutive points in time processors may test the same value for the limits test.

Redundancy may also involve the computation of values in parallel, when it is known that only one may be used. Indeterminancy in an algorithm occurs when the exact progression through the process is dependent upon the specific performance of a family of processors.

28.6. DYNAMIC ACQUISITION AND RELEASE

A problem arises in the design of a system as to exactly the meaning and use of the concept of minimum processors and the concept of independence of the number of available processors.

The concept of priority becomes involved here, and this necessarily invokes an image of the intent of the system. In many (perhaps most, for some time to come) multiprocessor situations, the computational power is acquired because a particularly large job, perhaps supporting the fundamental mission of the organization, is suitable for running in a parallel form.

This job, when it is on effectively, can saturate the machine, and the number of processors acquired (no one buys N processors, but a given number of processors) is a reflection of the appreciation of the processor usage capability of this single large problem, or a small group of large problems of the same magnitude.

It may or may not be realistic to consider that this problem, or any problem, can be effectively programmed for N processors, but that inherent in the design of the algorithm is the assumption of 5, 10, or 12 processors. In order to run, so many processors must be available. It is not unreasonable to expect that some processes cannot be processor-independent. Consider how few processes are independent of the number of tape drives or disc space.

It should be usually possible to simulate the presence of the specified number of processors, but this is not at all what the user has in mind for his job. Basically, he is not willing to tolerate the dynamic environment of FORK and JOIN, where processors become spontaneously available and the parallel effect is effectively one of staggering, dependent upon the rate of availability.

Further, even if the problem acquires a processor for each FORK, it is possible that it may not hold on to them, since the possibility of preemption might cause certain paths to requeue.

The use of priority alone will certainly build a tendency for this problem to acquire processors at a fast rate, and the use of priority preemption might allow him to take on the processors he requires. But he may have a process with a number of FORK and JOIN points, so that the number of processors in use is constantly varying, and he may face constant releasing and reacquiring of processors, which he does not wish to occur. Further, the reduction in the number of processors from his desired number to any less will certainly cause a disproportionate extension of his completion time.

As an added consideration, the overhead of using the scheduling mechanism for any path control appears to be very high, and the possibility of avoiding the scheduling mechanism of the system is almost a requirement for profitable cooperation. We have mentioned the "cost of parallelism" and "true parallel effect" earlier. The definition of small parallel elements is highly unprofitable if the burden of starting them is high. What characterizes all of these considerations is that the efficient performance of a given parallel process is more important than balanced effective utilization of the system.

We may look at Dennis's algorithm in this light. If there is a priority associated with this task, he will be subject to delay in acquiring processors, and he will be subject to loss of processors to higher-priority jobs. It is impossible to predict how many processors will work on the job, how many different ones, how many together, for how long.

If this uncertainty of processor allocation is unacceptable, two things may be done. First is simply to raise the priority; this will have the double effect of protecting the task from preemption and gaining processors for it. But this still leaves a considerable potential for stagger and delay. The other possibility is to separate the concept of number of processors from priority. This may be done so that priority represents the right of this program to call upon the system for any resource it desires and to preempt processors from lower-priority jobs. In order to sustain the number of acquired processors, each time a processor releases or an attempt at preemption is made, a check is undertaken to see whether the release of the processor will cause the algorithm to fall below its minimum number. If the release of the processor would do so, then it is not allowed to release, even if that means that it would do nothing but idle until the last JOIN was executed. If preemption occurred, the job would lose all its processors and be rolled (perhaps) out of core.

The REQUEST function is a form, then, of a static fork, which collects processors into a partition, preempting lower-priority processors when necessary. The "gang" of processors is then turned over to the job, which may then do its own scheduling, spawning, and collecting of tasks independent of the general scheduling mechanism, since there can be no contention from other tasks in the processor partition initially allocated.

This view of a multiprocessor is somewhat different from the very fluid and dynamic multiprocessor system we have been implying in order effectively to utilize equipment. The concept of FORK and JOIN as generally described is not adequate here, because they do not, as such, imply preemption and will not "collect" processors.

Dennis's example is much more sensitive to the number of processes than Anderson's because of its broad processor spread. This relates to a second conjectural principle of parallel programming.

There should be a reasonable bound on the performance of the parallel algorithm if less than its desired number of processors are available, and if only one is available, the cost of running on one should be within some reasonable approximation of the time of sequential processing. Dennis is considerably more vulnerable; on the other hand, if he gets his processors (not necessarily N, since some number less than N may be the balance between interference and efficiency), he has a potential for much quicker operation.

28.7. REDUCING QUEUE LENGTHS

Earlier we discussed the problem of queue overflow and a technique for reducing entry onto the queue. This technique guarantees staggered parallelism, and the programming to support it is very much oriented to an unpredictable number of processors available. In place of a mother task spawning daughters, we have essentially generations of tasks.

At an initializing phase a mother sets up counts of requests for processors, processors granted, and paths completed. The mother then FORKS a request to the CPU queue. It then continues to operate on the body of the computation. At the end of the computation it tests processor, request, and completion counts to determine whether it should release or continue with another iteration. If the requested second processor was never granted, then the first processor processes all paths; if the requested processor was granted, the path acquiring control of the second processor checks on activation as to whether it should request a third; the third, if activated, checks on whether it should request a fourth, etc.

The claim made is that at any time only one request for a path is on the queue. Consider, however, a system in which there is dynamic preemption. At any time the processor acquired may be lost and, if it is lost, each path must requeue for service. In the presence of a preemption, therefore, the claim made for the method cannot be sustained, and there is no general control on the size of the queue.

28.8. OTHER LANGUAGE FORMS

Extensions to ALGOL of the type reviewed here are opposed by Wirth. He feels strongly about the purity of ALGOL and contends that adding verbs of this type runs counter to the underlying structure of the language and the basic motives for block structure, which are to reduce the use of labels and jumps of any kind to these labels.

What he prefers is the replacement of the semicolon, which terminates ALGOL statements, with an AND whenever parallel execution is required. The AND mechanism is certainly more general, since it is evident that, unlike FORK and JOIN, AND can express parallelism at any level, whereas FORK and JOIN are seen to be restricted to no structure smaller than a compound statement. Wirth further recommends an avoidance of

LOCK and UNLOCK by the definition of CRITICAL paths in a procedure definition. The critical block is that part of the procedure which must be entered only by one processor at a time.

If the AND appears to be more general and powerful than other suggestions, however, the shared procedure would seem to be less so. It is certainly valid to restrict access to variables that are shared by prohibiting access to the code that accesses them, and, indeed, LOCK can be made to mean this. However, if one wishes to generalize sharing, one must accommodate the possibility that the access to the shared resource may, indeed, be through independent sections of coding and that only a lock on the variable or task itself can enforce sequentialization. It may be that all calls on a resource are not able to be embedded in a single outer block declared a shared procedure. This "outer block" may be the operating system or the I/O control package not programmer specified.

Another language extension that occurs in the literature is the parallel FORTRAN DOTOGETHER, suggested by Opler. This is a way of reducing a process into subprocesses executable in parallel. The DOTOGETHER statement names a list of statement numbers at each of which there is a statement to be executed. The last number in the DOTOGETHER list is a join point, which has a verb HOLD. The label of the hold is always in parentheses. The number of elements in the DOTOGETHER imply the number of processors to be involved.

The usage of the form is general and also usable as a parallel loop expression, since any process can be described as beginning at a specified label. The example of its usage in Opler's presentation, however, is as a cover for a decomposition of a matrix multiplication involving the DOTOGETHER reference of nested DO-loops, each nest of which is responsible for iterating through a portion of the multiplies to be performed.

A parallel FOR specification for ALGOL involving the use of AND has been made for SHARE ALGOL. In the suggested form the parallel FOR would appear as FOR condition AND condition. Two forms of parallel DO are suggested in the NYU report (see Sources and Further Readings, *Programming Considerations for Parallel Processing*). One form of DOP specifies two labels, a label designating a statement number for all CPU's where the initial index developed is less than the limit (i.e., $m_1 + m_3$ is less than m_2) and a label for the processors where the initial index is greater than the limit.

PL/I has been extended so that parallel tasking is possible in the language, and this has direct applicability to multiprocessing. The EVENT statement of the language may now be associated with a CALL statement, which can, in general form, be expanded to a call for the asynchronous operation of the called task. The task option merely names a subtask, and the only use of the name is to allow priority control. The EVENT option establishes the event name and associates it with the completion of the subtask. WAIT's can then use this name as a condition for proceeding.

PART FIVE:
SOURCES AND FURTHER READINGS

Ackerman, W.B., and Plummer, W.W., "An Implementation of a Multiprocessor Computer System," *ACM Symposium on Operating System Principles* (October, 1967).

Anderson, J.P., "Program Structures For Parallel Processing," *Communications of ACM* (December, 1965). (See also *CACM*, April, May 1966, Letters to Editor for additional comments and responses.)

Aschenbrenner, R.A., Flynn, M., and Robinson, G.A., "Intrinsic Multiprocessing," *AFIPS Proceedings,* Spring Joint Computer Conference (1967).

Ball, J.R., and others, "Use of Solomon Parallel Processing Computer," *AFIPS Proceedings,* Fall Joint Computer Conference (1962).

Barnes, G.H., Brown, R.M., Kato, M., Kuck, D.J., Slotnick, D.L., and Stokes, R.A., "Illiac IV Computer," *IEEE Transactions on Computers,* Vol. C-17, No. 8 (August, 1968).

Bauer, W.F., "Why Multi-Computers?" *Datamation* (September, 1962).

Bernstein, A.J., Analysis of Programs for Parallel Programming," *IEEE,* Vol. 15, No. 5 (October, 1966).

Bingham, H.W., Fisher, D.A., and Seward, J.W., "Detection of Implicit Computational Parallelism From Input/Output Sets," Clearinghouse for Federal Scientific and Technical Information AD 645438 (December, 1966).

Bingham, H.W., Fisher, D.A., and Seward, J.W., "Detection of Essential Ordering Implicit in Compiler Language Programs," Clearinghouse for Federal Scientific and Technical Information AD 650845 (February, 1967).

Bingham, H.W., Fisher, D.A., and Seward, J.W., "Plan for Detection of Parallelism In Computer Programs," Clearinghouse AD 655867 (June, 1967).

Bingham, H.W., et al., "Parallelism In Computer Programs and In Machines," Clearinghouse for Federal Scientific and Technical Information AD 667907 (April, 1968).

Blaauw, G.A., "The Structure of System/360. Part V: Multisystem Organization," *IBM Systems Journal,* Vol. 3, No. 2 (1964).

Chang, W., Paternot, Y.J., and Ray, J.A., "Throughput Analysis of Computer System. Multiprogramming vs. Multiprocessing," *IBM Simulation Symposium* (1969).

Constantine, L., "Control of Sequence and Parallelism In Modular Programs," *AFIPS Proceedings,* Spring Joint Computer Conference (1968).

Conway, M.E., "A Multiprocessor System Design," *Proceedings AFIPS,* 24 (1963).

Curtin, W.A., "Multiple Computer Systems," *Advances in Computers,* New York, Academic Press (1963).

Dijkstra, E.W., "Solution of a Problem in Concurrent Programming," *Communications of ACM* (September, 1965).

Draughton, E., Grishman, R., Schwartz, J., and Stein, A., "Programming Considerations for Parallel Computers," *AFIPS Proceedings,* Spring Joint Computer Conference (1968). Also Courant Institute of Mathematical Sciences IMM 362, November 1967.

Gonzalez, M.J., and Ramamoorthay, C.V., "A Survey of Techniques for Recognizing Parallel Processable Stream," *AFIPS,* 35 Fall Joint Computer Conference (1969).

Gosden, J.A., "Explicit Parallel Processing Description and Control in Programs for Multi- and Uni-Processor Computers," *AFIPS Proceedings,* Fall Joint Computer Conference (1966).

Goutanis, R.J., and Viss, N.L., "A Method of Processor Selection for Interrupt Handling In a Multiprocessor System," *Proceedings of the IEEE,* Vol. 34, No. 12 (December, 1966).

Keeley, J.F., and others, "An Application Oriented Multiprocessing System, I: Introduction, II: Design Characteristics of the 9020 System, III: Control Program Features," *IBM Systems Journal,* Vol. 6, No. 2 (1967).

Kuck, D.J., "ILLIAC IV Software and Application Programming," *IEEE Transactions on Computers,* Vol. C-17, No. 8 (August, 1968).

Lehman, M., "A Survey of Problems and Preliminary Results Concerning Parallel Processing and Parallel Processors," *Proceedings of the IEEE,* Vol. 34 (December, 1966).

Morenoff, E., and McLean, J.B., "Inter-Program Communications, Program String Structures and Buffer Files," *AFIPS Proceedings,* Spring Joint Computer Conference (1967).

Morenoff, E., "Job Linkages and Program Strings," Clearinghouse for Federal Scientific and Technical Information AD 651513 (April, 1966).

Mosier, R.A., "Multiprocessor Compendium," Clearinghouse for Federal Scientific and Technical Information AD 675937 (June, 1968).

Murtha, J.C., "Highly Parallel Information Processing Systems," *Advances In Computers* (1966) Academic Press.

Opler, A., "Procedure Oriented Language Statements to Facilitate Parallel Processing," *Communications of ACM* (May, 1965).

Perkins, R., and Mc Gee, W.C., "Programmed Control of Multi-Computer Systems," *Proceedings IFIPS* (1962).

Pariser, J.J., "Multiprocessing With Floating Executive Control," *IEEE International Convention Record* (1965).

Pomerene, J.H., "An Approach to Parallel Processing," *Proceedings IFIP Congress,* Vol. 2 (1965).

Schwartz, J., "Large Parallel Computers," *Journal of ACM* (January, 1966).

Slotnick, D.L., "Unconventional Systems," *AFIPS Proceedings,* Spring Joint Computer Conference (1967).

Slotnick, D.L., Bork, W.C., and McReynolds, R.C., "SOLOMON," *AFIPS Proceedings,* Fall Joint Computer Conference (1962).

Stanga, D.C., "UNIVAC 1108 Multiprocessor," *AFIPS Proceedings,* Spring Joint Computer Conference (1967).

Tesler, J.G., and Enea, H.J., "A Language Design for Concurrent Processes," *AFIPS Proceedings,* Spring Joint Computer Conference (1968).

West, G.P., "Best Approach To A Large Computing Capability," *AFIPS Proceedings,* Spring Joint Computer Conference (1967).

Witt, B.I., "IBM 360/65 Multiprocessor: An Experiment in OS/360 Multiprocessing," *Proceedings ACM National Conference* (1968).

Yarbrough, L.D., "Some Thoughts on Parallel Processing," *Communications of ACM* (October, 1960).

6

Multiprocess Scheduling (Multiprogramming)

Chapter 29

INTRODUCTION
TO THE
FUNCTION

29.1. ENVIRONMENT OF SCHEDULING ACTIVITY

The function of scheduling involves the determination of a set of tasks to be accomplished, the formulation of a sequence in which to perform them, and the allocation of resources required for their performance. A schedule represents a time mapping of the association of tasks and resources that reflects a strategic objective of the scheduler.

This strategic objective may be the guarantee of a task completion for a given task (or set of tasks) by a certain time, or, at the other extreme, the performance of the greatest amount of resource work regardless of the distribution of that work across named tasks or the impact of this on the rate of task completion. Between these extremes there is a continuum of strategies that stress service or utilization as a goal.

The formation of a schedule is primarily dependent upon the characteristics of the population of tasks to be scheduled. The critical population characteristics that affect a schedule are

1. The extent to which all tasks to be performed may be enumerated at some point prior to the formation of the schedule.
2. The extent to which the relative urgency and time requirements of each task can be known.
3. The extent to which the processing of tasks, if cyclic, involves identical resources in identical amounts for each processing event, or the extent to which resource requirements can be predicted.
4. The extent to which all relationships between tasks are known.

These population criteria affect scheduling in a number of ways. They affect the duration of a schedule, the completeness of a schedule, and the level at which scheduling is undertaken.

If population characteristics are perfectly known for the lifetime of the system, then a single schedule may be prepared which describes the relative sequence of jobs constrained by relative required completion times. This schedule may be constructed at the time the system is conceived, and indeed the determination of specific system resources may perfectly reflect the proposed schedule. The resource pool comes into being as a result of and to satisfy the known task schedule.

In computer scheduling perfect matching of work and capability is rarely perfectly achieved or approached. It is best approximated by "flow-shop" production organizations, where the computer system has been designed to support one or more production applications whose successful operation has justified the acquisition of the computer and perhaps the development of a specialized operating system. Among these are the real-time applications, which are tailored quite specifically for a given service to a given class of users.

The real-time dedicated system has many of the characteristics of a perfect scheduling environment, and schedule policy is characteristically determined at a high design level. Many imperfections of knowledge exist, however, and many variables are only statistically determined. For example, it is impossible to predict precisely the rate and frequency of operating certain modules, or the exact sequence of all modules operative in response to a given transaction.

In some processing systems, particularly the UNIVAC 494 and 1108, the concept of service to a population of "standard production runs" is embodied in the operating system. In such an environment certain jobs are registered with the system as "SPR's," and the system running on a clock initiates jobs at their proper initiation time, as recorded in the system.

Most systems are combinations of flow shop and job shop, with a given proportion of known and predictable jobs combined with the need to respond to a population of jobs that are generated during the lifetime of the system and whose arrival rate, resource demands, and urgency requirements are unknown.

It is because of these jobs that scheduling functions are partially transferred to an operational level, and "low-level" or short-range tactical decisions are passed from the policy-making level to the operating level. Great care must be taken in distributing the scheduling function. In many current systems certain ad hoc scheduling decisions are made by operations personnel who are not truly qualified to make them. The dynamics of "computer room" operation have, despite the stated goal of simplification by operating system, actually become more complex, and decisions by console operators about initiation, termination, and depth of multiprogramming constitute an intermediate level of scheduling and should be recognized as such.

29.2. AN EXERCISE IN ECONOMICS

Scheduling is essentially an exercise in economics. The formation of scheduling policy initially involves a determination of the economic utility of known tasks and/or a mechanism for describing this economic utility to some tactical agency. In basic form this economic utility is expressed as a "priority"; in more sophisticated form it is expressed as some point in time where the information produced by the task has economic value.

The concept of "external priority" is an imperfect reflection of the concept of economic utility. The conventional service bureau pricing policy, which associates a price augmentation with priority, is a reflection of the concept of economic utility. By acquiring "priority" one is implicitly saying to a service bureau that one is willing to pay an additional sum of money to defray the cost of potential misutilization of equipment caused by selecting and servicing a given job at a given service level. By assigning priority in one's own shop, one is stating that the economic utility of having this information before other information or by a certain time is larger than the loss due to potential equipment misuse.

Various kinds of information have various degrees of utility at various points in time. One can visualize a situation where the utility of the output of program 1 is highly valuable at 1 p.m., but of no use whatever at 1:01. The cost of delay of the completion of the program is its total value.

We may find a number of various utility value curves associated with a population of tasks. The ideal for multiprogram scheduling is a population whose value is equal and holds steady, since this then allows for schedule preparation and execution to be dependent solely on resource utilization bases.

A fundamental scheduling problem is that most frequently the economic utility of a task is only informally appreciated, and service priorities are accorded on an at best intuitive, and at worst political, basis. The classic struggle between production groups and development groups for computer time is an excellent example. Development people claim that the utility of quick development lies in the earlier availability of an application of high economic value; production people point to schedules of completion requirements based upon assumed information utility.

It is beyond the state of the computer or management art at this time to specify the dollar value of a piece of information to an organization at any time, but an appreciation of this must lie at the very root of the

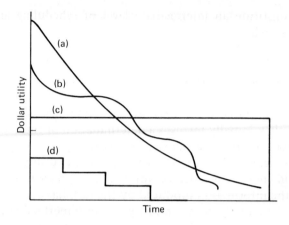

Various delay curves:

(a) The utility of this information decays continuously
    over time

(d) (b) The utility of this information decays continuously
    but holds within a series of defined ranges

(c) The utility of this information is insensitive to
    time (within some reasonable scheduling period
    or series of scheduling periods)

Fig. 29.1. The value of information.

process that establishes the "preconditions" of the scheduling process, i.e., the determination of when things must be completed. A proper schedule minimizes the cost of delay of information so that the utility of information made available to the organization is maximized.

The length of a schedule, the scheduling period, is sensitive to the relative time that tasks take upon resources, as well, as mentioned above, to the predictability of the set of tasks to be performed. The scheduler must know something about the rate of performance of his resources and the relative rates of resources. This is not absolutely essentially in terms of precise times, since schedules may be prepared in abstract time units, but it is necessary for a scheduler to have some initial idea of time frames, since scheduling over a shift, a week, or a month, is inseparable from a knowledge of stated deadline requirements.

29.3. FINENESS OF SCHEDULE

One final introductory point about scheduling is relevant to our earlier comments about the various levels of definition and perception of tasks and time. The more precisely we understand our resources and tasks, the better schedule we can make.

In early computer systems, we tended to be quite general about our appreciation of what we were scheduling. We allocated the entire computing system to a user and tended to schedule blocks of time in hours. This was partially because of the slower basic rates of our resources, but primarily because we did not distinguish between the various asynchronous parts of our machines and recognize them as being independently schedulable. (Often they were not.) Further, we scheduled jobs in bulk and did not perceive independently schedulable steps as finely as we have come to do.

One example of this growth in appreciation is in IBM's OS/360, where many of the scheduling and allocation options initially available only at the JOB level are being brought down to the STEP level. (A JOB is a collection of program "STEPS" and is primarily an accounting concept. The STEP is the actual processing unit.) The user may now associate priorities and times at STEP level. A current tendency to schedule in finer units means a great deal more flexibility in schedule formation, as well as an attendant complexity of systems.

29.4. BASIC PROCEDURE REVIEWED

Let us imagine that we wish to schedule one shift of a computer. We have a population of three jobs, each of which must be completed on the following day. Each job has a number of programs. We know the resource requirements for each program and the length of time it will require on the system. In addition, we know that there are time constraints on the completion of certain programs. A feasible schedule must accomplish the time demands while preserving the ordering between programs expressed in Fig. 29.2. It is manifestly possible to complete the work in a single shift, and manifestly impossible to achieve the stated time constraints if each job is looked upon as an inseparable unit of work.

Our first task is, then, to separate out the separately schedulable units and identify the precedence constraints that must be respected. The precedence relationships between programs may be represented as a graph, where each node represents a program and each line a path. We have seen this before in Part 1. We introduce it here again to show how exactly the same considerations apply in scheduling jobs as in scheduling instructions.

The complete graph represents a partially ordered set of programs. The set is partially ordered in that there is not a precedence relationship between each node. Each node is either a successor, a predecessor, or neither to other nodes. In our simple example 1A is a predecessor of 1B,

	Time	Precedence	Due At
Job 1			
Program 1A	30 MIN		
Program 1B	1 HR	1A	
Program 1C	2 HR	1B	
Program 1D	30 MIN		
	4 HR		
Job 2			
Program 2A	15 MIN		
Program 2B	30 MIN	A	10
Program 2C	15 MIN		3
	1 HR		
Job 3			
Program 3A	1 HR		11
Program 3B	1 HR		1
	2 HR		
Job 4			
Program 4A	30 MIN		5
	30 MIN		
Total	7 HR 3 MIN		

Fig. 29.2. A population of tasks whose attributes are known.

1B of 1C, etc. The graph in Fig. 29.3 may also be represented as a matrix called the "precedence matrix." This matrix, when read across, reveals for each program its immediate successors and when read down, reveals for each program its necessary immediate predecessors. All those columns that contain no indications of a predecessor identify the programs associated with those columns as being "unbounded," that is, available for execution immediately.

We see that the precedence relationship tends to restrict the choices that the scheduler has. Care should be taken in the establishment of precedence relationships for this reason, particularly in a multiprogramming environment, where systems performance in part depends on the broadest availability of programs capable of being selected for execution. As we have seen in instruction streams, it is possible to imply precedence where it truly does not exist. STEPS in OS/360 JCL have a false precedence relation.

On the precedence table we have shown the cumulative time for each predecessor in the column of its immediate successor (as the delay row). From this we can compute the minimum processing time of any program, calculated as its own processing time plus the delay caused by the necessity of running its predecessor(s). It is not necessary to represent the time values in the table in this fashion, nor is it truly necessary to introduce the concept of a table. Our motive in showing the pseudo matrix form is merely to indicate that there is a tabular form of showing these relationships that is suitable for internal representation in a computer and conse-

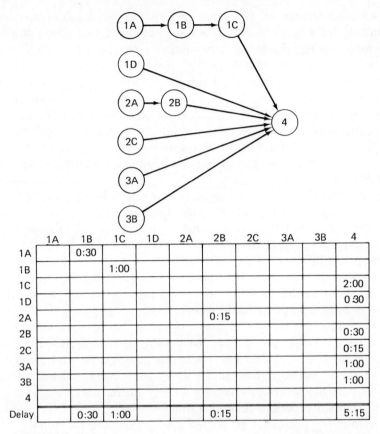

	1A	1B	1C	1D	2A	2B	2C	3A	3B	4
1A		0:30								
1B			1:00							
1C										2:00
1D										0 30
2A						0:15				
2B										0:30
2C										0:15
3A										1:00
3B										1:00
4										
Delay		0:30	1:00			0:15				5:15

Fig. 29.3. Graph and matrix of Fig. 29.2. Bounded and unbounded tasks.

quently to imply that various machine-scheduling schemes are possible and that these characteristically depend on the movement of the pro- grammed algorithm along the rows and columns of the matrix form. Schemes for determining schedules and for controlling resource alloca- tions are largely based on simple machinable matrix inspections.

From our figures we see that there still are a large number of accept- able schedules from which to choose, that is, a number of schedules that preserve the ordering relationships. The exact number depends upon cer- tain physical details. For example, it is possible to take the position that the scheduling of a predecessor node implies the scheduling of its succes- sor as an immediate adjacent operation. Such a position might be taken if there were a serious penalty involved in breaking the chain, such as dispro- portionate setup or breakdown time between runs. If this policy were taken, it would be superfluous to have represented so many separate programs. We shall consider that adjacency is not a constraint.

A schedule may be generated in a manner reminiscent of the code-optimizing techniques of Part 2. We shall consider that the required time of completion represents a priority number for those nodes having them, and we shall sort the initial list by priority for each deadline node. We have available to us, either explicitly or by computation, the duration and last possible starting time for each node. We may use the initial sorted list as an index to generate a precedence adjusted schedule, in essence associating with each predecessor the priority of the deadline's successor.

Starting with 2B in Fig. 29.3 we search down its column to find 2A as a predecessor. We now know that 2A must start at 9:15 if 2B is to be available on time. We may now determine if 2A has any predecessors; we see that it does not. We have adjusted our schedule from (1) in Fig. 29.4 to (2), adding 2A to the list. We now determine that 3A and 3B and 2C are unbounded, but that 4 has a direct predecessor, 1C. Knowing that 4 must start at 4:30, we determine that 1C must start by 2:30. We also discover that 1C has predecessor 1B, which must start by 1:30, and indeed 1B is "covered" by 1A, which must start at 1:00. Adding these elements to the list, we return to 4 to find its next predecessor, 2B; we continue in that way to find all other predecessors.

At completion we have a time-ordered schedule of all programs as in Fig. 29.4 (3). However, as our list developed we generated conflicting intervals between 1C, 2C, and 1D in the period 2:30 to 4:30. We must resolve these conflicts. We can do so by computing "slack," calculating the delay of the successor chain by a delay in the start time of the node and rearranging programs following some rule.

Among the rules that we follow are to minimize the number of jobs that are late (when any lateness is of the same cost), to minimize the total amount of late time, to minimize idle time, to minimize lateness in "hot" (deadline) jobs, which we are driving for here, and to minimize elapsed time for all jobs.

We see that among the three contending programs there is no time between the ending of 1C and the beginning of 4; consequently 1C cannot be delayed. 2C cannot be delayed because of its own deadline. There is no float in the system. However, since 2C and 1D are unbounded, it is possible to move them ahead into earlier time slots. We need a mapping of idle time to be processed against the requirements of the programs 2C and 1D to be moved. This map may be kept either in size or time order; in either order we shall find a spot for 2C at 9:00-9:15 and for 1D at 11:00-11:30, and we have an acceptable, executable schedule.

This is very much in the flavor of the critical path method techniques used for project scheduling. It is needless to point out that the schedule is only as good as its time estimates. There are, of course, various alternatives in representing the matrix of predecessors and successors. One could

(1) Initial ordering

Program	Complete	Latest Start
2B	10	9:30
3A	11	10:00
3B	1	12:00
2C	3	2:45
4A	5	4:30

(2) Intermediate at discovery of 2A precedence

Program	Complete	Latest Start
2A	9:30	9:15
2B	10	9:30
3A	11	10:00
3B	1	12:00
2C	3	2:45
4A	5	4:30

(3) Final list

Program	Complete	Latest Start	Idle Map
2A	9:30	9:15	9:00-9:15
2B	10:00	9:30	11:00-12:00
3A	11:00	10:00	
3B	1:00	12:00	
1A	1:30	1:00	
1B	2:30	1:30	
* 1C	4:30	2:30	
* 2C	3:00	2:45	
* 1D	4:30	4:00	
4	5:00	4:30	

Fig. 29.4. Schedule list.

list all predecessors (not only immediates) in each column; one could reduce to a binary matrix; one could represent the resources that are required for each program. But the fundamental determination of precedence and deadline is universal to all scheduling considerations when they intrude themselves into an environment.

This schedule formation is truly trivial compared to multiprogram scheduling, since we are mapping against a single resource. The complexities of multiprogram scheduling are introduced when we desire to develop a schedule for the system seen as a collection of resources, channels, devices, CPU's, storage media, etc.

Chapter 30

INTRODUCTION TO
MULTIPROGRAM SCHEDULING

30.1. SCHEDULING IN A MULTIPROGRAMMING ENVIRONMENT

The scheduling of a multiprogrammable machine involves the solution to a number of problems not dissimilar to the schedule of a machine shop. The fundamental nature of the resource to be scheduled is its parallel, asynchronous, separable componentry. The distinguishing feature of multiprogram scheduling lies in the essential detail that one of the resources to be scheduled is a virtual resource, which must be made to appear like a collection of machines. In essence, multiprogramming involves the simulation of a number of processors from one processor. Questions of priority and precedence, of course, remain, but in contrast to the scheduling of a sequential system, questions of effective usage and utilization of individual components are strongly present and may dominate policy. The utilization considerations are fundamentally identical to those introduced in the earlier chapters.

Multiprogram scheduling may be done in a static, preplanned way by either human beings or by a computer. If job population is stable and known, then preplanning may be quite useful. In any event, even in more fluid situations systems run under a general policy, which is enforced by the operating system through its appreciation of priorities and deadlines determined by the installation. There is more and more of a tendency to increase the intelligence of the scheduling function in the operating system and to develop dynamically adjustable schedules on-line with the scheduler operating as a program in a multiprogramming mix.

Whether human-planned, system-generated, long-range, or short-range, a multiprogram strategy must resolve questions of the following type:

1. How does one select programs that can be run together?
2. How does one determine how many programs to run at a time?
3. What is the nature of the interaction between programs running together?

We are going to introduce this area using concepts based on E. F. Codd's articles (see bibliography) as a structure on which we will hang ideas developed since his early work. Let us assume a simple situation where there is a population of programs, each of equal urgency and with no predecence relationships, which are candidates for a multiprogramming mix. We shall undertake to take a first look at the considerations involved in selecting which ones to run together.

We must know something about the characteristics of our program population. Figure 30.1 lists what we know. These characteristics, of course, relate to how the program behaves when it is running in a stand-alone environment. As we have seen from an earlier chapter, running characteristics naturally change in multiprogramming environments because of contention.

The scheduling task in the absence of priority and deadline and precedence, reduces to the determination of a parallel schedule, which will minimize the amount of time that it takes to run the known set of programs on the machine. The derivation of such a schedule involves the balancing of available resources on the machine so as to achieve desired utilization of asynchronous equipment. The "optimizing" of a mix is not unlike the optimizing of a program, where one attempts to achieve balance by buffering in order to achieve the theoretical minimum run times described in an earlier chapter.

30.2. LOAD ON A RESOURCE

From Fig. 30.1 we see that we wish to schedule five programs on a 32K machine with 10 tapes and two channels. Our first step is to determine which programs can feasibly run together. A first measure of feasibility is the load that the programs place upon the resources of the system. There are two measures of load that correspond to two species of resource, space and time (or passive and active).

The space load is a simple idea of whether there is or is not sufficient physical resource on the system to contain all the procedure and data required. In early approaches resources that are space-shared, such as core, discs, and tape populations, were considered to represent an inflexible

Program	Time (min)	Core	Tapes	Channel Utilization (expected over lifetime)	CPU (expected over lifetime)
P1	15	16K	3	.30	.70
P2	30	8K	5	.50	.50
P3	60	24K	2	.40	.60
P4	20	16K	6	.70	.30
P5	15	8K	1	.60	.40
SEQ	140				

MACHINE
32K, 10 TAPES, 2 CHANNELS

Excluded from same mix:
P1,P3:   Core
P3,P4:   Core
P2,P4:   Tape

Fig. 30.1. A list of programs to be performed.

limit on running programs together. It has been recognized that the space/time dichotomy, which we have used to influence the design of programs, applies as well to systems operations. Consequently, for example, and most dramatically in time-sharing systems, a load of greater than 100 percent on core is no longer necessarily considered an upper bound for a population of "active tasks" (although over a given interval the limit may be enforced over a subset of actives). A load of greater than 100 percent implies that core will be dynamically allocated, deallocated, and reallocated over time, increasing the time load on channels.

This tradeoff can be extended to any resource; that is, any resource can be considered deallocatable over the real life of a program in the system. Currently, however, because of the level of time penalty involved, most resource types (tape and disc devices) are considered undeallocatable unless the program specifically frees them.

The impact of dynamic core management is discussed in a later chapter. We shall here take the approach that 100 percent of core usage for a mix represents an upper bound for the use of this space resource. The number of the other space resources in the system, tapes and disc drives, is also a bound. Using these resources, we may begin to select programs that it is feasible to run together.

We may look for the programs that exclude each other. We determine that P1 and P3 cannot be run together because of core, and P2 and P4 cannot be run together because of tape.

Let us mention at this point that the time-shared facilities, the CPU and the channels, need not in any sense be thought of as having an upper bound of 100 percent loading; the establishment of loading levels on these resources is a matter of policy. For any given mix of programs the time utilization bound is the scheduling policy mechanism that reflects the user's throughput/turnaround policy for that mix.

Before continuing, let us look at Fig. 30.2, which represents the time to perform and the tape and core utilization of these programs in a uni-programming environment. What is shown here is the utilization of these two resources over the 140-minute interval required to run the five programs together.

Each program contributes for each resource a level of usage for the time that it is operating. Figure 30.2 computes the individual utilization of each program. We see, for example, that although program 1 uses 50 percent of core, it uses it only for 15 minutes, or 10.7 percent of the 140 elapsed minutes of the program population. Therefore, it contributes over this period only 5.3 percent usage of core. If, by some inclusion in some mix, the lifetime of program 1 were extended to 140 minutes by reducing the amount of core available to it or by alternating intervals when core was available to it, over 140 minutes it would use around 5 percent of all the space available over the longer period.

We are talking about available space over time. For example, we do not mean that the tape subsystem is delivering 32.1 percent of the data it can deliver in 140 minutes, but that 32.1 percent of total tape space is being used. What the table shows is that core utilization is the limiting passive resource but that it is sparsely used by the programs; it suggests that multiprogramming could be undertaken to increase the utilization of core and thereby to reduce elapsed time.

If we assume a perfect multiprogramming situation, the absence of

Over 140 minute time frame for sequential running of five programs

Core utilization

Program	Time	Time as % of 140 Min	% of Core Used	Utilization
1.	:15	10.7	.50	5.3
2.	:30	21.4	.25	5.4
3.	1:00	42.8	.75	31.1
4.	:20	14.2	.50	7.1
5.	:15	10.7	.25	2.6
				51.5

Tape utilization

Program	Time	Time %	% Tape	Utilization
1.	:15	10.7	.30	3.2
2.	:30	21.4	.50	10.7
3.	1:00	42.8	.20	8.6
4.	:20	14.2	.60	8.5
5.	:15	10.7	.10	1.1
				32.1

Fig. 30.2. Utilization by program of various resources over 140 minute time frame for sequential running of five programs.

precedence and priority, the goal we are after is to determine what is a minimum time and what utilization levels achieve that time for us.

Consider Fig. 30.3. We show here a perfect situation on two resources. We are assuming that there is simultaneous, constant usage of tape and core for each program. An allocation policy is implied by Fig. 30.3, which could contribute to the illusion it represents. If all the resources that a program might need during its lifetime are simultaneously awarded to it at initiation and kept by it until it completes, then any representation of the joint use of resources by a program will show identical start and release times for each resource and, necessarily, will show a uniform utilization of the resource over time. Figure 30.3 need not show the identical start and terminate times for each mix, but it does show a situation in which it was always possible to find a program whose use of core and tape so complemented the other programs to run in the interval that 100 percent utilization was achieved.

In this case there is (depending on one's viewpoint) either no limiting resource or multiple limiting resources. It is no longer usual for all re-

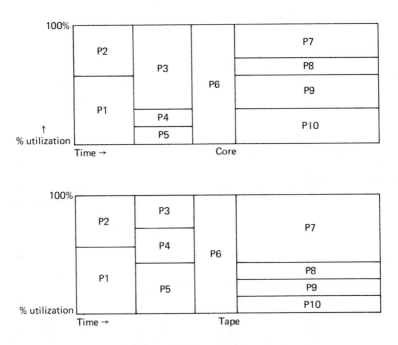

1. Simultaneous allocation of space resource
2. Uniform usage (no dynamic fluctuation of use)
3. Programs are found which run together so as to
   achieve a joint utilization of 100% on all resources

Fig. 30.3. Perfect situation.

sources to be acquired maximally and simultaneously (without some guilt on the part of system designers), and in fact increased flexibility in this area is a key to finer system scheduling.

Notice the programs of Fig. 30.2, however. If we disregard for the moment feasibility and allow loading on a resource which results in an overload on another, we shall see that there is a difference between the tape and core packing diagrams. This we expect because the utilization of the two components differs across time for the population of programs.

In Fig. 30.4 we show tape and core diagrams that really have nothing to do with each other, since programs are mapped into different mixes at different times. Notice that each program is represented by a rectangle, the height of which represents its utilization and the length of which represents its extent of time. The diagrams show that if one assumes that the utilization of core is held constant for a program, that is, it uses exactly .6 of core at all times, then all of the core space that is required to run this mix, if we consider no other constraints, particularly ignoring feasibility of joint acquisition and release of resource, can be granted in 80

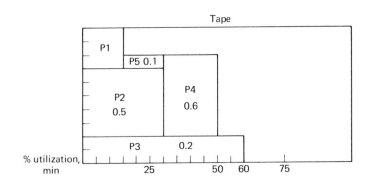

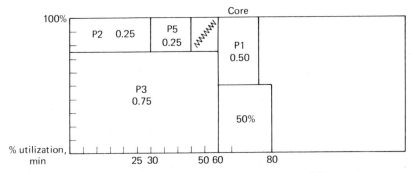

Fig. 30.4. Tape and core time maps for the PGM's of Fig. 30.2.

minutes of time. It further shows that this is done without achieving 100 percent utilization of core. The 80 minutes represents a possible mapping of programs over core. It is not the lowest possible time in which all required core space can be granted, since 100 percent of utilization is not achieved.

The tape diagram shows that all tape space can be granted in 60 minutes with a moderate utilization. We see that if the tape spaces are granted over the 80-minute interval required to grant all core, tape space utilization will further reduce.

We notice that if we take the sequential time of the programs that cannot run together because of tape loading we find a time of 50, not 60, minutes. That is, given that P2 and P4 must run together, a possible span time for the population is 50 minutes. This cannot be achieved because of the dominance of program 3, which requires 60 minutes to receive its percentage of tape space, if it is assumed that it uses it constantly at the 20 percent rate. The feasibility constraints on core are 75 minutes for the sequential running of P1 and P3 and 80 minutes for P3 and P4. Since there is no single program whose length exceeds these spans, these represent potential times for running these programs. The 80-minute limit coincides with the ability of the system to grant core, if constant use is assumed.

We immediately notice, of course, how decreases in span of time and increases in utilization go with each other. We must remember, however, that these utilizations are probably maximums, artificially inflated to represent maximum load, or possibly averages. In other words, we are either artificially restricting the schedule by inflating loading estimates or pretending to have information that we do not have.

Simultaneous allocation of resource implies a scheduling philosophy equally conservative. It restricts the addition of new programs to a mix to the points of termination of a program. In an operating system that consists of a "scheduler," this scheduler will operate to select a new program for the mix only when a program terminates.

An alternative strategy is to operate the scheduler whenever there is a change in the resource environment of the system. At this time the scheduler would be operated to see whether a program could be found to fit in the new resource environment. Release of any resource by a program would be sufficient to initiate a scheduling activity.

This "finer" scheduling approach has as its goal the enforcement of a more rigorous resource utilization policy. It still represents an intermediate policy relative to dynamic approaches, which we shall discuss below. It takes into account variances in the usage of resources over the life of a job.

Time loading is similarly treatable, but the derivation of meaningful loading levels for a program is considerably more complex than for space utilization. The statement that a given program will use 70 percent of available CPU cycles over a given period of time must be based upon a fairly detailed understanding of that program's logical characteristics and structure. At best, the statement is a very loose approximation. Further, the relative loadings of various time resources for a program are dynamically related to each other and suggest the compute/I/O balance of a program over its lifetime.

30.3. OVERLOADING AN ACTIVE RESOURCE

In Fig. 30.1 we simplified the time-utilization of our programs by showing channel utilization as if there were only one channel. For illustrative purposes it is adequate to show usage this way. We say that over 15 minutes, running alone, P1 will utilize 70 percent of the available CPU cycles and 30 percent of available channel time. It is a heavily compute-bound program. Similarly, P2 is balanced, P3 is compute-bound, and P4 and P5 are I/O-bound. In a uniprogrammed read/write/compute machine, much of the CPU and channel utilization will be overlapped if there is proper buffering.

However, buffering has been unable to achieve its ultimate goal with any of these programs, since the goal of buffering, to achieve the theoretical running time by smoothing and eliminating gaps, has not been perfectly achieved. This might be because buffering has not been fully applied or simply because the variations in processing time for records are so great that no reasonable level of buffering can achieve a smoothing.

Running in multiprogramming environment will have the minimum effect of redistributing the overlap between the asynchronous activities implied by separate time-shared facilities. In a system allowing no buffering, there would be effectively no overlap between the CPU and a channel for the same program.

We look at Fig. 30.5, the time diagrams for our population, and determine the minimum running times in terms of feasible time loading. We find that we have a CPU-bound mix that cannot complete before 105 minutes. This bound is so large because of the relatively sparse use of CPU cycles that we are able to make over long portions of the schedule. We are unable to make heavier use of CPU cycles, because we have no "light" CPU user to run with P1 or P2.

If we insist that 100 percent CPU cycle is a limit, we constrain ourselves to this schedule. By this limitation, we are insisting that no program be delayed due to multiprogramming; i.e., the elapsed time of a program may not be extended. This policy might be reasonable in an environment where particular programs were required to be run on a priority basis, but

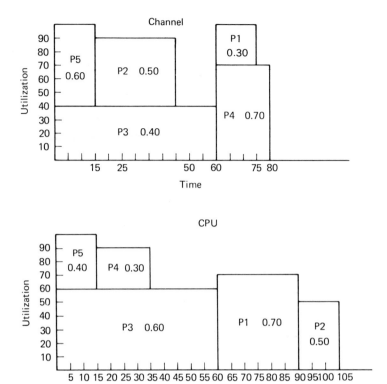

Fig. 30.5. Time (active) resource allocation. A feasible schedule with a unit limit (individual resource).

if we remember our earlier assumption in this section, we are now interested only in the shortest possible schedule, and we do not care about the finishing time of particular programs.

What would be the effect of loading time-shared resources beyond their capacity? The initial effect is to change the specific distribution of CPU or channel attention over the time interval. If we established 1.5 as a maximum time loading, the effect is as in Fig. 30.6. The running time of each program is extended by a third but the total running time of the mix is reduced to 97.5 minutes. This is because the higher time load limit allowed more active programs to be undertaken at the same time, representing a greater actual use of available CPU cycles. The contention between these programs caused a change in the service rate for each one, however, extending the period of time in which it received all the CPU cycles that it required for processing.

Program	Unity load run time	1.5 load run time
P1	15	22.5
P2	30	45
P3	60	90
P4	20	30
P5	15	22.5

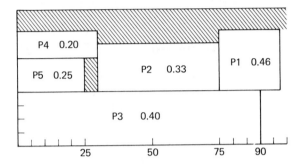

1. By extending span of each, reduce loading over time
2. Develop finer loading flexibility
3. Achieve higher utilization
4. Reduce mix time

Fig. 30.6. Impact of a loading policy of 1.5.

We are surely implying mechanisms for the distribution of CPU attention, "task-switching" mechanisms for multiplexing the CPU. We shall describe various task-switching considerations in a later section.

The question arises as to what represents a reasonable upper limit for the overloading of a time resource. Figure 30.7 shows the effect of assigning a loading limit of 2 to a time resource; in this mix it has a bad effect of extending the running time of P3 beyond the unit mix time. Notice the proportional effect of increasing the loading level until we run out of programs. The extension of completion time for programs in the mix varies directly with the variation in the load. The compensation for this lies in the ability to approach full loading at any given time more closely and to sustain high utilization of the resource by being able to schedule more active programs, since the contribution of each program to the load is smaller.

Whether or not the ability to run more programs in the mix has a payback is dependent upon a number of factors. First, of course, there must be programs to run. Second, the infeasibility constraints must not eliminate the possibility of running them together. Notice that Fig. 30.7 overlooks this and is artificially compressed. In order to achieve any

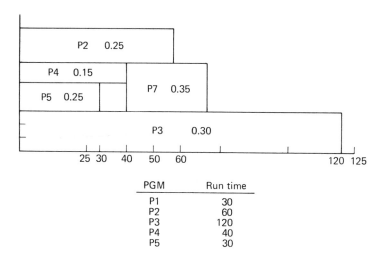

PGM	Run time
P1	30
P2	60
P3	120
P4	40
P5	30

1. Not enough work to sustain CPU
2. P3 overly extended

Fig. 30.7. Loading of two.

serious utilization of this resource, other feasible programs would have to be found. To the extent that feasibility constraints inhibit the mix, the increase in utilization and throughput possible with higher loading levels cannot be achieved.

30.4. HOW MANY?

The question of loading levels might roughly translate into the question of how many programs might be (or should be) run at the same time. We address this in part in the core management chapter, where we indicate that both core size and channel/device resources impose a limit. It is not bad, with any system, to start with the premise that we run as many as possible and work down to the number that is characteristically profitable.

The number possible with any system may be constrained by certain overridable hardware features. For example, on IBM 360 the number of protection keys [associating core space with privilege of access by comparing a value in the "stateword" (PSW) of a program with a value in core] effectively limits the number of programs in a mix to 15. On Honeywell 800, where multiprogramming is a built-in hardware feature, the number is limited to eight by the depth of the hardware program counters, which hold instruction addresses for eight programs.

The number of active programs in a mix may not actually limit the potential number of tasks actually competing for CPU or channel time, since the capability of "subtasking" (tasks spawning other tasks by their private use of the capability of the system to nominate a program as a candidate for resource usage, on OS/360 by the use of the ATTACH macro, on the U1108 by the internal START, for example) may proliferate this number. A constraint on subtasks may exist which limits them to use of resources granted to the mother. Usually, the vendor-provided operating system will contain a maximum number of programs that the system will put in the mix.

It is rarely profitable, because of resource paucity, to run the maximum number. In general, a higher number is preferable, because it increases the probability that a program will be ready to run at any particular time. However, heavy resource contention, which extends average device response times, has the countereffect of increasing the real-time interval within which an event occurs, extending event times and increasing event-dependent waits.

One strategy, which attempts to balance the benefit of broader selectivity with the resource contention problems, is to introduce a concept of admission on a contingency basis. A large number of programs are admitted, but only a subset of these is allowed to compete for resources at any time. All admitted programs are assigned resources (perhaps not core), and on the basis of some criteria some number of allocated programs compete for time-shared cycles over given intervals of time.

30.5. MINIMUM MIX TIME

With concepts of time and space load we may now determine what is a minimum time to run a given set of programs. We have seen that we have lower schedule limits imposed by infeasible combinations and by the loading levels of programs upon resources. The minimum schedule achievable is, therefore, the greatest bound. This is the theoretical run time of the mix.

Figure 30.8 shows all the bounds of our set of programs. These bounds are noticeably lower than the constrained limits previously shown in earlier figures. They are pure representations of the packing of programs on resources with no constraint. They represent the minimum amount of time that would be required if the resource were 100 percent utilized. In these figures we show the mix operating in minimum time by utilizing 100 percent of the resource.

It is interesting to note that we have introduced a new element here.

Minimum conceivable running time

(1) Infeasibility of core P1,P3	:75
(2) Infeasibility of core P3,P4	:80*
(3) Infeasibility of tape P2,P4	:50
(4) Delivery of all required core space at 100% utilization	:76:*
(5) Delivery of all required tape space at 100%	:45
(6) Delivery of all required Channel time at 100%	:66
(7) Delivery of all required CPU time at 100%	:73

*LIMIT USED IN TEXT IS :76 IN ORDER TO SHOW NEED FOR 100% UTILIZATION
OF A RESOURCE

Load levels

Utilization of Core				Utilization of Channels		
P1	.50 x 15 =	7.5		.30 x 15 =	4.5	
P2	.25 x 30 =	7.5	100%	.50 x 30 =	15.0	
P3	.75 x 60 =	45.00		.40 x 60 =	24.0	87%
P4	.50 x 20 =	10.00		.70 x 20 =	14.0	
P5	.25 x 15 =	6.25		.60 x 15 =	9.0	
		76.25			66.5	

Utilization of Tape				Utilization of CPU		
P1	.30 x 15 =	4.5		.70 x 15 =	10.5	
P2	.50 x 30 =	15.00	$\frac{45}{76}$ = 59%	.50 x 30 =	15.0	
P3	.20 x 60 =	12.00		.60 x 60 =	36.0	97%
P4	.60 x 20 =	12.00		.30 x 20 =	6.0	
P5	.10 x 15 =	1.5		.40 x 15 =	6.0	
		45.0			73.5	

Fig. 30.8. Minimum mix time.

This is the possibility that a program can vary over time in its use of the resource. This is manifestly true, and what it suggests is very important to us, because it implies a level of mix control just now becoming of interest to the field. The diagrams prior to this assumed a constant uniform distribution of resource utilization across the lifetime of a program and commitment to schedules based on uniformity. Much higher utilization can obviously be achieved by admitting programs to the mix that will vary and by observing and using the observations of variance. A true bound considers the possibility of local unevenness in the utilization of a resource, showing more clearly the dynamic nature of the utilization rate of any resource as locally dependent upon interaction with other resources.

The bounds of Fig. 30.8 are computed simply by multiplying the load of a program on a resource by the extent of time it places this load, the product being the amount of time it actually uses on that resource. The sum of all loads represents the true lower bound of the resource for the given set of programs. The nominal lower bound in our case is 76.25 minutes, the core space bound.

It is, perhaps, profitable to discuss exactly what this figure represents, for it is not obvious how a space resource load can be transformed into a time span. Our number represents the minimum number of minutes in which it is possible for the machine to grant a certain amount of space, if it is assumed that each location was used (occupied) every minute. We must, therefore, work with the concept of a "core-minute," or core-unit of time, which represents the amount of space occupied and the extent of time the occupancy lasted. The system is asked to grant 73,528 core locations and has 32,768 to give. Multiplying the length of occupancy by the core locations gives the total number of core-minutes for each program. The sum of these is the total core-minute demand on the system. Dividing this number by 32K gives the minimum amount of time, if we assume perfect packing, that it will take a 32K system to grant this amount of core/time. Avoiding the calculation of core minutes can be accomplished by replacing core requirements by the percentage of core they represent and deriving the core/time figure directly. Figure 30.8 shows both, and Fig. 30.9 shows perfect packing achieving the 76.25 minutes of elapsed time.

Since the minimum time in which we can run is determined by core, this is the limiting resource; we can determine what loadings are necessary on other resources by dividing the minimum time into the space/moments or event times associated with space and time resources, respectively. Figure 30.8 shows the utilization levels required for each resource to achieve a 76.25-minute schedule. We see that we must maintain very high utilization levels on CPU and channels and reasonably light utilization on tapes.

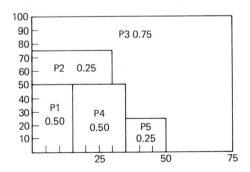

1. Assumes P3 can change rate of
accepting core usage over its life

Fig. 30.9. Use of core at 100 percent.

These figures represent the average loading over the interval that is necessary to support the schedule. Each program has an associated list of loadings that it will impose upon the resources; the job of scheduling is to admit programs into the mix so that they maintain utilization. We have previously discussed time resource loading limits as a policy decision that affects elapsed time of individual programs and the probability of achieving utilization by determining the fineness of the distribution of attention over an interval. The higher the limit, the greater the tendency to admit more programs to the mix to achieve the maximum loading, the greater the probability of achieving the desired load. Therefore, we may generate different schedules based upon variation in loading strategy.

In more advanced systems, the space/time tradeoff may be made through some levels of data staging (moving data forward through a hierarchy of data devices to devices of faster accessibility on the basis of rate of reference or need) of which the current example is core management paging. It is interesting to note that the assumption of flexible core usage turned our mix from a CPU-bound to a core-bound mix.

We have been assuming that we have knowledge of the existence (and characteristics) of all the programs we wish to run. The generation of a full schedule is possible by considering, through the use of various rules, what mixes approach the desired loadings most closely. This is accomplished by adding loading figures across the resources and selecting the schedule that maintains the highest loading levels. If there are time constraints, these can be used as guideposts to the schedule, and precedence relationships, if they exist, can also be taken into account.

Consider the situation in which we know less than we do here. A population of programs will be presented to us, dynamically arriving at some rate, and it is necessary to make decisions about whether or not to admit programs to the mix, and, indeed, when to consider programs for admission to the mix.

Chapter 31

OPERATING SYSTEM SCHEDULING: HIGH LEVEL

31.1. JOB OF AN OPERATING SYSTEM

The job of a multiprogramming operating system is to sustain a level of resource utilization within the constraints of deadlines, with a perhaps trivial subset of total job population to consider at any time and with no real knowledge of the utilization of resources these programs will make over time. The scheduling function properly includes the monitoring of actual resource utilization and the control of "short-range" tactics as well as the enforcement of broader strategies. A separation is often made between high-level and low-level schedulers ("coarse scheduling" and "fine scheduling") in contemporary operating systems.

31.2. HIGH-LEVEL SCHEDULING—WHEN SHOULD THE SCHEDULER RUN?

The high-level scheduler is the interface between the operating system and the operational environment of the computer center. This function has as input to it a list of programs that have entered the system and are candidates for activation. The output is a list of programs ready in some sense for operation. A number of questions surround this function, and the resolution of these questions determines to a large extent the "flavor" of a multiprogramming system. One question we have already touched upon: When should the scheduler operate? We noted that two possible operational times were when a program terminated or when the resource environment changed. A third possible choice is when a new program arrives in the system. This approach takes cognizance of the fact that in a changing population of jobs the dynamic arrival of new jobs may imply a change in the job-resource balance sufficiently significant to merit a reconsideration of the status of the system. This may be because the job has an urgency that requires its immediate activation or because it has resource usage characteristics that might just fit in and bring the system to its loading requirements.

Because of the burden of running the scheduler and reorganizing the active mix, it is common for urgent jobs to be submitted through special action by the operator and for whatever mix reorganization is required— rearrangement of priorities, premature termination, or preemptive allocation—to be undertaken by the operator. In systems organized this way, the actual activity of receiving jobs is separated from the scheduling function and split off into an asynchronous utility, which reads the "job stream" and develops an internal format on a system job queue. Entry of a job is possible at any time and does not imply scheduling of the job in any sense. The utility that reads the job stream may be operator-initiated or may be run on an interrupt basis when a ready signal is received from the job stream device.

The scheduler is a program like other programs in the system. Since it occupies space and time, it must compete with other programs for these resources. The question develops as to what level of resource the scheduler should be allocated, and on what basis. The answer to this lies in the operating system's approaches to this issue of mix stability, as is reflected in the decisions about when it should run.

In a conservative system, where job termination provides the sole occasion for running of the scheduler, it is reasonable to consider the scheduler as a "transient" utility of the operating system called into core at those relatively infrequent intervals when it must "replace" a job in the mix. It then is allocated the area just released or some dynamic overlayable area in the system residence portion of core, or is spread across both. Those portions of the scheduler that have direct responsibility for operator interaction may be made resident in core in the system area.

The scheduler, when summoned into core, may run at a fixed priority, or it may assume the priority of the job that it selects for entry into the mix at some point in its operation. Since the space for scheduler operation is often guaranteed, the priority is mainly for CPU and channel cycles. It is possible, however, to enqueue the scheduler on the core resource.

In a more dynamic system, for example, where the scheduler is run at resource release time, a number of factors become relevant. First, the relationship between high- and low-level scheduling begins to become more amorphous. The mix is less stable, and the need for scheduler becomes more frequent. In an extreme system, where variations in the level of resource utilization may cause the mix to change, the separation of high- and low-level scheduling may disappear entirely. The priority for resource of the scheduler depends upon the rate at which it will be expected that new jobs are admitted to the system and upon the distinction made between old and new jobs. In real-time systems, the scheduling function tends to be compressed, because there is no external job stream as such.

31.3. GETTING INTO THE MIX

A critical point in the decision process about the nature of a high-level scheduler lies in the criteria used to admit a new job to the mix. These criteria are basically of two types—resource-oriented and service-oriented—and themselves imply the relationships between high scheduling and allocation, as well as high and low scheduling. The service criteria are priority, time of arrival in system, deadline (if any), and time of job duration.

In a system where strict separation is made between high and low scheduling, the service criteria tend to dominate high-scheduler operation. Various queue disciplines are applied to job admission. A simple one is time since arrival within priority. The scheduler, when requested to run, investigates a queue of submitted available jobs ordered by priority and selects the longest-waiting job within the priority group. (Jobs are time-stamped at their time of arrival.) This process tends to maximize system responsiveness to the "purchased" priority represented, for example, by the external PRTY parameter on an OS/360 job card. No cognizance is taken of the resource implications of the selected job except to test the feasibility of selecting the job.

The scheduler, after locating a candidate, can undertake a number of approaches. It can attempt a "trial allocation" of resources. This involves determining whether the data files, devices, and core required to run this program are indeed available. To do so, the scheduler has access to the resource status tables. If all the resources are available, the scheduler may then itself make a bulk allocation. If all resources are not available, the job is rejected, and another job is considered for admission.

At the time of rejection the scheduler may or may not record the fact of rejection. If no recording is made, there is no protection against indefinite delay for this program. In rejection recording systems, the goal is to limit the amount of delay by reinforcing the time-in-system criterion. Each time a program is rejected, its priority is increased, so that it competes more strongly for resources each time it is considered.

A priority level may be defined such that one program has first call on all resources—no resource may be assigned to any program before this one. In effect, the system is allowed to "dry up" so that this program can get under way. Further, a priority level may be reached that will allow this program to preempt resources. An external preemption level may also be defined. Most pure batch systems attempt to avoid high-level schedule preemption unless it is operator-enforced. The cost of preemption is high. If no resume capability is built into the system, then the cost is all resource time since the initiation of the program string. If a resume capability is available (a checkpoint-restart of some sort), the cost lies in the allocatability/deallocatability characteristics of the resources. One approach is to hold all nonpreempted sources for the preempted program, in effect idling them for the period of suspension.

The scheduler may operate on a more discrete resource policy. There is usually a resource hierarchy defined in a system. The resource domain of a high-level scheduler may be constricted essentially to higher-level members of the hierarchy.

The hierarchy is formed by an appreciation of the expense of allocation and deallocation. For those resources within its domain, the high-level scheduler may allocate one by one. Initial data sets, devices, channel paths, and core may be allocated in some order with allocation firm. In an approach of this type, programs tend to be staged through the system, occupying a device queue, a core queue, etc. until they are ready to join the mix. Potentially, each queue may have its own discipline, and the program may acquire dynamic priority for a resource as it spends time on these queues. Each resource, however, is bound to it, so that all previously committed resources are idled through the progression interval.

If core management is such that frequent changes occur in core residence and the time that a program waits in a core queue can be expected to be small, then the previous allocation of devices can be expected to have a minimal cost. Whether or not one should allocate core or tape space first depends on the potential usability and paucity of the resource, and the rate of usage change perceived by the system.

Whether the allocation function is considered part of the high scheduler function or a separate activity, for any or all resources, depends upon the conceptual definition of the low-level scheduler and the attendant variations in resource usage perceived by the system during the life of a program.

Take, for example, the function of allocating core. If no changes in core allocation are allowed during the life of a program, then core allocation is a one-time function that could be considered a high-level scheduling function. If, however, extent of core is a variable allowed in the system, then core allocation mechanisms must be external to the scheduler, since changes in the core map will occur during a mix. This is also true if programs are allowed to change their usage within a predefined region. The user programs must have access to a core manager for the execution of get and free core operations associated with their spawning of tasks or the growth of tables within their area.

31.4. SETUP

In most systems that fall short of being true data base systems and in which "data-staging" is not implemented in any serious way, the allocation of data files represents a particular problem generally referred to as the "setup problem." This occurs when a path and a device are available, but the particular data file which is, by action of the scheduler, to be assigned that device must be physically mounted.

The introduction of device allocation in most systems makes it impossible for the operator to predict what data files will be assigned and require mounting on what devices. Specifying specific devices in the control stream has the undesirable effect of reducing the discretion of the allocation function and increasing the probability of conflict. Device independence is an efficiency- as well as a reliability-enhancing capability, since it increases the chance that a necessary resource can be found within a set of resources of the same type and that alternatives can be found. The undesirable effect, however, is to delay mounting until device assignment has actually been made, introducing a disproportionate delay between the selection and actual initiation of a job.

Various approaches to minimizing this delay exist. One is to allow an operator to premount and report his mounts to the system. This in effect extends the "data-base" catalogue to reflect transient data sets. The operator premounts on the basis of what he knows of the schedule over a small period of immediate time. Associated with premounting is an automatic recognition capability, where the scheduler reviews what is "on" the system and schedules to accommodate the data environment.

In the absence of a premounting feature, a system may enqueue a selected job on mounting, send out printed instructions to the operator, and permit the job to go forward when mounting is complete. The problem exists regardless of the allocation techniques of the scheduler. It is different in essence from other allocation problems in that the physical capability of mounting is already assured, and the delay is on the "conditioning" of a resource, not upon its availability.

The impact of setup leads us to a first consideration of a resource-oriented consideration on the part of a high-level scheduler. This leads to a classification of jobs as "setup" and "nonsetup" jobs, depending on whether or not the system anticipates a mounting delay. The object of this distinction is to allow nonsetup jobs to enter the system and utilize resources during the time that the setup job is waiting for its data files to be mounted. If, on the basis of priority and time in system, a setup job is selected and enqueued for mounting, the system attempts to find a nonsetup job to enter the system and use core and CPU cycles. The concept of a nonsetup job relates to the provision of certain basic system capabilities in the form of an input file, an output file, and some collection of "scratch" or utility files, which are always present and accessible. In OS/360, for example, there are defined SYSIN and SYSOUT files; for any program that can take its input from SYSIN, a stream of input records, and needs only one final output (SYSOUT), no setup is required. To the extent that the system maintains catalogues of resident data files, the population of setup jobs decreases.

31.5. A RESOURCE-ORIENTED CONCEPT

An extension of the presence of resource-oriented criteria in high-level scheduling comes in the concept of JOB CLASSES, also present explicitly in OS/360. The user of the system classifies jobs on the basis of their utilization of resource. Unlike many systems, which attempt to build scheduling criteria of this kind into the scheduler, OS/360 is completely open-ended about the definition of job "classes."

These classes are determined by the user on any basis that he chooses: high compute to low compute ratios, device requirements, or core occupancy. Each class is associated with a letter, and a separate queue is maintained for each class. The system has, therefore, at any time a preformed distribution of jobs of various resource patterns from which to choose for scheduling. External priority is defined within class of job.

The specific implementation of OS/360 varies from most other systems in that the high-level scheduling function is not represented in a single scheduler (as in EXEC VIII, for example) but in individual "initiators." These initiators may be associated with a class of jobs, and the breadth of flow from any class is, in part, determined by the number of initiators associated with that class.

Initiators may be associated with more than one class, in which case they attempt initiation from all classes with which they are associated in a

predetermined order. The number of active initiators is an operator decision, so resource balancing in part depends upon the acuity of the operator. Local extensions to OS/360 provide for the classification of jobs by
the control stream reader on the basis of the preformed rules for job
classes and the resource characteristics of the job as reflected in its description in the control stream.

31.6. CENTRALIZED VS. DISTRIBUTED CONTROL

The presence of separate initiators in OS/360 is reminiscent of our image
of the virtual machine. One may think of each initiator as representing a
single thread uniprogrammed machine, a virtual partition, competing for
resources in the resource pool. Initiators run with the priority of the job
that they are "running." As opposed to a centralized scheduling function,
they would seem to have no inherent drawbacks in the presence of the job
class concept. In a highly dynamic mix optimizing environment, however,
the relative independence of the programs would make it more difficult
truly to balance an entire system.

In an environment where discrete job steps representing program executions form sequences of operations, a question arises as to the degree of
commitment the system makes when it selects a particular job. This exists
in all batch operating systems of the current generation. We acknowledged
the problem of precedence and adjacency in the opening paragraphs of
this chapter.

In operating system scheduling the point resolves itself as to the preferability of attempting to proceed onto a new program of the same job or
opening a new job. Most systems are biased toward continuing down an
open thread, but a centralized high-level scheduler has the potential option of opening a new thread if it cannot continue. Independent initiators
do not have this option and, in case of a resource conflict, will tend to
"dry" up in enqueueing on a resource between program initiations. This
drying up reduces the effective level of multiprogramming.

Whether this is a good or a bad thing depends very much upon the
specifics of the system and the mix. If the goal of the system is to
maintain a certain level of throughput in a given time frame, it is not clear
that there is a manifestly proper way to do this. The opening of a new job
thread is efficient if a job can be found that will use the resources that are
available effectively.

For example, let us say that the fifth program of a string in a job
requires 84K of core. In an independent scheduler environment, if this
core were not available, the string would dry until it was available. If 16K
were available and a 16K job could be found (perhaps a nonsetup job)

that could utilize the system effectively, then the level of work could be maintained. The vulnerability of independent high-level scheduling lies not only in the inability of the system to determine freely what a proper depth of multiprogramming should be, but in its inability to determine the best participants in the mix in a sufficiently discrete way.

The counterargument, however, is also easy to see. If the level of multiprogramming is such that there are resource interlocks at the STEP level, then it is indeed proper temporarily to dry the level. The probability of finding an adequate replacement is sufficiently low so as not to justify the effort. Further, it is truly impossible at high-level scheduling time to make such a subtle resource decision in any event.

Finally, the temporary drying of a thread may well lead to the faster performance of the active programs by reducing contention for CPU cycles and channel cycles and consequently allowing programs to complete before the time they would complete if the level of multiprogramming were maintained. The faster completion will free resources more quickly, enabling the interlocked program to obtain its core more quickly than if a substitute had been thrown into the mix. Proper loading balancing, this argument concludes, is not a high-level scheduling function.

This is an interesting argument, with empirical evidence currently on the side of the conservatives. This may be because of the current state of the art of appreciating the resource usage of programs. In any event, the validity of either argument is constrained to examples. It is easy to see how substitution could be fruitless, but it is also easy to see that the drying up of a level of multiprogramming will not enable the remaining jobs to proceed more quickly, because of their own interference with each other or because of their inherent operational characteristics. If jobs are to be "dried," they should be dried by an active decision to balance the load and not by default.

31.7. CHANNEL ALLOCATION POLICIES

We might make some observations here about a problem of allocation policies on a machine with multiple asynchronous channels. In a system where there are multiple channels, each asynchronous but each associated with a particular set of devices, one is interested in achieving overlap not only between the CPU and the channel population, but between channels themselves. The system is more productive as the amount of time that the channels are running together is increased.

The ability of channels to be concurrently active is a function of the distribution of data across those channels. Operating systems make attempts to equalize the assignment of data files across a population of channels so that for any program concurrent channel usage is maximized within the limit of the appreciation the system has for what true utilization might be. A minimal approach is to balance by the number of data files assigned to devices on a channel. The allocation algorithm, assigning tapes, for example, attempts to assign an equal number of tapes of each channel. In OS/360 assignment may be further honed by a capability for the user to specify some of the balancing information he may have about his data structures without actually being forced to specify specific channels or devices. He indicates a preference for having certain files associated with or separated from the channel to which another file has been assigned.

These channeling considerations are, of course, more critical for machines of the IBM 360 channel logic than they are, for example, on a machine like the Burroughs B5500. The organization of the 360 is such that devices have specified paths to memory (although provision of alternate paths is possible in planning configurations, as well as the possibility for some devices to be manually switched from one path to another; this allows for some increase in scheduling flexibility as well as for an increase in reliability) specified at the time the configuration is developed. The B5500 allows for any device to be reached by any path through a channel exchange unit. The presence of an "open" network reduces the importance of channel balancing, because there is a preponderant probability that a path to a required device can be found, due to dynamic pooling. The allocatability of channel time is, therefore, dynamic and schedulable in time units related to the average time required to complete a particular I/O activity.

The presence in a system, however, of space-shared devices like discs and large drums reintroduces the problem of balance even for "open" systems. If data files are distributed across a disc file in such a way that a reference to a file may find the file unavailable because the device is busy, then effective channel utilization will fall off, because data transfer cannot occur in the referenced file.

An interesting distinction may be made at this point about "nominal" and actual utilization. The channel referencing the busy device may be nominally busy during the time it is waiting for the device to become available. It may "lock in" to the device. An example of this was shown in the description of the logic of IBM's channel-to-channel adapter unit.

Similarly, because of the distribution of intelligence between control units, devices, and channels, there may be times, during the active interaction of these components of a path, when a channel is nominally utilized but when it is not truly performing its function—transferring data between memory and a device. The difference between actual and nominal utilization is especially clearly seen in the core situation, where blocks of core may be nominally "utilized" because they have been allocated to a process, but are receiving little real usage because the process is not characteristically referencing addresses in their locality.

Effective resource scheduling often involves a finer definition of what constitutes true utilization. A further example of nominal as opposed to actual utilization lies in the relationship of a CPU to its memory. In most systems CPU utilization figures include the delays caused to the CPU because of memory contention. Often the differences between nominal and real utilization are functions of hardware design and cannot be ameliorated in any way of software strategies or design; as we shall see in the core management chapter, an appreciation between real and nominal space usage is central to core management approaches.

The shared-device balancing and channel balancing problems reflect upon our observation that time-shared resource loadings are related to each other, as well as upon our earlier remarks about the limiting device in a uniprogrammed system. We saw that the theoretical minimum run time for a program was the rate at which the slowest device could deliver service cycles to the program, and we saw how this limiting rate affected the service of other units to the program. It determined to a large extent the rate at which inherently faster units could be employed by limiting the rate at which service requests could be made of them. We noticed that sustaining a high utilization of a limiting device still left us with serious utilization gaps throughout the system.

Our response to this was to go to a faster device. We attempted to balance our system by removing the printer. With channels of nominally equal rated speeds attached to devices, the equivalent action is to spread the time load across channels by distribution of data files. One can appreciate the effect of this by observing a system in which one channel is heavily overloaded. This has the effect of increasing the average response time of that channel because of the development of chronic queues.

It is necessary in most programs to synchronize channel activity; the ability to write a record is partially determined by a previous read; the read of a record may be dependent upon a previous read, etc. The heavily disproportionate use of a single channel has the effect of converting that channel to a limiting resource in precisely the same way that a slow peripheral is a limiting resource. Gaps in the service request rate to other channels begin to appear and become more serious, because the time to complete activity on the overloaded channel increases.

As the proportion of service requests going to a single channel increases, the average queue delay increases on that channel, and utilization of other channels decreases. In addition, the amount of overlap between channels naturally decreases in cases of imbalanced reference. This situation leads to extended running times. This deterioration of performance exists independent of the level of load placed upon the channels. Further, conditions of imbalance are obviously not discernable from gross utilization figures, since utilization of channels over a given time frame at any level need not be concurrent.

Some considerations reminiscent of our remarks about buffer balancing are relevant here as well. The general problem of which both are part is the problem of the optimization of individual programs in a multiprogram environment, and whether this is a uniformly good or bad thing.

The polar approaches to resource allocation with regard to channels and devices in an attempt to optimize mix performance are, on the one hand, attempts to balance each program as well as possible, making the widest distribution of loads, and, on the other, attempts to constrain each program as much as possible. The latter strategy is based upon the premise that when the mix is organized so as to make minimum interference between programs performance decay can be contained. When active, each program will make maximum efficient utilization of its resources when it requires them. An ideal situation is to have target resource loading on a resource from a single program.

Although it is feasible dynamically to allocate and deallocate buffers, it is not so feasible to allocate and deallocate channels and space on devices, and channel-balancing algorithms are more limited than buffer- or core-balancing algorithms. The appreciation of time loads across channels is quite critical to a scheduling/allocation scheme. The great problem is the determination of what service rates (references) to a given file are actually going to be. Numeric equality is not sufficient to guarantee balance. The size of records, the relative activity of the file, and the interrelationships of reads and writes from various files all affect the concept of balance. The adequacy of a static allocation technique is dependent upon the adequacy of information about behavior and variations in behavior.

31.8. WHEN TO ALLOCATE DEVICES

The extent of system commitment to a job also involves the timing of device allocation. To what extent should the total population of devices be assigned to the entire string when the string is selected? To what extent should it be delayed so that allocation is made only for the program in the string? Certainly, the capability of delaying acquisition is the more powerful and flexible approach. It introduces, however, the problem of what to do with a program if it cannot get allocation. One approach is to enqueue it on the resources, and this is certainly acceptable.

Whether or not devices should be immediately assigned is completely

402 Multiprocess Scheduling (Multiprogramming) Sec. 31.9

a function of the time extents of predecessor programs and the variation in device utilization between programs. Manifestly, if the first step takes one half hour and uses 8K of core and one device and a successor takes up 16 devices, it is noneconomic to allocate all devices at job initiation time. The broad availability of core makes it reasonable to assume that other programs that can profitably use the free devices can be found during the half-hour interval.

On the other hand, if the initial program is short and large and users of the devices may not be found, it is profitable to allocate immediately. The general criterion would seem to be that the potential delay due to setup of successor runs must be tolerable because of the effective utilization of those devices during the duration of the early program. Immediate allocation would, of course, permit mounting of successor data files during the running of the initial program. A system that allocates immediately has an insurmountable bias toward string following, in the sense that finding a substitute from another thread becomes much more difficult because of the devices that are prematurely allocated. In systems where program level allocation is built into the system, it is possible to "fool" the system by overstating device requirements. The penalty for this may be delayed start of the string; the reward early mounting.

31.9. SOME SERVICE-ORIENTED SELECTION SCHEMES

Before going on to low-level scheduling, we should look at some popular service-oriented selection schemes other than age within priority.

The absolute simplest selection mechanism from a waiting line of jobs is FIFO or FCFS (first in, first out or first come, first served). This mechanism accords most directly with our intuitive notions of "fairness" in according service. The basic representation of a waiting line for service is an ordered list, which orders itself naturally by the dynamic appendaging of later arrivals. The scheduler in selecting from the list merely selects the top (front) of the line.

If for resource reasons this member of the waiting line cannot be serviced, then the techniques discussed above become relevant. That is, the rejected member may be accorded a special status in some manner, and the pure FIFO nature of the system breaks down, since a measure of priority accrues to the rejectees. This situation is found in human servicing when someone reaches a bank teller and finds he has improperly filled out a form, is sent away, but is allowed to return to the head of the line.

The staging of a program through local resource queues in a system is fundamentally a way of granting service, though delayed, to someone at the head of the line. Alternatively, the program may be dropped to the back of the line. This is a primitive form of cyclic reenqueueing that we will discuss later. This would be an extreme penalty for the rejected program, and would, in effect, replace the criterion of arrival time in system with the criterion of time of last look by the scheduler. A more common approach is to leave the rejected program at the top of the queue for the next inspection.

The FIFO-priority scheme, which we described as our first scheduling technique, is an urgency-oriented modification of FIFO in which an external or "purchased" priority is considered. In scheduling systems in which priority is used as a high-level scheduling criterion and which tend to make a strong distinction between high- and low-level scheduling, the most often found application of priority is for mix admission. A higher-priority job will be preferred for entry to the mix. It is possible, however, to apply a preemptive priority scheme, where a high-priority job may preempt lower-priority jobs. Real-time systems commonly contain priority preemption mechanisms with a resume capability. Preemptive priority would seem to be consistent with a scheduler that runs whenever a new job arrives rather than with a scheduler that considers applicants for service only at the time of termination of a job.

Another service criterion used by high-level schedulers is the shortest job in either its preemptive or nonpreemptive forms. The concept of shortest job presents some difficulties in remaining nonresource-oriented. Surely the concept implies that this is a short job based upon some assumed level of resources. We have previously seen that we can manipulate elapsed time by manipulating access to resources. Unhappily, the job time usually taken for consideration in an SJF scheduling scheme is a programmer's estimate, independent of load considerations. The goal of an SJF discipline is to increase the number of total jobs processed by the system in a given period of time and by consequence to offer a particularly fast turnaround time to small jobs. The effect is to reduce waiting time for fast jobs and cause delays to long jobs.

There are a vast number of total system strategies that can be built around the concepts of FIFO, priority, and shortest job first. Shortest job first may be constrained within priority level, allowing external urgency to be expressed. The concept of standard jobs or job classes may be invoked in connection with the service disciplines. For example, a schedule may be built around a dominant job, perhaps one of the set of jobs that justified acquisition of the machine, and all other jobs may be scheduled around it on a priority basis or on an SJF basis.

The difficulty with all non-FIFO schemes is that no guarantee of service is made to anyone; that is, there is no constraint on the amount of time a low-priority or a long job may be asked to wait for service. This is a function of how fast high-priority or short jobs can arrive during the scheduling interval. In very complex scheduling situations one often finds a constraint that attempts to guarantee some minimum service rate to low-caste jobs.

A service criterion of recent interest to high-level schedulers is the deadline. The description of a job contains an estimate of the execution time for the job (a program) and a time by which the job must be completed. An example of an operating system primarily batch-oriented which allows deadline scheduling is the UNIVAC 1108 EXEC VIII. In this system a list of deadline jobs is maintained, and jobs are admitted to the mix as the time approaches where the true day time and deadline time difference approach the expected execution time of the run. After admission, a low-level scheduler attempts to keep these runs on schedule. The concept of on-schedule also exists in TSS 67.

Deadline scheduling will be discussed more fully in the low-level scheduling section. Manifestly, however, in a deadline scheduling system, the high-level scheduler must have some appreciation of time in order to activate the program so as to make it possible to meet its deadline. The deadline is a form of purchased priority; it is an absolute external priority replacing the vaguer notions of relative priority, which exists in most systems. Proper high-level scheduling beyond the limits of a high-level scheduler in an operating system is, of course, necessary to guarantee that a system is not asked to accomplish impossible tasks.

31.10.   FINAL REMARKS

A few concluding characterizing observations on high-level batch scheduling are appropriate here. Characteristically, high-level schedulers treat the jobs that they schedule (or the programs within that job) only once. There is no requeueing a job after it has been admitted to the mix. There are exceptions, where "suspended" jobs may be returned to a high-level scheduler if they are preempted, but this is not true in general. This is partly because most systems do not allow preemption or consider resumption as the domain of the low-level scheduler. The high-level scheduler behaves as if a job once commited to the system will receive uninterrupted service by the system. In this sense, the scheduler is a sequential scheduler, behaving as if jobs ran sequentially, once released to the system.

There is a serious question as to whether the distinction between high- and low-level scheduling must be made. The question, in part, resolves into whether or not the door between jobs waiting to be admitted to the mix and jobs in the mix should allow passage in one or in two directions, and whether or not those resource-oriented criteria or more subtle service-oriented criteria, which characterize finer scheduling, should not be applied to admissions policy as well.

The separation is also based upon an appreciation of the allocatability and deallocatability of resource. Some systems attempt to marry the concepts. This is most easily done in real-time systems, where waiting jobs and active jobs exist in a resource environment that is preplanned and where allocation plays a minimum role. If one considers the essential discontinuity of service that truly exists in multiprogramming, the question is raised as to whether, except for allocation complexity, it is a truly different thing to schedule the first slice of service than it is to schedule subsequent slices of service.

One concept connecting the two in batch processing is to allow the fine scheduler, during its mix control activities, to go out to the gross scheduler to request admission to the mix of programs of various resource usage characteristics.

Chapter 32

OPERATING SYSTEM
SCHEDULING:
LOW LEVEL

32.1. LOW-LEVEL SCHEDULING

The low-level scheduler (LLS) or "fine scheduler" is primarily responsible
for the dynamic allocation of time resources among a mix of "active"
programs, programs which have been passed to it by the high-level scheduler, which have been spawned by active tasks, or which have become
active from a time-sharing terminal. The job of the low-level scheduler is
to ensure that active jobs receive services at a rate consistent with the
intent of external priorities and/or with the proper use of resources available to the mix. In addition to allocating time, the low-level scheduler
may be empowered to make core residence decisions (especially in time-
sharing systems) dynamically as it monitors system activity.

Low-level scheduling may not truly exist in a system, or may be
reduced to such trivial functions as truly to be more properly considered a
dispatcher—a mechanism charged with controlling queue selection but
with no decision or policy capability of its own. This is characteristic of a
conservative batch operating system oriented toward smaller machines.

When terminal or real-time support is added, the first function that is
added to the basic operating system is a subsystem, which provides the
low-level scheduling function. The history of many current operating
systems is a history of the grafting of special-purpose low-level schedulers
onto basic OS services in order to support more complex environments.

Low-level scheduling is traditionally more characteristic of real time
and time-sharing than of batch operations. In mixed systems one often
finds a batch and a real-time or terminal domain (partition). The LLS

operates on a set of programs in the terminal partition and a batch partition runs essentially independently.

The reason for the low level scheduler lies in the essentially cyclic nature of time-sharing, where a mechanism must exist to sustain response rates and multiplex CPU attention at a microsecond or millisecond level, and in the time-critical nature of real time, where critical tasks must be guaranteed finish times and where this guarantee can be enforced only by careful monitoring of an active task list.

By the nature of terminal service, the scheduling criteria tend traditionally to be service-oriented and concerned with response (turnaround) time rather than with load balancing or maintaining loading levels. The current interest in developing general operating systems that can run programs together in batch or time-sharing modes is leading to an interest in the extension of the low-level scheduling concept to batch programs, since these must be part of the multiplexed CPU's attention in a conglomerate mix of terminal and batch jobs.

Further, the developing interest in measuring performance of systems is leading to an interest in the techniques that might, in a batch environment, be associated with a low-level scheduler to achieve throughput—that is, properly to utilize resources in the machine.

There are two fundamental approaches to the comprehensive operating system. One is to incorporate the scheduling and monitoring functions in a central facility that controls time and core allocation to all resources; the other is to continue to define separate environments with separate scheduling rules in partitions of the resource base. The arguments for the extension of separation or the collapse into common control are not entirely related to scheduling and performance, but involve considerations of the special nature of services required by different environments, the ease of moving from one operating system to another, the nature of interfaces, etc.

So far as our interest here is concerned, we shall explore the general nature of low-level scheduling, the various rules that may be applied for task-switching, and the implications for performance that lie therein.

32.2. IN THE MIX

When a task is a member of a mix, it joins a queue of tasks, which one may call the CPU ready queue. This queue in basic form represents all tasks which have the resources they require and can perform if given control over the CPU.

The low-level scheduler determines which task is next to be given to the CPU, or to a CPU in the more general environment. During the course of its operations, a program may find itself unable to continue the use of the CPU because of a requirement for I/O. At this point the program enqueues a request for an I/O operation on a queue associated with (most commonly) the channel or device on which the operation will occur.

The modification associated with the readiness of a program for the CPU may take two forms. It may be placed upon a waiting list of programs that cannot take CPU cycles, or it may remain on the CPU queue with a flag set that acknowledges his unselectability. At the time that the program "releases" control of the CPU and enqueues on an I/O operation, the low-level scheduler runs to select another program from the list of CPU-ready programs on the CPU queue. This event is called "task-switching," and the balance of task-switching—flexibility and intelligence on one side, and the "overhead" of task-switching on the other—is the central problem of low-level scheduling in all environments.

In the simplest batch environment task-switching occurs only when a program informs the system, using a system macro, that it cannot continue its use of the CPU. Inability to continue use in a simple batch environment is usually associated with a need for I/O; however, in more complex environments a program may be dependent upon the completion of asynchronous tasks other than an I/O server. Further, and we remember our discussion of buffering, an I/O request may not be in and of itself the cause of inability to continue.

The probability of suspending CPU utilization is a function of the probability that an I/O record may not be available from a buffer. In a batch environment, where CPU/I/O utilization ratios are critical to performance, buffering depth on any given file will have an effect upon the need to discontinue CPU control. Given sufficient back-up (as in a large-capacity core box given over to buffering), a CPU "bound" program can be smoothed into a balanced or I/O-bound program for substantial periods of its system life.

The most primitive means of task dispatching is to continue all active programs, whether they are awaiting an I/O completion or not on the CPU queue. The event (I/O task completion or other) that will enable a program to continue or resume CPU usage is not made known to the system. Event completions are posted in a central communications area. When a program reports inability to continue, the dispatcher (the vestigial, or more properly, embryonic, low-level scheduler) transfers control to the entry at the top of the CPU list. This program determines for itself whether or not it can continue; if it cannot, it reports inability to the dispatcher, which then attempts the next program, etc., cycling around the queue.

The foregoing is a software variation of cyclic multiplexing. Its disadvantage is the time spent in determining for each program on each entry whether or not it can continue. As we earlier saw with fixed time division multiplexing on a channel, there is nonproductive time associated with the visiting of "posts" where there is no need for servicing. A nontrivial time is the restoration and subsequent saving of the statewords of tasks that cannot, indeed, run. Even in such a system, of course, queue structure may be made to reflect priorities so that higher-priority programs have more frequent opportunities to run than lower-priority programs.

More commonly in contemporary operating systems, the dispatcher is aware, either by absence of the entry on the CPU queue, or by a flag on that queue, of the programs that can run on the CPU. Associated with a request for I/O service (or for the running of any asynchronous task) is sufficient information that the system can relate the completion of an event to the program that requested it.

32.3. TASK SWITCH AND I/O

In conservative systems the task-switching function has been intimately associated with the I/O interrupt. The reason for the I/O interrupt on batch machines was initially to allow for closer synchronization of a program with its I/O support (IOCS). Since CPU/I/O balance in a multiprogramming system was the dominating concept of multiprogramming utilization, it became quite natural to use the I/O interrupt as the central mechanism for task-switching.

The interrupt provided a natural point in time for the review of CPU status and the enforcement of scheduling strategy. This enforcement implies the involuntary release of control by a program. We have two occasions when a program loses the CPU: on a self-generated release and on a system-perceived inability to continue.

If the system attempts to fill a request for I/O from a buffer and finds that buffer to be empty (or, more generally in a true pooled buffer environment, finds a record from the referenced file to be unavailable), then the system places the program on a wait list and selects another program to run.

Whether or not it is "proper" for the system to do this is a point of some argument. For example, it is possible for a program to have quite complex I/O patterns involving multiple reads and multiple writes. One postulates the situation in which an input record is not available for a given file; the program will lose control because it cannot proceed, the system thinks, until the record is provided. However, there may be a number of reads and writes that could be undertaken independent of the response from a single file. If the program loses control, these submittable

I/O operations will never be enqueued. The system will look for another program and may or may not find one that will initiate or request activities on the channels that the quiesced program would have used. Potential channel overlap will be lost to the system because of premature abruption of a program interval. This, of course, relates to the general problem of program optimization in a multiprogram environment.

32.4. PRIORITY ENFORCEMENT AND RESOURCE USAGE

Consider an environment in which there is "purchased" or external priority; this external priority is to be enforced not only by the high-level scheduler in admitting jobs but by the low-level scheduler in its enforcement of multiplexing rules. Observation of the priority requires that the high-priority job be running whenever it can be running.

The I/O interrupt that brings news of the completion of an I/O event is a point at which a higher-priority program may be switchable from waiting to runnable status. At interrupt time the system not only switches that status of the program, but investigates whether the program that last held the CPU should be "preempted" by a now available higher-priority program. This is an OS/360 feature.

The intermittent discontinuance of programs due to I/O interruption is a basic method of multiprogramming. A recommended approach to such a system is to grant priority to highly I/O-bound jobs, and low priority to compute-bound jobs. The idea here is to allow I/O-bound jobs to drive the system (keep the channels and devices going) while compute-bound jobs "soak" up remaining CPU cycles when I/O active programs are not ready to run.

The concept has no basic flaw, but it encounters a number of operational problems. Primarily it encounters the problem that many installations are unwilling to grant priority on the basis of assumed operational characteristics of a program. They feel that priority is more properly a statement of urgency and economic utility and should not be used to describe job characteristics.

The development of job classes in OS/360 is a result of the inability of a system to rely upon priority for its mix optimization. Oddly, the effect of classes is more strongly represented in the lower-level version (MFT) than the more advanced version (MVT) of OS/360. In MVT jobs compete for the CPU by their purchased priority after they have been selected for the mix. The class concept is used truly only as an admission rule in this system. In MFT classes are associated with particular partitions of the system, and each partition has an associated internal priority. This "internal priority" is used as a dispatching priority by the dispatcher, and consequently class definitions and the mix-balancing properties of class definitions are brought forward to low-level scheduling.

A further limitation on the effective use of priority to express preferences for dispatching comes from the fact that even when priority is associated with balancing attempts, it can only reflect an appreciation of balance for the program in general—that is, the general tendency of the program over its entire lifetime for all of its runnings. The variations that may occur from running to running can be reflected by assigning a different priority for each running—a possible but not especially realistic approach.

Variations in balance during the program cannot be reflected at all. The classic example of this is the FORTRAN program, which reads in a large amount of data initially, then computes with little or no I/O for a long period of time, and then writes an enormous amount of output. What is the proper priority of this program? Indeed, what is its proper class?

The problem is extremely severe if priorities or classes are associated only with the job in which a program is a member. By allowing programs in a job to have individual priorities and individual class designations, the problem is somewhat relieved. At least in this way programs of grossly different characteristics may be treated differently when they are members of the same stream or thread. The problem still remains, however, that during the course of its life in the system a single program may vary in its I/O-CPU balance, and no age-based (FIFO) or purchased-priority-based dispatching scheme can adjust the dispatching priority of a program to reflect these changes.

In our discussion of loading and resource utilization, we saw that the maintenance of "target" loadings was the means by which to execute a mix of programs in the minimum period of time. The difference between assuming a constant rate of utilization and allowing for variations in the rate was shown in the differences between Fig. 30.4 and Fig. 30.8. If programs are admitted to a schedule on the basis of average load upon a resource, utilization figures go up, and the schedule contracts, as it does not do when we add to a schedule on the basis of a constant rate. Since the job of multiprogramming is to maintain resource loadings, it follows that the submission of jobs for CPU cycles should be sensitive to the resource load contribution that the program is making at that particular point in time. It is only that "instantaneous" load that will truly have an impact upon resource loadings. Misappreciations of load in an instantaneous sense will tend to unbalance the system and to cause glutting and consequent underutilization.

The fundamental idea is that of the buffering and back-up, "look-behind," "look-aside" techniques. The recent history of a system is valuable in predicting its immediate future, and scheduling (by look-ahead instruction buffering, which assumes localized addressing, by associative memories holding last referenced pages, by recent resource loadings for tasks) on this basis is an effective control technique. More will be said about this later in this chapter. The concept is introduced here to show the relative weakness of priority as a technique for balancing mixes.

32.5. TIME-SLICING

The I/O dispatch rule for multiprogramming has one basic flaw. Consider a system in which there is, indeed, a mix of I/O-and CPU-bound jobs. When the CPU-bound program receives control, there is no way for the system to retrieve control from it unless there is a priority preempt dispatching rule, and I/O-bound programs take control when their I/O is ready. If priority is not truly assigned on the basis of I/O-CPU ratio, or if it is incorrectly assigned, then there is no way to dismiss the compute-bound program from the CPU.

In this instance channel and device utilization will "dry," because those programs that load the I/O subsystem will be unable to take the CPU to generate I/O requests. In order to protect the system from being taken over in this way by the compute-bound runs, many current batch-oriented multiprogramming low-level schedulers interrupt programs at given intervals of time.

This "time-slicing" is traditionally a feature of time-sharing systems. Its use in the batch environment is to provide a fixed point in time where the status of the system may be reviewed and some resource load policy enforced. The sophistication of the review is, of course, a function of the information available to the reviewer. As with all techniques, there are a number of approaches to time-slicing that determine its impact on the system.

The most straightforward approach is to time-slice all active jobs. In a nonpriority environment, the system may simply develop two internal classes of jobs—those that finish before they are interrupted and those that do not. Those jobs that do not use their full time slice are characteristically I/O-bound jobs, since they have been unable to continue because of an I/O dependency, and are unable to use their full time allotments. Those jobs that complete their time-slices are computer-bound jobs—jobs that can make extensive use of computer time between I/O requests.

The concept of I/O- or compute-bound is not truly mapped here, but is roughly approximated by a measure of a program's capability to use certain amounts of CPU time. When a program is interrupted, it must be a high CPU utilizer (or it would not have run to its time-slice end). It is placed on the high CPU queue, and an attempt is made to give control to the "low CPU" queue, where programs that have not completed their slices reside. If a program of this class is ready, it receives control; otherwise the high CPU queue in inspected for a successor program.

This scheme may be constrained by priority rules, which apply to whichever member of the low or high CPU queue is chosen. This simple approach to time-slicing has a real advantage in that it truly relieves the user from knowing which jobs are CPU-bound and which I/O-bound, and introduces a level of dynamic observation into the low-level scheduler. The UNIVAC 1108 EXEC VIII contains a mechanism similar in flavor to this. Programs in the mix drift downward in priority when they complete their time slice and are maintained at priority when they do not.

Other implementations of time-slicing do not go as far to relieve the user of his need to know which of his jobs are I/O- and which compute-bound. OS/360 relieves the user of the need to manipulate his priority but maintains his need to express an opinion about the job. In OS/360 MVT priority levels are designated as "time-slice" levels. If a program has one of the time-sliced priorities, it becomes a member of a "time-slice group." Members of a time-slice group are time-sliced together, each program receiving a guarantee of a minimum slice of time.

Judicious use of the system dictates that I/O-bound and CPU-bound jobs are given the same priorities in order to force the attention of the CPU to the I/O bound programs. The possibility of giving CPU-bound jobs lower priorities, of course, still exists. If CPU-bound jobs are given higher priority, however, this priority is still enforced in a preemptive way. Time-slicing occurs only within the priority level designated as a time-slice group, and continues so long as no higher-priority task is ready to go or until no member of this group can proceed.

One can visualize a number of alternative strategies associated with this system. One can achieve balanced channel utilization by making assignments to a time-slice group on the basis of channel contention. One can time-slice among all I/O-bound jobs with CPU-bound jobs at a lower priority "soaking" up cycles, etc. However, the OS system still falls short in this regard of the concept of "dynamic running" priority or "deserved" priority achieved by "good" resource utilization by a running program. No system goes very far in this regard, but there is a developing interest in the technique of enforcing dynamic running priorities.

Among systems that have time-slicing capabilities are GECOS II and III for the GE 635, EXEC VIII, Chippewa for the CDC 6600 and, of course, OS/360. Surely there is a "burden" in time-slicing. The burden occurs in two forms. The first and obvious form is the amount of CPU time lost to the process of interrupting a running program; the secondary form is the loss to the system in computing potential when no suitable job may be found to run. If "low CPU" tasks are ready to go, all analysis time, as it were, is completely lost. Care must be taken to set the time-slice period at a proper level so that CPU-bound programs are not allowed to "dry channels," and so that the probability is high that an I/O-bound program will indeed be ready to run when a search for one is made by the system.

The significance of "quantum" size for a system was revealed by GE in a comprehensive statistical analysis of the GECOS II operating system for the GE 635. Initially a time slice of 62.5 ms was allowed. Under study this time was found to be too long and was reduced to 15 ms with a nontrivial attendant increase in effective utilization for the system.

A question that arises in connection with time-slicing is what to do with I/O interruptions. Does one want an "event"-driven system, a "time"-driven system, or a mixture of both? The I/O-interrupt system is an event-driven system, in the sense that the status of the system is (to some extent) analyzed upon the occurrence of events that occur at irregular intervals. The clock-interrupt system is purely time-driven, in the sense that system status is sampled on a regular basis, regardless of external events.

One could visualize a system with no I/O interrupts. Each I/O completion would set a bit. At various timer interrupts, I/O status would be sampled "en masse," and dispatching would be done on the basis of whose I/O events had completed in the time interval. Current time-slicing systems (with the exception of the 6600) are mixed systems. The mixed system is more efficient if strong preemptive priority is to be enforced.

The introduction of time-slicing to the batch environment brings it closer to the time-sharing environment in the possibilities that are available to a low-level scheduler. Now both systems face the phenomenon of a regular enqueueing of jobs asking for service, the enforced partitioning of the service interval of a job, and the constant dynamic decision making about who deserves or requires the next chance at the machine.

The goals of the two systems are not necessarily the same, one having a primary interest in resource utilization, the other in maintaining a service rate (response time) to conversational users. However, they intersect in the area of the basic mechanics of handling queues that characteristically and cyclically readmit members. Before going on to describe some resource-oriented approaches for a batch low-level scheduler, we shall investigate some basic techniques for handling queues of this type.

32.6. ROUND ROBIN

The simplest scheme for handling the dispatching of jobs that are going to be serviced in discontinuous intervals is the basic round robin scheduling rule. This approach is analogous to the FIFO rule for high-level scheduling, in that it accords most closely with our notions of "fair play." In a basic round robin queue each member of the mix is represented. Starting from the top of the queue, the scheduling mechanism grants a fixed interval of time to each member. At the end of the list the scheduler returns to the top of the list. Each task is served in order of the time it last received service. Time is divided equally between tasks so long as they can use the time allotted to them. If all tasks are represented on the queue, it is possible for certain tasks to be unable to use the CPU at any given time, and this fact can be simply represented by a bit in the entry for that task, which causes the scheduler to bypass it. This cyclic inspection of a queue of tasks is an analog of fixed time multiplexing. The technique guarantees service for all members of the mix in an equable fashion.

In a time-sharing system, where the preemption of programs in time is a fundamental technique for balancing service to different users, a round robin or a variation of round robin is a reasonable approach to scheduling.

If the shortest job first feature is desired in order to reflect a large number of job completions, one can adjust the time slice so as to maximize the number of preemptions for all jobs, thereby reducing their waiting time and the consequent elapsed time to their completion. As the time slice becomes short, short jobs will tend to be serviced more frequently and therefore to spend less time waiting for their next turn. Of course, as the time slice becomes larger, the technique begins to resemble FIFO more and more, since jobs will tend to run to completion.

The great difficulty with round robin scheduling comes from its implications for resource utilization, especially core. Since the technique is so purely service-oriented, it is insensitive to resource loadings. One effect of this in a limited core environment is to maximize the need for "paging in" the procedure and data of a given program. This can be partially ameliorated by partitioning core into areas so that the swap-in of one program can be overlapped with the processing of another. Since the system operates in a cyclic fashion, the selection of a given program for execution always implies its successor, and the core requirements of this successor can be established. Quicktran on the IBM 7044 used a scheme of this general nature to maintain an overlap of input and execution. If core can be made available for all members of the mix, of course, the problem disappears.

External priority may be applied to round robin scheduling by associating the size of a time-slice with different priority levels. A given program will receive its selected activations in a cyclic fashion but will retain control of the computing system longer each time it does so. To the extent, however, that the process is limited by I/O, it may not be able to take advantage of its slices, and the priority may not be effective.

Round robin scheduling has no particular interest for the batch low-level scheduler for a number of reasons, primarily because it is a resource load independent scheme based upon a concept of fair service that is of no particular interest to a batch scheduler with no illusions to sustain. By its very nature the technique denies dynamic running priorities, since it is cyclic in nature and takes cognizance of neither resource nor time left to use resource information.

32.7. LEVELLED QUEUES

Various elaborations of round robin, however, allow it to become more resource-sensitive. One common technique is to develop a dynamic running priority for active tasks based upon their age in the system. Levels of queues are formed to which programs gravitate on the basis of their usage of computer cycles. The high-priority queue contains jobs that have received less than a given amount of time (been given a number of activations). This queue receives frequent service of time slices that are of relatively short duration. Whenever a member of this queue is capable of using the CPU, it receives an opportunity to do so.

There is a strong "shortest job first" flavor to the exclusive servicing of this high-priority queue. "Short jobs" tend to finish their function while they are members of this queue. Time is apportioned equitably between all members of the queue. Since short jobs tend to stay on this queue for all or most of their time in the system, they will be given service at the best possible rate.

After having the CPU for a given number of slices, however, a job loses its claim to being a "short job" and it is moved to a lower-level queue. This queue receives service only when the first queue contains no element that can receive useful CPU time. Attention then transfers to the second-level queue. Elements on this queue receive fewer "shots" on the CPU, but they receive longer shots when they are activated. After a certain number of time slices on this queue, a program may be made to descend yet another level, etc.

Because the size of the time-slice given at various levels is often considered to be a power of two greater than the time at a higher level, this technique is usually called "exponential scheduling." It is not at all necessary that the slices be defined in this way, however, and it is not even usual in existing systems. The SDC time-sharing system using a three-level queue does not exponentially relate the time slice; EXEC VIII for the UNIVAC 1108 allows such flexibility of time-slice definition regardless of level that it is not effectively an exponential scheme.

The intent of the scheme is to overcome the need constantly to establish and reestablish environments in core for all active programs. The concept of the "dynamic priority" is invoked by assuming that one can predict further demand on a resource by a program by observing its demand history. If a program establishes itself as a "long program" with large CPU cycle requirements, the technique of allowing it to run less frequently for longer periods of time attempts to reduce the load on the system associated with running that program while maintaining a high rate of service to short jobs. The number of program loads and swaps is reduced across the lifetime of the longer job.

The scheme, of course, can be extended in a number of ways. In its basic form it is only minimally resource-oriented. One problem that would develop from the pure use of this scheme is channel underutilization for those channels associated with I/O not related to paging and swapping.

We may take cognizance here of a tendency for time-sharing systems to overlook the traditional batch concern in this area. To the extent that conversational systems tend to deal with programs that make very little or light use of the general I/O subsystem, the time-sharing systems which support them are concerned with a narrower usage of this problem, which might be called the CPU-paging balance of the system.

The general balancing of the system is left to "background" processes (if they exist), which have as their major function the soaking up of unused system cycles. In a mixed conversational/batch environment with batch jobs running from the same queue, the batch jobs would fall very quickly to the lowest internal priority levels. This would have the good effect of indeed reducing them to background cycle "soakers." However I/O-bound and CPU-bound jobs would compete on the same basis. The CPU-bound jobs would make good use of their extended time-slices; however, the extended time-slice would be of no use to an I/O-bound run, since it would characteristically release the CPU before the end of the slice. The period of time between activation of an I/O-bound job would be long at lowest priority level; as longer jobs drifted to the lowest queue, the interval between I/O-bound activations would become longer and longer, and use of the I/O subsystem consequently more and more sparse.

If high channel utilization, in general, is still meaningful to the system, it is necessary to modify the scheduler to give priority to the I/O-bound jobs. This can be done by recognizing them (as we have described before, by observing which jobs do not complete the time slice) and dynamically adjusting their priority within the lowest-level queue.

A scheme described in EXEC VIII that uses a levelled queue scheme of this type for control of batch (in connection with deadline scheduling, described below) is to reduce the queue level of a program when it completes its time slice, but to maintain it at a level for so long as it does not complete its slices at that level. This leads to a nice distribution of I/O-bound and CPU-bound programs across quantum levels, and it tends to seek out the I/O-bound programs.

Since a program may appear to be compute-bound (finishes its time-slice) at high levels, where slices are shorter, it drops in priority to the point where its processing time is such that it cannot continue without requiring I/O, and it does not fall past that point. True "number crunchers" (seriously compute-bound jobs) fall to the bottom levels. Within each level of queue other elements, including external priority, determine the specific size of the time slice a program will get.

An even more flexible approach to levelled queues is present in IBM's TSS 360. The "table driven scheduler" provided with this system allows the user to specify what constitutes a reason for a level change for an active program.

The resource insensitivity to levelled round robin comes from its implicit assumption that a long user of the system is a "large" user of the system's resources. One technique for making the scheme more sensitive is to impose the concept of job classes upon it. Programs may be classified by core size or device utilization; movement from level to level in the queues may be a function of a dynamic request for core which exceeds some system limit. By dropping large core users to the bottom level, the system truly guarantees a relief of load for task-switching based upon its use of the resource (core) that most significantly contributes to switch load and to (perhaps) channel load.

32.8. DEADLINE SCHEDULING

The levelled queue technique is also used in connection with a highly service-oriented discipline called deadline scheduling. This is a technique associated with both time-sharing and batch systems. In a time-sharing system the concept associated itself with notions of when a response must be made to a terminal in order to sustain the service goals of the system. In batch system the concept of deadline associates itself with the ideas relevant to economic utility of information.

The deadline is a time after which the system may be said to have failed if it does not deliver a result. It is "priority," a relative concept, in absolute form.

Deadline is also relevant to another economic concept of computer usage. In selling service to users (internal or external users) a service agency must make some statement about what level of service is associated with the fee it wishes to charge. It is not truly acceptable to declare to a customer that he will get your "best" effort for $900 an hour, a little less attention for $750, etc. This has no specific useful meaning for a customer who does not know what "best" effort means. It will mean something different if he is sharing the machine with a number of others also receiving best effort than it will if he is the highest-priority job at the time it is running.

What the customer really wishes to know is what is the cost of time. He would like to buy "10 minute turnaround," "30 minute turnaround," "3 hour turnaround," etc., and he would like his prices to be set in that context. The deadline is absolute in this sense. Of course, no system, no operating algorithms, no high- or low-level scheduler can produce results in a chronically overloaded system. It must be possible to give service in time and space to those users who are scheduled to complete by a certain time.

Deadline scheduling is a member of a class of time-dependent scheduling techniques that are variations of service-oriented scheduling. Deadline represents the "time by which" rule. Other rules are "time left to go" and "time already received." When a "time to go" rule is used, CPU cycles are always given to the task that can run which has the least or the most time to go. "Time already" gives CPU to the task having the most or least time already received.

In attempting to distribute time equitably among tasks, different tasks will fall behind because of inability to use the CPU, and in a dynamic environment different tasks will arrive later than others. A constant attempt is made to enable late and delayed tasks to catch up. If, in a simple levelled queue system, priority is always given in a preemptive way to a new arrival, we have a form of the catch-up rule. The system is driving to complete all jobs at approximately the same time.

Deadline differs fundamentally in that it is not attempting to sustain equitable service, but to gather preferred service for a task in order to keep it "on schedule." The system is basically dependent upon an estimate of how long the task will take and on record keeping of how long the task has run. The deadline may be expressed as length and ending time. The high-level scheduler computes a start time (or the time may be directly expressed). The low-level scheduler constantly monitors the progress through the system of the deadline job so as to develop for it the dynamic priority that it requires in order to get sufficient service to proceed on schedule. If it falls behind schedule, its priority is increased; if it gets ahead of schedule, its priority may be reduced.

The monitoring process involves a computation that develops a ratio of the service required to the service available. Let us visualize a program that must be finished by 3:00 p.m. and which will run for ten minutes on the CPU. For simplicity, let us say that the CPU delivers 1000 cycles per minute, so 10,000 CPU service cycles are required for this program. Let us further postulate that the high-level scheduler admits this job at 2:30 and there are consequently 30,000 CPU cycles available.

Initially the program will receive a priority intended to allow it around 33 percent of the CPU cycles available. The program will be run at that priority for a period of time, and its behavior will be observed. Let us postulate that it receives only 20 percent of the CPU cycles, and it has been run for one minute. It has received 20 percent of 1000 cycles, 200 CPU cycles. It now has 9800 CPU service cycles remaining to go, but it must acquire these in 29 minutes, or during 29,000 maximum possible cycles which could be made available to it.

A recalculation of the program's priority (by dividing time remaining into time available) will cause it to be run at a higher-priority level. Let us postulate that it receives 25 percent of available cycles over the next minute, or 250 cycles. It has now received 450 cycles and requires 9550 more. These must be received from 28,000 remaining available cycles over the next 28 minutes. A higher priority will be developed by a recalculation of the ratio of "time to go" to "time remaining." Eventually a priority will be achieved that will get it sufficient cycles to complete on deadline.

The problems associated with the technique are numerous indeed. Some of the difficulties are procedural, in the sense that they may be solved (or responded to) by a choice of a technique among available techniques; other difficulties lie in the mapping of the scheme upon a system of the nature of an asynchronous digital computer system.

The initial problem lies in the determination that a deadline program must start. This is a problem only for a batch system, of course, since start time for a time-sharing task is determined by system log-on.

The batch problem lies in the fact that the high-level scheduler has the deadline and estimated run time of a program (or its start and deadline times) but has rarely any appreciation of system load. The times it has available to it are in all probability based upon some anticipation of stand-alone running time. The probability, therefore, that a job with a deadline will be admitted to the system in such a way as to cause a minimal disruption of system operation is rather low, since the admission will tend to occur at the latest possible time that the job may be activated and still complete on schedule. Since the program will develop "acquired" priority in terms of its behind- or ahead-of-schedule status, the closer to deadline the program is admitted, the more demanding of service it will be, and the tendency to "dry" the system will be greater and greater. Resource load balancing will tend to fall by the wayside and levels of multiprogramming to contract as the deadline job becomes more and more insistent.

A further problem lies in the allocation of resources to deadline jobs once selected for admission. If resources are not available, they must be preempted. The technique of staging is used by the UNIVAC 1108 EXEC VIII system to acquire resources for all jobs, and it is particularly relevant to supporting deadline jobs. Selected programs are phased through a "facilities wait" and a core wait status. Facilities are acquired through priority competition (an external priority); core is acquired on a "best-fit-in-core-within-priority-group" basis in order to minimize core repacking requirements.

Deadline jobs, as they become more critical, may interlock and then preempt resources in order to get under way. A deadline job may cause the roll/out of a program in order to get a needed core resource. As a job becomes more critical, it has a higher and higher call on required core. The system allows a number of deadline jobs to be active at any time. Care must, of course, be taken to preserve feasibility. Since deadline jobs have this strong disruptive power, the system attempts to "qualify" jobs by requiring that they be registered by name to the system in order to avoid whimsical abuse of the deadline privilege.

The central problems of deadline scheduling are the determination of what "time left to go" and "time still required" ratios demand what priorities and the determination of the interval at which the dynamic priorities should be calculated. There are two fundamental environments in which deadline may run. The EXEC VIII environment supposes a mixture of deadline and nondeadline jobs; a pure time-sharing environment might well suppose all deadline jobs (with perhaps a background job).

In either case the levelled queuing mechanism may support the deadline concept. The UNIVAC system uses levelled queuing with a pure exponential rule to determine when a program is to run.

As we described before, existence on a level is acquired by age in system and I/O-CPU balance. Deadline programs move back and forth between levels, depending upon their progress along their schedule. At a given interval the low-level scheduler (dynamic allocator) computes the progress of a job and determines on what level queue it should be placed in order to meet its schedule. The low-level scheduler plans the work of the system for the next interval of time, as it were, and prepares the dispatching list for the dispatcher, which executes this list on a priority preempt basis until the next activation of the low-level scheduler. Activation may occur because of a change in core usage or on the time cycle.

A number of schemes for determining proper priority level have been studied. One scheme involves partitioning the jobs by their ratios and finding the proper queue within ratio intervals. For example, those jobs requiring more than 50 percent of CPU cycles would reside on one queue; those requiring 25 to 50 percent on a lower-level queue, those requiring 10 to 25 percent on a third level, and less than 10 percent on a last level.

The specific partitioning must come from an appreciation of what CPU cycles are represented by priority levels.

The queues may be treated in a manner equivalent to normal exponential scheduling; that is, the top priority is serviced so long as there is an active member, and lower priorities are serviced only when higher members are quiescent. This has the effect of time-sharing equitably among programs at a given "critical" level. The extension of the time slice at the lower levels would not seem to be appropriate, however, since the extended slice to noncritical jobs would tend to interfere with the ability frequently to recalculate priorities and guarantee percentages of service over given intervals.

The EXEC VIII approach, therefore, of using queue level to determine when, but not how long, seems appropriate to deadline scheduling. In EXEC VIII the size of a quantum is determined by a number of elements, of which only one is the queue level. A basic allocation time slice is determined at system generation time, or at system initiation time. This basic slice is multiplied by the queue level to give a basic priority slice. This number is then further modified by the program's external priority.

The impact of external priority is controlled by a division of a system parameter (F), which is divided into the priority of a program. There are 26 priorities—A to Z. If F is small, then external priority will have a large impact on quantum size; if F is large, (and as F increases), the impact on quantum size of external priority becomes less and less until at $F = 26$ it has no impact at all, the level basic allocation unit being multiplied by one.

In EXEC VIII, therefore, external priority is a strong factor for the high-level scheduler and for acquiring resources. The internal priority is a "deserved" or dynamic priority, and the usage of the CPU is a number derived from the "deserved" priority and the regulated adjustment of this number by the "purchased" priority.

In OS/360 a rough equivalent is the partition priority by class in MFT. Notice that EXEC VIII attempts to impose the concepts of resource load and deadline through its use of queue selection on the basis of time-slice completion and deadline. Again, TSS 360 provides an even more general capability for moving from level to level from its table.

32.9. RECALCULATION OF PRIORITY

The question arises as to when the priorities should be recalculated. This is a question appropriate not only to deadline scheduling, but to all dynamic internal priority systems whether service- or resource-load-oriented. The low-level scheduler has an opportunity to observe the system whenever an interrupt occurs, either time slice or I/O. Dynamic priorities may be calculated with this frequency, and there is no reason on principle why they should not be. If the "dispatching" function is intimately connected with the low-level scheduling function in situations where there is effectively no true low-level scheduling (except the enforcement of preemptive external priority), then in environments of highly dynamic scheduling the close connection may be unchanged. In such an environment each interrupt would involve an analysis of the behavior of all programs and a dynamic selection of the next to run.

Such an (extreme) dynamism would be consistent with the general philosophy of monitoring a system in the smallest possible scale. As we saw with our earlier discussion of loading levels on time-shared facilities, the more discrete the elements of load that we can manipulate, the greater the flexibility that exists in the system and the greater the potential for achieving our objectives.

The constraint exists in any dynamic scheduling system, however, that the scheduling interval must be long enough so that significant new information may be acquired about the processes involved; further, in computing systems that use schedulable resources to compute a schedule, the load of schedule analysis must be kept proportional to the expected increase in system effectiveness. The "burden" of the operating system in terms of the percentage of time and space that it requires must be reasonably bounded.

In the time between interrupts in a computer system, it is unreasonable to expect that significant new information will have been acquired to justify a reanalysis of the set of dynamic priorities in the system. Therefore, it seems reasonable to constrain the occurrence of the operation of the low-level scheduler and to separate the scheduling from the dispatching function.

In this context the low-level scheduler prepares a schedule for an interval of time (perhaps a few seconds) based upon its appreciation of the status of the mix and adjusts that schedule upon notification of a change in resources or on a fixed time frame. The penalty for this is the need to record information about the performance of the system so that it will be available to the low-level scheduler when it operates. The collection of information (CPU cycles used, I/O requests initiated, etc.) is input to the low-level scheduler.

In deadline scheduling it would be possible to compute the new ratios for a program whenever it lost control [gave up because of a resource delay (I/O)] or when its time slice ended. This, in effect, would maintain a constantly current list. The order of the list would change with each program release. Alternatively, it is possible to choose more extended intervals for analysis, which would allow a program more "shots" at the machine to collect more significant data about its progress.

Since it is desirable to recalculate priorities for those jobs most critical in the system more frequently than for those jobs less critical in the system, one scheme involves a different rate of calculation based upon what level queue the job is in. Those jobs in the "critical" queues (depending on ratios) are investigated at each interval for the possibility of moving them down (reducing their demand on the system); those in lower-level queues are investigated less frequently (every two intervals, every four, every eight, etc., perhaps using an exponential rule).

This scheme is based upon the assumption that the better the ratio is, the less urgent it is to inspect a queue for upgrading. If this is not so, under conditions of heavy system load, for example, where jobs are characteristically tending to fall behind, the infrequency of upgrading can result in a lower percentage of deadlines achieved.

An alternative is simply to recalculate the priorities for all deadline jobs at some given period of time, and to recalculate for any one job as frequently as for any other, regardless of its current position in the queues. If queue levels are not merely a function of deadline, but tend to be composite (as in 1108), this might be a more reasonable approach, since competition for the machine from any queue is not only a function of time to deadline, but of age in system and I/O-CPU balance.

32.10. PROBLEM WITH DEADLINE SCHEDULING

The critical environmental problem for deadline scheduling comes from our earlier appreciation of the interaction of loads on resources. These tend to be minimum in a time-sharing environment (and those perhaps hidden behind staging mechanisms so that all references to data are to a "virtual memory") but of a high degree of importance in batch systems.

The difficulty lies in the use of CPU cycles in determining the priority of the program vis-à-vis its deadline. If we compute a priority based upon required CPU cycles and basically posit that we require 3000 of the next 5000 cycles, we are overlooking the fact that at any priority this program may be unable to use 3000 cycles because of the set of loadings it makes across all resources and the constraints they place on its use of the CPU. What we really wish is a modification of the priority by this program's demonstrated ability to use CPU cycles. In a sense, the recalculation of ratios provides us with a picture of its ability to use cycles. A program that is falling behind at a given priority level may be doing so because competition at that level is too high, because it is not getting the "shots," or because it is not able to use the shots that it gets. Moving it to a higher priority will alleviate the first condition but not the second.

If the "limiting" resource is not the CPU (the lower bound), then any appreciation of being "on schedule" based upon CPU progress must be naive, since it does not truly reflect the progress of the job through the system. This problem leads us to a discussion of our next area—dynamic priorities based upon resource utilizations and loadings and what might be done to reflect the dynamic nature of programs as they are running.

32.11. RESOURCE-ORIENTED DYNAMIC RUNNING PRIORITIES

We have seen before that the job of a multiprogram scheduler is to maintain resource loadings so as to complete a schedule in a minimum time; more generally, its job is to sustain the throughput of a system by making optimal usage of the asynchronous resources available to it. We have seen that minimum times are achieved when resources are scheduled on smaller rather than grosser scales and that allowing variations in the use of a resource over time is necessary to achieve minimum bounds. We have also seen techniques for the monitoring of progress on a system—basically deadline scheduling, most to go, and least executed—in order either to maintain an equitable distribution of system attention or to grant special service to classes of users.

So far we have not seen the use of monitoring techniques to achieve the basic goal of multiprogramming—enforcing resource load. Programs will vary over their lifetime in the level and pattern of usage for CPU and I/O. The concept of deserved priority is present in the 1108, and the concept of dynamic recalculation also exists. The thesis of this section is that dynamic running priorities that are calculated on the basis of the recent history of a program's behavior across its resources are a meaningful way to increasing system throughput.

The first constraint in such an approach is external priority. To the extent that this "purchased" priority is allowed to dominate a system, one may expect poor general system performance and efficiency. In a system where high- and low-level scheduling exist as separated functions, it is possible to restrain the effect of priority to admission and initial allocating and to modify the effect or ignore it for low-level scheduling.

In a dynamic resource-oriented system, however, allocation balancing is an important element, and it should not be ignored even at the mix admission level. The high-level scheduler should, in addition to finding resources for a job, attempt to find those resources in such a way as to minimize system interference and contention.

In a minimal way such things are currently considered. In OS/360, for example, allocation strategy attempts to balance data files across channels, and the system is sensitive to directions by a programmer that will aid in optimizing the channel balance for a given program.

High-level schedulers, however, are generally not sensitive to the utilization status of resources and do not admit programs to the mix on the basis of their contribution to system performance. The interface between low- and high-level scheduling might be softened here, even to the extent of (as discussed below) jobs moving in and out of the mix in order to achieve resource target loads.

The concept of resource-oriented, dynamic running priority is quite simple. Since the assignment of high priority to I/O-bound jobs has the effect of maximizing system effectiveness by providing for higher levels of overlap, and since increased utilization is achieved for all resources by giving priority to I/O-bound jobs, some scheme is needed to allow the system to observe what the I/O-bound jobs are and to grant them priority. The observation may take a number of forms, and it can be run with or without "time-slicing" on a millisecond level.

One scheme is to monitor every few seconds the amount of time a job has spent using the CPU. The ratio of total CPU time and time waiting to time waiting gives a rough balance of I/O boundedness. A job, for example, which during a previous 6 second interval had spent two seconds waiting and had used the CPU for four seconds, would have a using to CPU/wait ratio of 2/6; a job that had spent one second waiting and used the CPU for five seconds a ratio of 1/6. The second job, more compute-oriented, would have a lower dispatching priority, based upon these ratios.

Another scheme is to count the number of I/O requests submitted by a job during an interval of time and to grant highest priority to the greatest submission of requests. These schemes have been attempted in user-modified versions of the Chippewa operating system for the CDC 6600 and for OS/360 MVT. In both cases the use of dynamic running resource-oriented priority increased the throughput of the system in a nontrivial way.

Both schemes depend upon fixed-time investigations of the recent history of a running program (not unlike the conceptual basis of look aside) as a means of predicting its immediate future performance. Alternatively, a new dispatching priority could be calculated for every program at the time of its release of control. Both the 6600 and 360 schemes conceptually separate low-level scheduling from dispatching and preplan a time period for the dispatcher.

A number of questions arise in connection with dynamic priorities. As we shall see in the later chapters on core management (Part 7), the critical resource of a system is often core. The free rearrangement in core determines the number of programs that can be active in a mix and, consequently, the population of programs that can be considered for internal priorities. If the number of programs is small, the breadth of possible choices is restricted, and the use of dynamic priorities will tend to be lessened, although it is interesting to see that the OS/360 implementation achieved its result (up to 23 percent more throughput) in the conservative core environment and with the shallow level of multiprogramming characteristic of OS/360.

Beyond this, the number of channels is a serious consideration since (as we show later) the more programs and the more I/O activity there are, the greater the average device response time on a system of few channels, the greater the probability of any program's becoming I/O-bound, and the smaller the probability of an I/O-bound program being truly ready for CPU service. It is here that more help from the high-level scheduler might be useful. In addition to CPU-I/O ratios, priority might include such elements as breadth of channel distribution, or dynamically channel queue lengths might be watched and a low priority given to programs that are glutting a channel.

A further consideration must involve buffer space. A fundamental assumption of dynamic running priority is the variation during short intervals of a program's I/O boundedness and its rate of submission of I/O requests. Current operating systems will currently force a program to "wait" if it refers to a file that has an "empty" buffer. Thus the program is unable to continue submission of any other I/O. The variation in the number of requests is constrained by the artificial "waiting" of a program. This can be overcome by allowing a program to request return, regardless of buffer condition, and thereby to allow him to submit his further I/O if it exists.

In general, buffering is a critical system concept too loosely regarded by most programmers. System control of buffer depths for a mix of programs is a (potential) form of mix-optimization. Given any mix of I/O-bound and CPU-bound programs, it is possible to change the mix by dynamic acquisition and release of buffers by the system on behalf of programs, rather than by the programs themselves. In the presence of large-capacity core units (as on CDC 6600 and IBM 360) many I/O-bound programs can be turned into compute-bound programs by affording buffering of sufficient depth. The ability to change buffers by the system is dependent upon sufficient space (of course) and also upon a system observation that the mix is currently all I/O-bound or all compute-bound and that something should be done. Changing buffering levels is one way to balance a mix when the system is constrained to a given set of active programs.

A "balanced" mix is a mix that is sustaining the loading that the system requires. It consists of I/O-bound and CPU-bound programs; the maintenance of mix balance begins by knowing what programs are placing what load currently on the complex of resources. What is a proper strategy when the active members of a mix cannot be balanced? One technique is to attempt to avoid this by a further classification of program states. The active mix is defined to be greater than the dispatching list, so that at any point in its planning the low-level scheduler selects some subset of possible active programs for the dispatcher for the next interval. The scheduler artificially constrains the number of programs that will receive cycles during the next interval.

An instance of this is found in TSS 360. The dispatchable programs are those whose scheduled next running times are less than a given system clock time. The TSS scheme is based upon the addition to a running "next" execution time of a "delta to run" time, which derives, at the end of a service interval, the next time to run this program. If that next time is less than clock time (the run is behind schedule), the program stays on the dispatchable list; if it is greater, the program falls behind onto the eligible list. Only a fixed number of tasks receive service in any interval. TSS uses a queue level technique in which queue length and time of time slice, as well as "delta to run" (scheduling interval), are represented as system parameters. In addition, the maximum number of pages a program may have in core (its permitted load on core) is associated with each level.

The intent of TSS is not dynamic load levelling, as we are discussing, but the constraint of submissions for dispatching from a population of eligible jobs shows one potential technique of mix balancing. As programs become I/O- or CPU-bound they could move forward or backward from the eligible to dispatchable list, the dispatchable list containing those programs imposing the best load in the system.

The high-level scheduler would overallocate core and CPU cycles. One effect of the constrained size of the dispatching list is to minimize core relocation. The dispatching group that maintains balance tends to balance their core utilization also, and hold the core environment stable for the dispatching period. Periodically and/or on the change in performance of a program from I/O- to compute-bound, a replacement would be found for it in the mix. One disadvantage to this scheme would be the idling of assigned devices over the period of time that a program was eligible but not on the dispatching list.

One other scheme involving resource appreciation involves the use of resources as a criterion for program switching. If a program A, using one level of resources, is running, and a program B, using another level of resources (3K vs. 10K core, 5 vs. 10 tape, etc.), has an I/O completion, and if both programs have the same priority, is there any reason to preempt one for another? If preference were always given to the larger user of resources, then he would tend to finish more quickly and release his resources to the system so that more jobs might be gotten underway. This is a nondynamic appreciation of resource use that would tend to increase nominal job completions over time.

What would be an ideal environment for resource-oriented, dynamic dispatching is to have as part of the "stateword" for every program a number representing its load on each resource to which it is associated. The act of scheduling would be the act of submitting the loading vector of a program to a "black box" (perhaps a sorting associative memory), which would compute the system load over the next interval if that program were to be added to the dispatchable list. The dispatchable programs would be the set that achieved target loadings for each resource. At the end of the interval the new loadings would be available, and a new dispatchable set would be created. The system performance activities that are currently under way might well lead to the availability of such numbers in an active useful way (rather than out on tape). Room in the stateword could be found by eliminating those program addressable registers for which there is so little real use at this time.

32.12. FINAL COMMENTS

Some final remarks are still to be made in this complex area of scheduling and control. One current dispute is the "level" of task-switching burden that a system must be asked to absorb. Within the constraint of reason a system should switch tasks at every possible opportunity. It should switch a task whenever it becomes apparent that by such a switch better utilization of equipment will be achieved. Time-slice switching is already established as a batch requirement by most vendors.

The true question is the balance of burden vs. increased throughput. The statement "10 or 20 percent burden or overhead" is not a meaningful statement, since it gives only half of the picture, the other half of which is increased throughput. Care must be taken in the design of future systems to aid in dynamic task-switching. Smaller statewords are one requirement; duplicate registers to enable the supervisor to field interrupts without status saving and restoring are another potential aid. One problem in the design of cache memories is the extreme penalty paid for task-switching at a high level on such systems.

A more thorough investigation of the relationship between high- and low-level scheduling might be due. Currently the relationship is seen in two forms: One is the way we have described it here; another way of viewing the relationship is more similar to the relationship between the low-level scheduler and the dispatcher. In mixed real-time, time-sharing, batched systems, each program type may be viewed as a subsystem. The high-level scheduler hands off periods of time to each subsystem, which is then free to impose whatever local scheduling rules it requires during its interval. Coordinating, if any, is effected by a guarantee of percentage of time across subsystems.

This implies that the two- or three-level concept of scheduling is not sufficient and that more elaborate scheduling networks may be required. One must never, in any case, forget that there is always a hidden level of scheduling, priority rules, etc.—that of the hardware itself, which gives priority to processors, channel positions, etc.

One nice expansion in the relationship between high- and low-level scheduling would be for the low-level scheduler to request of the high-level scheduler the submission of programs of certain given characteristics. This would come closer to making the posited "infinite number of jobs" a reality and would enable more program mix balancing from that number, rather than from the limited number named at any time to the low-level scheduler. Conversely, the low-level scheduler should be able to "ship back" any programs which it cannot use, or which are behaving badly relative to the mix.

Now certainly objectives like these are not consistent with the concept of rush jobs, jobs of special economic value, etc. Such needs must be enforced on a system, of course, but the inclusion of such mechanisms as deadline scheduling or priority protection to some degree should be extensions of a resource control system and not at its heart.

PART SIX:

SOURCES AND FURTHER READINGS

Cantrell, H.N., and Ellison, A.L., "Multiprogramming System Performance Measurement and Analysis," *AFIPS Proceedings,* Spring Joint Computer Conference (1968).

Codd, E.F., "Multiprogram Scheduling," *Communications of ACM,* Parts 1 and 2 (June, 1960); Parts 3 and 4 (July, 1960).

Coffman, E.G., and Kleinrock, L., "Computer Scheduling Methods and Countermeasures," *AFIPS Proceedings,* Spring Joint Computer Conference (1968).

Corbato, F.J., and Vyssotsky, V.A., "Introduction and Overview of Multics," *AFIPS Proceedings,* Fall Joint Computer Conference (1965).

Dahm, D.M., Gerbstadt, F.H., and Pacelli, M.M., "System Organization for Resource Allocation," *Communications of ACM* (December, 1967).

Franklin, E.S., "Job Stream Manager, OS/MVT Release 16," *Contributed Program Library* (IBM) 3600-03.6.010.

Gaver, D.P., "Probability Models for Multiprogramming Computer Systems," *Journal of ACM* (July, 1967).

Greenberger, M., *The Priority Problem.* Clearinghouse for Federal Scientific and Technical Information, Department of Commerce. AD 625728 (November, 1965).

Haberman, A.N., "Preventing System Deadlocks," *Communications of ACM* (July, 1969).

Havender, J.W., "Avoiding Deadlock in Multitasking Systems," *IBM Systems Journal,* Vol. 7, No. 2 (1968).

Heller, J., "Sequencing Aspects of Multiprogramming," *Journal of ACM* (July, 1961).

Hellerman, H., "Time Sharing Scheduling Strategies," *IBM Systems Journal,* Vol. 8, No. 2 (1969).

Howarth, D.J., Jones, P.D., and Wyld, M.T., "Atlas Scheduler," *Computer Journal* (October, 1962).

Lampson, B.W., "Scheduling Philosophy for Multiprocessing Systems," *Communications of ACM* (May, 1968).

Lonergan, W., and King, P., "Design of B5000," *Datamation* (May, 1961).

Manacher, G.K., "Production and Stabilization of Real Time Task Schedules," *Journal of ACM* (July, 1967).

Marshall, B.S., "Dynamic Calculation of Dispatching Priorities," *Datamation* (August, 1969).

Oestreicher, M.D., Bailey, M.J., and Strauss, J.I., "George 3-A General Purpose Time Sharing and Operating System," *Communications of ACM* (November, 1967).

Ramamoorthy, G.V., "Analytic Design of A Dynamic Look Ahead and Program Segmenting System for Multiprogrammed Computers," *Proceedings of ACM* (1966).

Ramsay, K., and Strauss, J.C., "A Real Time Priority Scheduler," *Proceedings of ACM* (1966).

Schwartz, G.S., "An Automatic Sequencing Procedure With Application to Parallel Programming," *Journal of ACM* (April, 1961).

Sperry Rand Corporation, "1108 Multi-Processor System Executive, Programmers Reference Manual," UP-4144 (1966).

Sperry Rand Corporation, "494 Real Time Operating System General Description," UP-4055.

Stefferud, E.A., *Environment of Computer System Scheduling: Toward An Understanding.* Clearinghouse for Federal Scientific and Technical Information, AD 661666 (November, 1967).

Stevens, D., "Overcoming High Priority Paralysis: A Case Study," *Communications of ACM* (August, 1968).

7

The
Management
of
Primary Store

Chapter 33

DEVELOPMENT OF
CONCEPTS

33.1. INTRODUCTION AND HISTORICAL NOTES

The concept of "storage management" in its broadest sense includes the
entire set of ideas relating to the physical devices on which data is represented, the cost and speed of such devices, and their relationships with
each other; it includes the entire set of ideas relating to the formation,
representation, initiation, and execution of programs, particularly including the functions of compilers, loaders, and monitors; it includes the set of
ideas relating to operating system schedulers, I/O support, and data
management. In this broad view the "storage management" problem is the
intersection of all aspects of the computing problem.

We shall be primarily interested in a much more modest concept of
storage management: the motives, strategies, and techniques relevant to
the allocation of elements of "primary store" to elements of a program
competing with other programs or with other elements of itself for the use
of this resource. We are particularly interested in those techniques because
of the implications they hold for throughput in a multiprogramming system or a multiprocessing system.

Consider the program represented in Fig. 33.1. This program is represented in what has been called "unfolded" form. Each and every required
information element—procedure, constants, subroutines, temporary storages, and input and output files—is represented as being part of a single
linear list that would be 2900 words long if placed in the "primary" store
of a machine. If we had a machine with 2900 words of this "primary"
store, it would be possible to execute this program directly; the only ex-

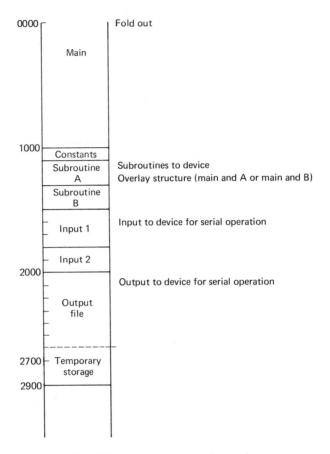

Fig. 33.1. A program in unfolded form.

ternal device required would be some mechanism for loading the program.

Instances of pure unfolded programs are rare in the real world. Usually some form of representation on a device other than primary store is required for the output of a program. Almost as commonly, the quantity of input data required by a program is too large to be represented totally in memory, and the data are allocated to another device, commonly tape or disc.

The program is aware of this secondary allocation because it is now required to establish a means of accessing the data that is, by convention, quite different from referencing it in store. Input/output operations in which logical partitions of input and output data files are dynamically overlaid by successors are the fundamental form of "folding" a program to fit into available core.

We are so used to I/O overlays that we tend in the current state of the art to accept it as a necessity without question as a fundamental of programming. We may now think of those areas in primary store associated with the files as buffer areas used at the interface between I/O devices and the program.

Whether or not this traditional form of dynamic overlay and the distinct conventions for referencing I/O will long continue in the industry is now becoming a point of contention. Concepts of data staging and I/O transparency are rapidly developing in connection with concepts of "virtual storage," a mapping of the unfolded form of a program into a conceptually large address space, which allows the programmer to reference all procedure and data uniformly as if they were in core. Hardware mapping devices, discussed below, transparently handle the transfer from virtual (disc or similar device based memory) to real store.

Currently, fundamental folding introduces the separation between the concepts of "device" allocation and "storage" allocation, with drum and disc straddling the problems of both. The concept of hierarchy is also introduced, as well as those questions relating to programmer control and knowledgeability.

One interesting and positive by-product of the separation comes in the "shareability" of the data allocated to a secondary device. This is achieved by the recognition of the data file as a named element, schedulable and assignable in its own right, independent in space and time of the particular program. This capability would not exist if data were "buried" in the unfolded form of a program.

I/O impact on program formation has, traditionally, been straightforward. If the program is written in machine language (or a pseudo machine language, like a basic assembler) or a higher-level language, the program creator is aware of what is expected in primary storage in the way of procedure and data and what he expects of I/O. The program achieves "executable" form when it is "bound" to certain specific resources of a machine. For elements represented as residents of core storage this "binding" is the process of associating a machine specific address with each element. For elements associated with I/O, the binding is the process of associating with each "file" a specific device address.

In early systems the "binding" of primary storage was accomplished by the programmer writing in true machine language, or by a compiler (or assembler), which mapped symbolic references into true machine addresses. Address assignment was static and completely defined before presentation of procedure for execution.

The "binding" of I/O was also initially directly represented in coding. UNIVAC I programmers, writing in C-10 machine language, wrote direct device addresses into coding. The undesirability of this in terms of availability was overcome by providing a patch panel that allowed different physical tape drives (UNISERVO's) to be addressed by device addresses as required by the program. The concept of "logical SERVO 1" and "physical SERVO 1" etc. is, therefore, a concept associated with the first commercially available machine.

IBM went further and associated address selection switches with its tape drives. The selection of devices, therefore, was very early recognized to be a process most expeditiously executed at program initiation and not at program creation time. The "binding" was delayed one level. Very early operating systems concentrated on formalizing the device assignment procedure at initiation time. This limited form of "device independence" was an absolute operational necessity.

Another "folding" practice is the partitioning of procedure into formal or informal segments that follow each other in some predefined or random sequence into core. By a "formal segment" we mean a sequence of code which is named and which characteristically has some "linkage" procedure associated with it. It is a subroutine, a subtask, an inner block of ALGOL or PL 1, a paragraph of COBOL, etc. An "informal" segment is a sequence of code that is part of a larger sequence but one that, for reasons of physical space, cannot coexist with other subsequences in core and during execution must overlie other sections of coding.

The reason for informal overlay planning is simply that a given program is too large to fit into the core of the machine. Before there were operating systems, "overlays" tended to be informal, and each programmer established his own techniques for reading in a "block" of code. Since binding to machine addresses was already a fait accompli, special areas of core were set aside to hold "overlays" addressed to them.

Operating systems have tended to formalize linkages and conventions for overlay structures, but to leave the basic concept untouched. Programmers, whether tending to the mechanisms themselves or linking with an operating system, are explicitly aware of the request and satisfaction of an overlay activity.

Conceptually, an "informal" segment is truly a logical and combined part of a single program. It has been coded or compiled and associated with machine addresses at the time the entire program was put into executable form. Necessarily, in the overlay environment there exists another section of the program with the same addresses, and these cannot coexist in primary store.

33.2. CONCEPT OF RELOCATABILITY EMERGES

Formal segments came into being with advances in operational concepts. Probably fundamental is the concept of the library-provided subroutine. The UNIVAC I (R-U Pair), IBM 704 (TSX), and most other early computers of the one-address type had instructions that facilitated the recording of an address in a specific place in the machine and the transfer to a routine that would use this address to branch back to the calling routine. The efficiency in storage space achieved by allowing easy access from different points to common procedure was early recognized.

The provision of these procedures from a common library was an early development associated with FORTRAN, COBOL, and assemblers. Since the subroutines were written to be used by a population of diverse calling programs, it was necessary to establish conventions for their use— for calling them, for passing parameters to them, and for returning from them. Initially references to formal subroutines were handled by the programmer or compiler, both of which had access to the library. Before initiation time these subroutines became integrated into the program and lost, except for linkage conventions, their formal nature.

Certain of them, however, were not integrated into the calling program. These were the I/O support subroutines in slightly more advanced systems. These were initially coded into fixed locations and provided at initiation time into those locations.

To accommodate this memory assignment, statements were added to languages in order to instruct the compiler/assembler to avoid using the locations needed by I/O support. The subroutine library did for procedure what data file separation did for data; it externalized and formalized the existence of a code capability which thereby became "sharable."

A formal segment has as one of its characteristics an identity of its own independent of the calling program. It may be copied and "bound" to a given program, in which case the incarnation becomes private, but the concept of independent functional modules gives rise to the possibility of other techniques and raises questions about when things should be bound to each other. We introduce here another form of "binding," that of associating one program with another as distinct from associating a program with machine addresses.

Consider the FORTRAN environment, where a large number of system functions and subroutines are available and where it certainly seems redundant and inefficient to recompile these routines for each use. Since compilation time is often a major consideration, it seems useful to represent these functions in a machine form that could be later associated with a using program.

Unlike an I/O support environment for an early machine with relatively uniform and shallow I/O resources, it is impossible reasonably to represent all library routines in fixed core locations or to predict which will be used; therefore, unless one represents all possible routines for all possible address assignments, it is impossible to code to machine addresses.

Therefore, the recompilation for each instance of usage can be avoided only if one conceives of a machine address independent, "relocatable" form and another memory address "binding" time for these routines. The defined "bind" time became "load" time and a new system function, "loader," came to be recognized.

Further motivation for the development of the "loader" function came from the development of large programming projects, where it became convenient to partition program creation and "modularize" system organization. This led to a need for an ability to link independently compiled programs together in various ways for various debugging and integration activities. Since it was not possible to predict exactly the environment in which a module would run, it was convenient to delay "binding" until load time. Compiler output became "relocatable" program elements. The compiler produced code relative to the origin of the element with which it was working for all symbols "internal" to that element. For references outside the element, the name of the symbol was passed forward. At load time the "loader" resolved external references with the address assigned to the symbol at the time of load.

The basic function of the loader was to transform the "relocatable" form into an absolute form by adding to each "relocatable" instruction the actual address assigned to the origin word of the element. Special information bits were associated with each instruction, indicating whether or not the operand address was "relocatable," i.e., required loader augmentation at load time. As each element was allocated and external references were resolved, real addresses were written into reference points to effect linkage.

This process of binding modules to each other caused them to be concatenated, to "unfold" in an executable core image form. The development of this "collection" function of a loader was enhanced by the availability of block addressable "direct" access storage devices. By the very early 1960's systems based upon the residence on drum or disc of relocatable program elements available for "collection" from a system or programmer library were operative.

The overlay requirement for a loader brought into being assembly or control language statements directing the allocation of defined segments into the same address space. At the present time loaders accept quite extensive directives regarding the inclusion of segments into the executable core image form of a program.

The impact on languages of this allocation environment was to introduce conventions and directives for the identification of symbols externally referenced, and symbols externally referenceable. Directives like ENTRY, EXTERNAL, etc. or the convention of U1107 SLEUTH II, which showed by the number of preceding asterisks the levels of availability of a symbol, were added to assembly languages.

33.3. MORE DYNAMIC BINDING

The concept of formal segments became more formalized in languages like ALGOL and PL/I, where the very structure of the language tends to reduce "go to" programming and encourage nesting of procedures. In the late 1950's the concurrent development of ALGOL (and the concepts surrounding the language) and the development of multi- and real-time programming systems gave rise to an interest in the possibility of delaying binding even further.

In the presence of large complex programs, where it is not possible to predict storage requirements or segment usage even at load time, and where complex processing requirements make it impossible to predict who may or must coexist with whom and where, the total storage requirement may exceed physical space binding at load time, fixing addresses and overlay structures may be impracticable, and the requirement may exist for the dynamic allocation of segments on an as-called basis.

Segments are never truly bound to each other, but transiently to core locations with a mechanism in the system to provide addressability of modules to each other. The functions of location and linkage are redistributed in program time so that they constantly intersperse periods of computation. The overhead for this interspersion and later resolution repeated for every call on a segment is the price of flexibility, minimization of "dead" core space devoted to dormant routines, and the increase in system throughput due to maximum use of "active core."

The ALGOL influence on allocation comes from the concept of allocating space for local variables as the blocks that reference them are invoked. ALGOL, its variants JOVIAL and MAD, and its descendant, PL/I are "block-structure" languages with very precise rules for the "scope of names"—the conditions under which a symbol is defined and to whom it is defined.

In the program in Fig. 33.2, all declared variables in the outer block are available to all inner blocks. E is not available to block 1 and F only to 3, and the local declaration of A and B gives rise to the unavailability of the outer A and B during control of the locally declaring blocks. The A of block 2 is a local A, not the A of the outer block.

In ALGOL space for the data declared in 1 would be acquired at the time of its invocation, and released at the time of its completion. ALGOL is implicit in its dynamic storage conventions; PL/I extends to the explicit

```
BEGIN 1
   DECLARE  A,B,C,D
BEGIN 2
   DECLARE  A,E
BEGIN 3
   DECLARE  B,F
END 3
END 2
END 1
```

          1 A,B,C,D
          2 A',B,C,D,E   A' IS NOT A
          3 A',B',C,D,E,F   3 IS AN INNER OF 2

Fig. 33.2. Block structure scope of names.

language level the capability of controlling storage through various features of the language, which allow a programmer to specify when he wishes to override the basic ALGOL-like scope of name rules.

The implementers of ALGOL made large contributions to the concept of separating pure procedure (nonwritable segments of code) from data segments. The data segment for a block is dynamically established and released as a separate formal segment, and the block must be given addressability to its data on each invocation.

Whether or not this dynamic allocation is truly meaningful to a system depends in some measure on the interface between the operating system and the compiler and on the basic operating system core management philosophy. If segments truly acquire space from a system pool and release to a pool, and if that pool is available to a large number of users (say in a multiprogramming mix), then the free use of core may benefit all users. In some systems, however, the dynamic GETMAIN, FREEMAIN macros operate only within a predefined region associated with the job. This would make all dynamic memory allocation private to the region and unavailable, regardless of usage, to other jobs.

In any case, the dynamic loading of unbound program segments and the dynamic allocation of data spaces have produced a number of problems and a significant overhead. In the face of these problems a number of attempts have been made to provide hardware mechanisms to support some of the work necessary to support dynamic schemes.

The hardware support is by no means complete, and much work must still be done by programming systems, but there seems to be a general tendency, despite some bad surprises, to continue in the direction of partial hardware support for dynamic allocation. Many manufacturers have already built such machines, and we may expect more and better machines under the pressures of "time-sharing" conversational systems, where dynamic environment establishment for a terminal is an inherent operational requirement.

Chapter 34

DYNAMIC
STORAGE MANAGEMENT

34.1. IMPACT OF MULTIPROGRAMMING

A fundamental goal of store management is to achieve a balance between supply and demand for core and to provide mechanisms to place things in core when they are needed there.

We are classically concerned with a "two-level" storage system, where one has a backing device—a disc or drum (or, more recently, a large, slower core) associated with the system, which is used to hold elements when they are not in core.

Reflecting for a moment on the larger view, we may in future expect much deeper hierarchies of storage. In Fig. 34.1 we have indeed seven levels of storage associated with the system, and the total management system is responsible for the movement forward or backward in the hierarchy of program and data elements based upon frequency of use, expected next reference, etc.

Since associated speeds and costs vary widely across the gaps, it is no trivial problem to balance data flow across such a system. We are content here to consider what has been done to address the relationship between a primary store, directly word-addressable by the CPU, and a backing device, characteristically block-addressable over a channel.

The multiprogramming environment has, by its very nature, contributed to the interest in dynamic storage allocation. The effectiveness of multiprogramming as a technique for increasing the effective utilization of components of a computer system in part depends upon the range of choice of programs that may be run together and upon the overhead the running together of programs implies.

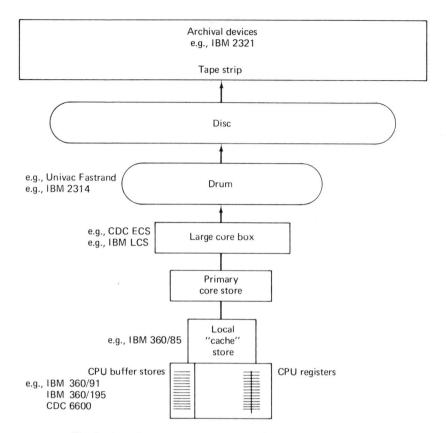

Fig. 34.1. A fully evolved hierarchy of store media. The data-schedule problem to be solved.

Consider two independently compiled programs. Because one of these is classed as I/O-bound and the other as computer-bound, it becomes desirable to run them together. If we for the moment assume that there is enough physical core space for both to coexist, the only constraint on their multiprogram operation is that the two programs have overlapping "address" space.

They have been compiled to the same absolute locations and consequently must operate from those locations. It would, of course, be possible to operate in such a manner as to bring in the core image of PGM 1, operate, and then bring in the image of PGM 2. Each incidence of task-switching would require a memory roll out/roll in. This has a number of undesirable effects. The assumptions concerning I/O-computer utilization that lead to the multiprogramming decisions would be entirely undermined by the channel burden involved in establishing programs in core. Further roll in/out time is entirely nonproductive.

Possibly some technique might be accomplished to multiprogram the read in of the remainder of a core image with early execution of already transferred portions, but unless the program was formally structured to a great degree, this would be difficult to achieve without complex and expensive monitoring. Finally, an additional mechanism would be required to allow I/O to proceed for PGM 1 while PGM 2 was in operation. System buffering could do this; the definition of all I/O buffer areas as being outside PGM space would provide areas for the deposit of I/O while an alternative program was running. Roll in/roll out and the elaborations and variations of the concept comprise a very real and interesting technique, but to rely on it as the only means of achieving multiprogramming implies an overhead unpopular with system designers.

A quick alternative, of course, would be to select groups of programs that can be run together and plan for doing so, allocation control statements being used in the assembler (ORIGIN = X) or compiler. There is no real reason why this is not an acceptable solution in some environments. Surely it is a reasonable technique in those real-time system developments where the population of programs associated with the application is known and coexistence may be predictable.

In job-shop operations, however, we must have flexibility in choosing a mix in the face of dynamically changing job populations; and in real-time applications, where processing chains are nonpredictable, preplanning is not an acceptable solution. The first basic approach to multiprogramming with regard to core management strategy involves a basic relocatability, which provides an ability to switch programs with "minimal" overhead and a wide flexibility in choosing the breadth and members of a mix. The responsibility for allocating space resides with a module of the system monitor, which maintains records of available storage and operates a relocating loader when it initiates a program. The number of programs in the mix may be predetermined, and definite "partitions" of core may be preassigned to programs of various types ("foreground" and "background"), or programs may be assigned to partitions on an as-available basis. The relocating loader may actually modify all relocatable addresses on the load, or it may only resolve unresolved "external addresses," binding a program as it is placed into core.

If the load function performs only resolution, then it depends upon some form of register to achieve the mapping of the program onto physical locations. The simplest hardware register support is to use an index register to hold the address at which the program starts (origin) and dynamically to augment addresses by the amount in the register. This had led, in machines like UNIVAC III and IBM 360, to conventions in the assembler for specifying which index register is to be used as a "cover" register for a given portion of program. In both these machines it is impossible to address the full range of memory without register cover, and a requirement for conscious program segmentation (SEG's in UIII, CSECTS in 360) at the assembly level consequently exists. In the 360 references outside of the segment are resolved by the loader writing absolute addresses (ADCONs) into procedure space.

34.2. RELOCATION REGISTERS AND FRAGMENTATION

The concept of a separate "relocation" register, however, developed early in the field in machines that did have an ability to address full core. The IBM 7090 could have, as a special feature, a seven-bit relocation register, which would add its contents to the high-order bits of the address register to form an effective absolute address. The system effectively defined 256-word segments allocatable in contiguous core. The relocation register would be loaded with the "segment" address, which represented the base of some collection of segments allocated to this program when it ran. Task-switching involved the switching of contents of the relocation register.

Such hardware support is perfectly adequate to support a conservative core management strategy. Since "relocatable" elements are bound to each other by the collector function, the total amount of space required for a program is known at allocation time, and all elements used by a program are also known. There is no reason not to be content with contiguous allocation from a base address.

A number of machines, including the UNIVAC 1108, the GE 635, and the CDC 6600, are relocation register machines. Characteristically, protection is afforded by associating with the relocation register the number of words or the highest address that can be addressed by a given program. References beyond this limit cause interrupts.

This basic hardware assist to relocation leaves a number of questions unsolved, however, if finer utilization or core is desired. Since each program must be assigned contiguous locations of core and cannot dynamically acquire more core after operation, it is reasonable to allocate the maximum amount of core that a program may require. Therefore, if a program has dynamic properties, expanding or contracting during processing, the core that is not used at any time is not available to the system, nor is it available to other elements of the program.

Since data requirements as well as storage requirements for procedure may change, more ambitious systems attempt more flexible data management with or without hardware assist. IBM's OS/360 MVT allows more dynamic management within a partition (a region) in two ways. Fundamentally it provides for the specification of overlays within a bound program. Beyond this, it allows for the dynamic acquisition and release of core space within a region. Program elements may issue GETMAIN and FREEMAIN macros, to acquire more working space. Further, it is possible dynamically to invoke procedures through an ATTACH macro. Using the ATTACH, a program may call upon a "subtask." Space for this task will be found in the region, and a relocating loader brings in the task, assigns it to found core, and provides addressability.

A fundamental problem associated with storage management of this level of dynamism is the fragmentation problem. The essential goal of the OS/360 definition of regions was to constrain the scope of the fragmentation problem to the area of a program rather than to allow it to propogate through core. Fragmentation occurs when space is dynamically allocated in various sizes to elements of various size, which themselves may grow and require more storage.

Consider Fig. 34.2. We have initially allocated space for five "ele-

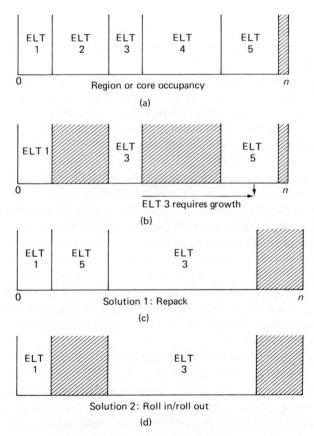

Fig. 34.2. Map of storage allocation.

ments." We may think of these as formal segments (subtasks) of a single program or as five programs in a multiprogram environment. In the time A-B we have quiesced element 2, freeing its store, and element 4, freeing its store. We now find that element 3 requests 1000 additional locations. We see that there is no possibility of granting 1000 additional locations at the end edge of 3.

We may take a number of approaches. We may abort 3. This is a real possibility for an unconditional GETMAIN in an OS/360, which would cause an exceed of REGION size. As a scheduling refinement, we might only suspend 3, releasing its storage space and enqueueing it for "reawakening" (as in EXEC VIII) when sufficient contiguous core becomes available. This is, in effect, a resource-oriented preemptive-resume strategy. The suspension of 3 might, in fact, be undertaken even if there were sufficient core for a contiguous extension. System strategy might well impose a limit on the amount of store an element may occupy. One might visualize a time-sharing system where, as a program expands in core utilization over time, it receives a lower and lower priority until it is subject, not only to refusal for further core, but to loss of its own. The intent of the system would be to push "large" users into the background and to maximize service to small users.

The suspension approach has some merits and demerits. As merit, it preserves the work that has been accomplished for element 3, system effort that would otherwise be lost time. As demerit, it introduces additional overhead for suspension and resume, and it makes completely unpredictable when 3 will complete.

If the suspension strategy is used, it is subject to a number of refinements and variations. When element 3 is enqueued on a core request, it may or may not lose the core it has. If it loses the core it has, the old image of 3 may be "rolled out," so that 3 may be brought back only to the area from which it was removed, so that its request for additional core is always to specific locations at the end of present allocation. Alternatively, it may be resumable at any expanded available area in the system. On the "core queue" 3 may compete with all jobs waiting for core—both those waiting for initial operation and those suspended—or it may compete only with those suspended. It may compete within a priority class; it may have its own priority incremented over time, so that it may be able to preempt another element for its core.

In a priority-sensitive environment the decision may be made to grant element 3 core immediately. If this decision is made, the problem of how to do so arises. We can surely remove element 5, making room for the contiguous extension of 3. This is the nicest thing for 3, of course. In a relocation register machine, we merely increase its limits entry and achieve addressability to more store. If the system has written absolute addresses into 3 at some binding time, these may remain unchanged.

Notice that on the removal of 5 we have the option of fitting it into two open spots immediately. If we elect to do so, we face the general problem of what to do if we had written absolute addresses into 5. These absolute addresses arose because of binding between independently created segments. In various systems they may be distributed throughout coding or collected together in a linkage section. If they are distributed through code, then the job of "relocating" 5 involves an inspection of the entire code to adjust references on physical movement. If a linkage "preamble" is maintained, then the job of moving 5 is somewhat simplified.

The "preamble" linkage section might be an area to which all unresolved memory references are relatively referred at binding time. In a sense, the elements are never fully bound. This technique is often used in the very simple environment of a one-pass assembler to resolve "forward jumps"—references to transfer points not yet defined. In a machine with indirect addressing, references to external symbols would be made through indirection at the linkage section. Regardless of the difficulty involved in relocating 5, it is absolute system overhead physically to move 5 from its current area to its new area, just as enqueueing 3 was overhead. Another possibility for getting 3 its space is to move it; exactly the same considerations apply as for moving 5.

34.3. A FIRST CAPABILITY FOR NON CONTIGUITY

We avoid the problems of the previous section by developing an ability to acquire noncontiguous pieces of storage for element 3's growth. In a relocation register machine this introduces a serious problem in addressing the space that is newly acquired. If protection is enforced through a limits register, it is impossible legitimately to address the noncontiguous space, or it is necessary to abandon protection all together. This difficulty is not so critical if elements of a single job are associated together in a region. Whenever any element of a region is operative, then the entire region is addressable. This makes it impossible for a calling routine to constrain the addressing power of a subroutine, but it is a reasonable situation. It is reasonable, therefore, on a relocation register machine to expect noninteractive regions with dynamic allocation limited to operating within a region.

The UNIVAC 1108 attempts to ameliorate this problem by using two limits registers—one intended to cover "procedure" and one for "data." This allows easy discontinuity between I (instruction) and D (data) elements of a program. The system contains in its basic stateword unit, the PSR (program status register), nine-bit base addresses for the I and D allocations of the running program. BI and BD are essentially segment selectors. In addition, there is a 36-bit storage limits register, which defines the legally addressable area of the program in terms of independent upper and lower bounds. These representations are nine-bit representations, of which eight are used to separate memory into 512-word segments. Within each defined area there is no protection; in the area excluded from the ranges there may be total exclusion or write only exclusion, depending on the setting of the protection/mode (F) bits of the PSR.

Whether an address is formed by using BI or BD depends upon a value in BS, the switching field. The relative magnitude of BS vis-à-vis the sum of the unmodified address selects which bank will be addressed.

To form a full memory address the upper seven bits of the 16-bit address from the instruction is added (in parallel) to both the contents of the BI and BD registers. The value in the BS register is the integer power of 2 higher than the greatest relative address that the I portion can generate. A third add, of the instruction address plus its specified index register, is also performed. This add develops the full 18-bit address before relocation; it is the full relative address of an operand. If this address is greater than BS, then the reference is considered to be a reference to the D segment area, and the BD relocation address (modified by the index register) is used. Before memory is referenced, a limits check is made against the bounds in the storage limits register.

The basic scheme allows for two discontinuous areas to be allocated and protected. The I and D (pure procedure and data) implications are not truly reflected in the machine, since it is not possible to protect I from writes, because the protection feature is restricted to those areas outside the bounds of the storage limits registers.

34.4. TRUE DISCONTINUITY

To achieve noncontiguous, usable space for element 3, we require some scheme whereby we may achieve the appearance of contiguity without its actual presence. We require a scheme under which the addresses in 3 are usable regardless of the actual physical locations that are to be referenced.

This is the basic relocatability problem. If all the elements 1 through 5 are truly independent programs in a multiprogramming mix, if there is to be no sharing, and if there are no formal segments to be dynamically acquired, that is, if 3 is considered to be basically unfolded except for I/O, then a simple "paging" mechanism can achieve the desired result.

Storage limits register

Program status register

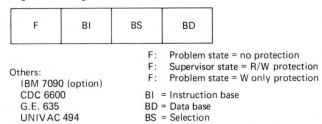

F:	Problem state = no protection
F:	Supervisor state = R/W protection
F:	Problem state = W only protection

Others:
  IBM 7090 (option)
  CDC 6600
  G.E. 635
  UNIVAC 494

BI = Instruction base
BD = Data base
BS = Selection

Fig. 34.3. UNIVAC 1108. An example of a bounds machine.

If elements 1 through 5 are not so, are interrelated named formal segments with complex desires to share code and data, then simple paging relocation will not suffice, and a more "complex" scheme is required.

Various machines have undertaken to be "paging," "segment," or combination machines. The ATLAS is the simple paging machine; the IBM 360/67 is a paging machine with some segmentation overtones; the GE 645 provides both; the Burroughs B5500 is a segment machine without paging relocatability.

34.5. PAGING

Since paging is the simplest concept, let us undertake to describe its underlying mechanisms first. The relocation register machines, in effect, defined fixed blocks of core addressable by a two-part address consisting of the block address and the position in block. The size of a block was limited by the number of bits that were not to participate in the relocation register addition.

In effect, the "granularity" of memory was defined by the number of bits in the relocation register. This was effectively the determinant of the minimum space that might be allocated to a program. The selection of this size is of some importance to a system using dynamic relocation. Since the motive for relocation schemes is to maximize effective use of core, too-large blocks mean sparse usage of a block and a low usage of words relative to the number of words read in per fetch of a piece of program. On simple bounds machines this is not a dominant consideration, since storage management strategies tend to be conservative.

A paging machine employs the fundamental concept of fixed partitions of memory and the two-part addressing structure. The mapping of program addresses, however, is supported by a list of addresses in real machine space, rather than by a single relocation register. Programs are informally partitioned into a number of blocks of contiguous space, in essentially unfolded form.

The concept of "virtual memory" relieves a programmer from any consideration of the relationship of his addresses to the addresses that will be used on the machine. This concept is associated with all paging systems but is not a requirement. Its use is basically to allow the programmer to write unfolded programs without planning overlay structures. The absolute size of physical core in "virtual memory" environments is smaller than the addressable space that a programmer or compiler may use. The concept is not fundamental to paging, however, since the fragmentation problem to which paging fundamentally addresses itself occurs, as we have seen, whether or not real core is greater or smaller than the space of any given program or combination of programs.

Consider a program A to have 4000 words, and consider a machine with a 100,000-word memory and an addressing structure where the first two digits (represented in decimals for convenience) are page or block addresses. This machine has 100 1000-word pages [see Fig. 34.4]. Let us first postulate that program A is read into real core at load time. The allocator assigns locations as available for every 1000 words of program A. As it does so, it records in a "page table" the base address of the 1000-word area that it has just contributed to program A.

Whenever program A makes a reference to memory, the first two digits of the address specify a position in the page table where the origin address of the block of core to which that page has been allocated may be found. That original address, when added to the low-order (page relative) digits of the address, defines an absolute core location for the reference.

We see from Fig. 34.4 that the physical contiguity of the 4000 words of program A is necessary only in 1000-word "pages," and that any distribution of real addresses over program A is as good as any other. We have effectively allowed program A to fragment over core, finding space where it can.

There are various ways to implement the hardware and software of a paging system. We shall describe two implementations briefly, the Ferranti ATLAS and the IBM 360 Model 67.

34.6. ATLAS

The ATLAS developed its paging scheme largely to support the virtual memory concept. Addressing on the machine allowed for reference to

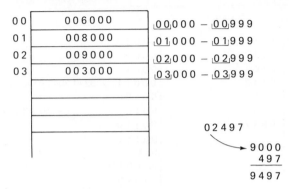

Fig. 34.4. Basic mapping mechanism.

2048 blocks of 512 48-bit words (1,048,576 storage locations) across 20 bits of the 1024K addressable locations.

The basic core storage was 16,384 words. The remaining addresses were mapped onto a magnetic drum, which was word addressable. Each block (32 in the case of a 16K machine) has a "page address register" associated with it. The PAR for any block contains the 11-digit block number as it is represented in the upper 11 bits of a 20-bit word address. Protection is provided through a high-order twelfth bit, which can be set to lock out by the ATLAS Supervisor. Since all resident block numbers are represented in the 32 PAR positions, those blocks which belong to nonactive programs must be locked out by the supervisor.

The 32 positions of PAR are implemented as an associative array which can be searched in parallel. If the 11-bit high-order pattern is found in a PAR, a five-bit core partition number is concatenated with the nine low-order bits of the address to form a 14-bit full storage address. See Fig. 34.5. Here the page 0347 is found in position 5 of the associative memory, causing a 00101 to be placed as the high-order five bits to form an effective address.

Each program in the system is allocated a block directory large enough to accommodate the number of pages that it will require. The directory contains the block number and the position in the system where this block exists on drum. When a "no match" is found in the PAR's, the computer takes an interrupt, and the block directory is scanned for the address on the drum of the desired block. When this is found, the block is read into store, and its number is placed in the PAR. When this occurs, another block is selected for roll out so that there is always an empty position in PAR.

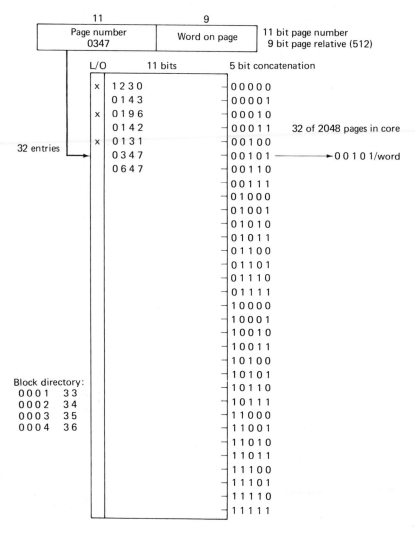

Fig. 34.5. ATLAS mapping mechanism.

34.7. INDEX PAGING AND IBM 360/67

A variation in one-level page systems associates with each program an
allocation map (page table) in which each position contains the real
address in core of the indexed page. A presence or absence bit indicates
whether the referenced page is truly in core. (A further variant might
contain in the page table the drum or disc address of the page when the
presence bit is off).

When a program is running, a system register is set to point to the base of the page table. Program switches are accomplished by changing the pointer to the page table of the program being undertaken. On a find the base address found in the indexed page table is added to the low-order position of the instruction address to form a full address. No associative memory is required for the operation of such a variation. Additional memory cycles are required for the computation of the index and formation of the address. A look-aside mechanism in the form of an associative memory might be used to provide performance improvement for such a system by holding in associative store the segment numbers recently referred to by the system.

The IBM System 360/67 is fundamentally a two-level paging system, which seeks to overcome some of the limitations of one-level schemes. These limitations derive basically from the fact that simple paging schemes primarily support unfolded images of programs.

Essentially all the functions of binding and relocation are preaccomplished into "virtual storage" space. The dynamic acquisition of addressability to external routines is not necessarily facilitated by one-level paging schemes. The dynamic address translation provided by the hardware is an approach to the fragmentation problem but not to the problem of general memory utilization. Since binding has been effectively accomplished, it is not possible without some artifice for programs to share the same coding in core. Further, the dynamic growth of structures is not provided for in the hardware system. The user must provide for sufficient space to hold the maximum size of an array or list. This leads to the definition of pages with very sparse actual usage and to consequent expansion in the size of page tables.

What is desired is the ability dynamically to acquire addressability to code external to the program and to acquire space in a way that permits the orderly growth of data elements. The two-level paging scheme of the 67 and of the segmentation machines, B5500 and GE 645, attempts to give serious hardware support.

The 67 scheme was designed to support a time-sharing environment. However, it has been successfully used in a standard multiprogramming environment with impressive throughput results. The basic mechanism involves a two- rather than a one-level paging scheme. Addresses assigned to programs are assigned in "virtual" space, and functions of binding and relocation operate in this virtual space, which represents the full addressability of the machine.

Virtual space addresses are 24-bit streams (a 32-bit feature is available) divided into three sections—a four-bit segment selector (allowing 16 segments), an eight-bit page selector, and a 12-bit byte selector. When a

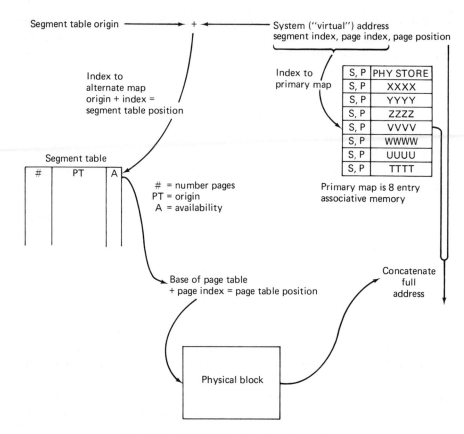

Fig. 34.6. IBM 360/67. Two-level mapping.

program gains control of the system, a segment table register is set to to the program's segment table, which must exist in real core storage. The 16 segments of 256 pages each account for the total permissible address space that a user may have in virtual storage (16 million bytes).

The segment table register containing the origin of the segment table is added to the segment portion of the address to develop an index to the segment table. At this location the physical address of the origin of the page table associated with that segment is found. This origin is added to the page selector portion of the address to form an index on the page table, which will locate the physical address of the page. This address becomes the upper 12 bits of a concatenation of which the lower 12 bits are the byte selector portion of the address.

This mapping process takes time to perform. A reference to memory is required to achieve the page table origin from the segment table, and a reference to memory is required to achieve the physical location from the page table. Since storage referencing constitutes (as we have seen in earlier chapters) a serious bottleneck to performance, the system attempts to reduce the amount of address translation that must actually be performed by using an associative memory and by providing an instruction counter in relocated form. Instructions in sequence may be fetched without reference to address translation or to the associative array.

34.8. A FIRST LOOK AT REPLACEMENT

The associative memory has room for the representation of eight segment/page pairs. Each entry has the segment number, page number, and associated physical storage address (upper 12 address bits) recorded at the time of first reference to a given page. All subsequent references to that page bypass the segment and page table look-ups by finding their high-order address in associative memory.

The device is a look-aside mechanism that, on the basis of recent history of references, makes a prediction about future references (in the sense that it assumes the efficiency of holding recently referenced pages) in order to reduce the burden on the system of dynamic address translation.

Residence in the array is dynamically controlled by bits associated with each entry. When an entry is made (on first reference), the entry is set to a "loaded, referenced" status. (Whenever program switch occurs, all entries are set to "unloaded, unreferenced.") While there are less than eight separate pages referenced in a program, the referenced status of all pages represented in associative store remains unchanged. New page references are added to the store in sequence as they occur. When an eighth page is brought in, then competition begins for space in the array.

The resolution of this competition is a microcosmic instance of the general competition for space in a system by all competing elements. A fundamental part of core management is the schemes used to select which element is most overlayable at a time when space must be made free. We shall discuss various approaches to this later in the chapter.

The fundamental algorithm used by the Model 67 for page representation in associative store is to set all entries open and available for replacement, that is, to change their status to "nonreferenced" at the time a competition for space must begin. Subsequent references to a register will reset the status to "referenced." If a new page representation then must be put into the array, the first (in a sequential top-down scan) unreferenced entry is replaced. (This has no connection with the residence of that page in real storage. This determination is left to another mechanism in the system. We are describing here only the representation of a page address in associative store.) When all entries have again been referenced, the set is again set to "unreferenced."

The general effect of this scheme is to bias the residence in the store toward those page representations most frequently referenced, since these will have the highest probability of being in referenced status after a resetting when an open space is being looked for.

There will be times when the replacement of an entry in the associative memory does imply a corresponding activity in core. This will occur when, on access to the segment table, the referenced segment is not found to be in storage, or when, on reference to a page table, the page is not found to be in storage. On such conditions interrupts are caused, and the operating system (normally TSS 360) undertakes to bring the required page to core. This operation is a complex one, involving the determination of the location on disc or drum where the page is being held, the determination of a free area in core where the page may be put, or (in other systems) possibly the determination of a page to be rolled out. These functions receive no hardware assist and represent serious overhead on all paging systems.

Neither the ATLAS nor the MOD 67 hardware, nor indeed any hardware implement what might be considered an allocation strategy in either a uni- or multiprogrammed environment. Much of the conversation about the effectiveness of paging (or segment) machines is really about the strategies that surround the machine in terms of the contention algorithms and degree of allocation fluidity allowed, and the resulting paging rates and burden of overhead associated with various strategies. Paging is simply a way to make certain strategies feasible in terms of time-economics. The software allocators and schedulers associated with these systems and the concepts therein embodied determine performance characteristics.

34.9. SHARING ON A ONE-LEVEL PAGING SYSTEM

The existence of the second level of structure (the "segment") permits a number of features that we mentioned as desirable and that are awkward in a one-level scheme. Primarily it introduces the possibility of structures of various sizes. Segments, unlike pages, may be of different lengths, depending on the number of active pages required at any time. Consequently, dynamic growth and contraction may be provided for segments. The size of page tables is also compressed. Sharing is accomplished by having segment table entries point to the same page tables.

The claim, however, that sharing is per se "impossible" on one-level paging machines is not true. This claim is made on the basis of the IBM

360/67 form of one-level page table, where the contents of the table are a memory address, and the page reference is a relative position in the table. Surely, sharing is impossible here, since page location is a function of position in the table. If any other one-level scheme is used, however, sharing is possible, if awkward. On ATLAS, or in any system where pages have names or numbers, or unique identification as such on the mapping mechanism, it is possible to share at a single level.

One can envision a system in which memory is partitioned into fixed-size blocks such that there exists a mapping mechanism entry for each block of memory. Memory blocks, rather than pages, are positionally indexed. Each page is named for each program with certain registered system programs having special system names (reference numbers). By placing these pages into the page directory of each program wishing to access them, it is possible for programs to share system subroutine or system functions, or any external object whose name is known, on a one-level paging mechanism basis. The trick to sharing is not levels, but names.

One must, of course, take care to understand exactly what one means by "sharing." Sharing the capability of using a function does not necessarily imply actual use of the same physical representation of that function's code in core.

We may share at the "virtual memory" level—the level of the space that is apparently available to the programmer, his "name space" only. This can be accomplished with a one-level paging scheme by assigning functions to be shared to the virtual memory of the program wishing to use them. If, for example, user A wishes the square root routine and user B wishes the square root routine, it is not difficult for the system to place in the map of virtual memory, at whatever address is available, the location in auxiliary storage of square root function. A may have the function at memory location 6000, B at 9000, in their virtual memories. The virtual memory map associates the virtual memory address with the auxiliary storage address. [In TSS 360 this map is called the external page table (XPT).] The function is shared, but no sharing of the physical representation of code in executable core is implied.

34.10. SEGMENTATION, WITH AND WITHOUT PAGES

Paging perceives memory as essentially a linear space of contiguous addresses. All addresses in a program are perceived to run linearly from 0 to maximum. The addition of one to the highest-order address of a page brings one to the first address of the next page. This one-dimensional

Linear space

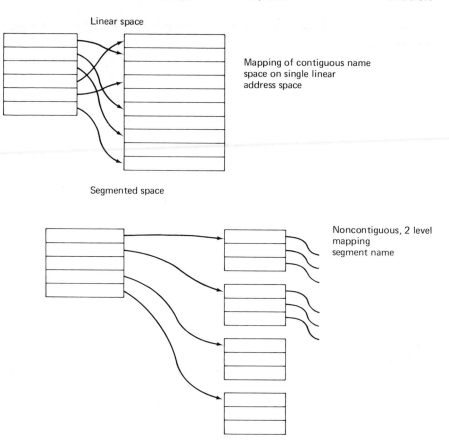

Mapping of contiguous name
space on single linear
address space

Segmented space

Noncontiguous, 2 level
mapping
segment name

Fig. 34.7. Two space concepts.

concept of memory is essentially why simple paging has no necessary
impact on the binding conventions of a system.

The concept of segment is fundamentally two-dimensional, in that
one perceives an address of the form S, W (segment, word) and memory
as organized into partitions S of space W. These partitions may be
thought of as being absolutely independent, as on Burroughs equipment,
or as being related as on the Model 67. In full realization a segment
address is treated by the system as a symbolic name. In a segment system,
the concept of a page may or may not exist. If it exists, it has the quality
of a Model 67 page—an informal partition of a segment—and is of fixed
size corresponding to the allocatable size of real store. In the GE 645 two
page sizes were planned for, a 64-word page and a 1024-word page. This
was done further to increase the flexibility of core assignment and segment growth.

The determination of a page size is a critical decision. It is equivalent to determining the granularity of storage in a relocation machine. Too-large pages will cause the fragmentation problem to reoccur within a page, as it will become necessary to "pack" code or data in order effectively to utilize core and to attempt to maximize usage for a page when it is referenced. Further, the penalty for having to replace a page is high. Too-small pages, however, increase the probability of page movement, since less effective data can be held on a page and consequently the number of individual swaps is greater.

In the Burroughs B5500 segmentation is provided without paging. As in other systems, a distinction is made between "pure procedure" and data in a program, where relocatable addresses and linkages are isolated (as in TSS, where they are put into PSECTS) with data into an allocation-sensitive structure called a PRT (program reference table).

Procedures reference elements on the PRT, where a PRT element may be a data word, a data descriptor, or a program descriptor. A data or program descriptor represents formal segments at the ALGOL block level. (The system can be programmed only in ALGOL, COBOL, or FORTRAN. No assembler is provided.) These descriptors contain absolute addresses of the objects that they represent, or indications that the objects are not in storage. If they are not in storage, an interrupt is taken, and allocation to bring them into storage must occur. The allocation and referencing mechanism is software-only procedure, supported only by an instruction that makes it more convenient to search a linked list of available areas.

Chapter 35

POLICIES,
STRATEGIES, AND
TECHNIQUES

35.1. FLEXIBILITY IN STORAGE MANAGEMENT

The availability of dynamic storage systems in a multiprogram environment leads to the opportunity to consider more flexible storage allocation than would be feasible without them. This means that storage utilization will be higher, and potentially more programs can run in the same storage space. With more programs active, the effective throughput of the system potentially increases, and the goal of multiprogramming is achieved. This increase is due to the increased probability that a program in the mix is ready to run.

We have already seen a conservative allocation strategy, where each program in the mix was allocated to a fixed region of store and where that region represented the maximum space available to the program throughout its life in the system. We introduced paging and segmentation to support more flexible strategies, particularly the dynamic allocation of storage to program elements as they are needed, and an expanded capability to expand sharability in the system.

The fundamental difference between flexibility levels in storage management is the amount of storage that is left effectively unused for periods of time. Small amounts of storage may be left unused for long periods of time, because they are caught in the fragments of private regions, or large amounts of storage may be left briefly unused, because they are not large enough to accommodate the needs of a potential new member of the mix; either effectively reduces the performance of a system.

The question of how much for how long is a function of various strategies. We have discussed these in terms of when programs are bound to storage, and what the basic allocation unit of the system is. In the most conservative multiprogramming system, the time frame may extend beyond the existence of a program in the system.

In OS/360 MFT, for example, fixed-size partitions are created into which all programs must fit; any unused space is entirely lost. In OS/360 MVT space is allocated for the duration of a job step (COMPILE, LOAD, and GO are job steps; any named procedure on an execute card is a job step. These are distinct from tasks within a step for which storage is dynamically allocated from the steps region. In addition, certain system functions, such as device allocation, operate out of the region of a using program set, establishing a system minimum allocatable space.)

In a more dynamic system, space is allocated as a response to explicit requests from a common pool of available memory, or perhaps from preemptible store. In paging systems these allocations are in the fixed page size of the system; in a segmentation system like the B5500 space allocation will be the specific area requested.

A final step in allocation is to provide space implicitly upon recognition of a need. Release of space follows the acquisition pattern, only at the end of a job, at the end of a step, at explicit demand on or implicit recognition of usage.

35.2. CORE AS A LIMIT ON MULTIPROGRAMMING

The breadth of multiprogramming in a machine is limited by the availability of all resources; however, the core requirement has been traditionally the most constraining element. The relief of this constraint promises the greatest improvement in utilization balancing.

An interesting implication in this area concerns the relationship between the scheduler and the resource allocator in a multiprogramming system. Traditionally resource allocation has been subordinated to scheduling. In the traditional literature on scheduling certain concepts are postulated that seem almost independent of resources, although they are surely resource-constrained. A primary example of this was the concept of shortest job first. Schedulers that select programs for running not on the basis of "best use of resources" but on external criteria or arbitrary standards related to service standards will not achieve utilization.

An example of a resource-oriented scheduling function is the "best fit in core" criterion of the EXEC VIII system. A similar and more dynamic approach consistent with the tendency to more dynamic scheduling of

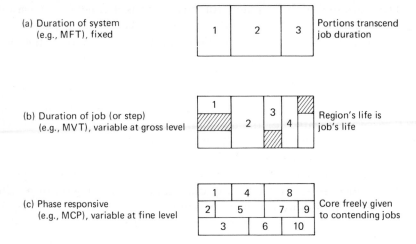

(a) Duration of system (e.g., MFT), fixed — Portions transcend job duration

(b) Duration of job (or step) (e.g., MVT), variable at gross level — Region's life is job's life

(c) Phase responsive (e.g., MCP), variable at fine level — Core freely given to contending jobs

Fig. 35.1. Levels of management.

smaller elements is represented in the experimental IBM system. This system monitors the paging level of the system, and if the swapping rate becomes too high, a member of the mix is quiesced.

In the dynamic management of memory there is a need for policy in two major areas: (1) how much space a program may occupy, and (2) how it gets its space. The question of "how much" relates to a continuing conversation in the field about the dangers of overpaging and the relative effectiveness of "demand paging." The question of "how" concerns the strategies by which core is made available to a program, the placement and replacement rules of allocation.

35.3. LIMIT OF A MIX

The ideal multiprogramming situation is to have as many programs available for execution as there are independent paths to I/O devices and to assign each program to independent data paths. This maximizes utilization of I/O without causing delays in programs because of queueing on channel contention. In the first part we discussed the effect of resource sharing at different levels. In systems where paths to devices are limited by channel logic, delays to programs are caused because of the interference between programs for access to a device through that channel. The probability of a program's being ready to take CPU control is, therefore, reduced, since the expected response time from an I/O service request is necessarily higher. One would expect that the effective level of multiprogramming would be sensitive to the level of resource contention, the "loading" of various resources.

In Fig. 35.2, if each program has access to devices independently of the other, no common queue forms; then the expected response time from a device is the physical minimum one can achieve. We have come as close as possible to simulating six independent systems. At any time a job is (1) waiting for a CPU, (2) waiting for I/O response, (3) engaging in I/O, (4) being processed by the CPU, or (5) engaging in I/O and being processed by a CPU. The memory associated with a job is, therefore, active in cases 3, 4, and 5 and passive in cases 1 or 2.

For each job there will be periods of time when a particular locality of its storage will be active; for example, when it enters a loop during which it is cyclicly executing instructions and advancing sequentially through a data structure, an array, or a vector of some sort. During this period of time, portions of memory are unused.

The fundamental problem is to what extent the utilization of this

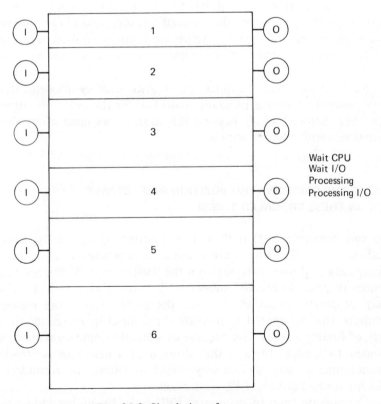

Fig. 35.2. Simulation of resource.

memory could increase the productivity of the system. This question must be resolved within the framework of other considerations:

1. The number of physical devices imposes an upper limit on the number of jobs it is possible to run. The number of paths to devices imposes a limit on the number of jobs it is profitable to run.

2. The overhead associated with levels of fluidity in storage management are chargeable to the CPU in terms of the activity that it must undertake to implement the strategy and in terms of the memory cycles that it must give up to support "paging." Charges accrue to I/O subsystems in terms of channel and device utilization, and finally contribute to lowering the probability that a program will be able to use the CPU. If we remember our first reference to the rolling in and out of entire programs, we there described a worst possible case, where by design the probability of finding a program ready to run was reduced to 0.

The storage strategy, then, has two goals, not necessarily consistent: (1) to optimize space-time utilization of memory, and (2) to minimize the burden placed on the system. Achieving them is equivalent to ensuring that the choices it makes about what is to reside in core are the right choices.

We have seen that the most conservative strategy allocates the maximum amount of core a program might use for the entire duration of its active life. Before looking beyond this strategy, we must surely show that there is something wrong with it.

35.4. IS THE CONSERVATIVE POSITION RESPECTABLE IN THESE TROUBLED TIMES?

One can demonstrate that there is something right with the strategy of maximum allocation when core is not a limiting resource and where channel capacity or device distribution is the limiting path of the system. If the number of paths in the I/O subsystem is small, then a low upper limit of ready programs would characterize the system. Any core management technique that attempted to increase throughput by increasing the probability of finding a ready task because of a greater population might well be doomed to failure. In fact, the addition of a new task contending for channel time in any serious way might well have the countereffect of reducing throughput and CPU utilization.

The phenomenon of decreasing CPU usage by adding additional tasks is infrequently observed but is nevertheless a real possibility. What develops is that under conditions of heavy channel contention the entire mix becomes I/O-bound. The period of time between the submission and the satisfaction of an I/O request is extended to the point that programs have a high probability of waiting for service from I/O at any particular time. The addition of another program has not only had the effect of increasing the waiting time for a job because it is giving up intervals of CPU service to more contenders, but has had the effect of increasing nonready to go time. CPU utilization will fall off when the mix becomes so dominantly I/O-bound that there are times when no one is ready to go.

A better use of core would certainly be to open it up as much as possible for the largest possible blocking or the deepest possible buffering, maximum space being allocated for each member of a reduced mix, the hope being that by liberal allotment of core a program would finish quickly and release its resources to other programs.

Of course, if a program can be found for the mix whose utilization of the channel capability is so low as to be inconsequential, there might well be some profit in making room for it in core.

Conservative strategy becomes inadequate when we can observe a system with both CPU and channel utilization low and realize that another program could be run if only there were room for it in core. That is when core is truly the constraining resource. Unhappily, configuration decisions are often informally made, and more core is acquired in order to increase CPU utilization when what is really required is more channels, and vice versa.

It is interesting to observe that the acquisition of more core may improve the throughput of a system as nicely as any sophisticated memory management strategy. Much of the interest in memory management developed when it was truly a dominating cost consideration for a system.

The time-sharing equivalent of the conservative batch strategy is the "all or nothing" approach in which an environment for a terminal is either entirely in or entirely not in core. The total environment is in for the duration of every service interval that the terminal receives. A difficulty here is that some systems impose a fixed size of work space for a terminal regardless of need. That fixed size must be reasonably large, so that the environment tends to be sparsely used by smallish problems and incapable of extending itself to accommodate larger ones.

Fixed systems of this kind are really prepaging systems that always nominally guess right and never really know how much of what they have brought in is actually used. In the absence of paging support, however, such design might well be more realistic than attempts dynamically to prefer certain elements of the environment of a terminal or set of currently preferred terminals.

Insofar as utilization balancing is concerned, we see that it is a subtle art. We shall indulge in one further reflection: that the task is hopeless in the absence of the dynamic monitoring schemes suggested in the last chapter. The exposure to the danger of actually reducing CPU utilization by our attempts to increase it is reason enough to take such things seriously. The availability of a new active program is in and of itself a good thing, unless the increase in contention actually reduces the availability of a process that is ready to go.

We have already described the difficulty in judging what an I/O-bound or CPU-bound program is because of the dynamic qualities of program performance. Beyond this, in order to handle multiprogram mix determination in the presence of constrained core (any core where the maximum size of a mix is greater than available core), we discover that there is no way of predicting on the basis of size the I/O or compute character of a program. Surely. a conservative storage policy is justified in the presence of narrow I/O paths. No overhead for roll in/roll out, paging, swapping, etc. could be justified. The effect would only be further to I/O-bind a seriously I/O-bound situation.

One might soften this conclusion with one observation. If dynamic management of core allowed for deeper buffering so that a program might be rolled out in part to make buffer space available to another, or if procedure were rolled out in part for its own buffering, then the average response time for devices would tend to go down (there being time when it was effectively 0, being available from the buffer), and deeper buffering would tend to make programs compute-bound. In this situation it might be conceivable for a management scheme that replaced sections of infrequently used code with buffers to be productive.

The constraint on number of active programs being partially a system balance problem independent of core storage indicates that no core management strategy or tactics will per se provide for increasing performance by increasing the number of programs active at any time.

In time-sharing systems where channel burden is particularly constraining—since all I/O activity (other than terminal input, which is relatively trivial) is confined to a channel (or a pair of channels in a dual path system)—the "paging" or roll-in delays tend to be the major factor in limiting the number of active users of a system.

Still the goal of maximizing core utilization is present in system design, since the number of active users can increase if the space that each one requires at any time can be made minimal. In batch or mixed systems, an increase in the number of active programs, as we have said before, increases the probability that a ready run can be found and increases the flexibility under dynamic "deserved priority" scheduling of finding a mix balance.

A goal of dynamic core management is to grant to each active program as much space as it requires for any period of execution and only that much space. If priorities are to be respected, the contents of core at any time should reflect the priorities established by the dynamic scheduler; if priorities are not respected, the contents in core should represent those blocks of code and data with the highest probability of usage over a time frame set by the system. The goal is to balance and synchronize CPU and core utilization.

We refer again to the relationship between core allocation and scheduling. In an earlier section we described paging as a response to the fragmentation problem. An alternative solution to this problem lies in constraining the scheduler so that the number of programs admitted to the mix is low enough so as to minimize the probability or the cost of fragmentation.

A form of the fragmentation problem is the "thrashing" problem of paging machines, where the level of page recall and roll out is so high as to seriously impede performance on the system.

A solution is to constrain the number of active programs so that memory load is sufficiently low that larger numbers of pages may reside for each program, and the need for paging is reduced. The art is to find a balance between scheduler conservatism and good core utilization so as to increase the number of active programs without serious thrashing.

35.5. MINIMUM SPACE

What is the most liberal policy regarding storage allocation? To give to a program the amount of core that it requires to operate in the system. The absolute minimum of core is a system variable related to the execution rate of the machine, channel interference, I/O speeds, and the way programs are structured by assemblers and compilers. In a paging machine a single page would be a reasonable minimum allocation. In a segment machine a segment would be reasonable; in a segment-paging machine, a page of a segment.

If, as in B5000 and its descendants, there exists a conceptual "root segment," a section of the code that contains directories and virtual memory maps, etc., then that section must be in core for the program to operate. A necessary condition for transferring control to the program is the residence of that root segment.

In the B5000 MCP system (master control program) the high-level scheduler (selection procedure) reads in a directory segment of a selected task, from which it finds the location on disc of the segment dictionary (containing disc location and size of every program segment developed by the compiler) and the PRT. PRT contains scalers, program descriptors, and data descriptors, describing the size and presence in core of procedure and data blocks. All references to memory are through the PRT. Addressing to PRT is relative, assisted by a base register so that PRT's can initially be located anywhere in core. Descriptors, however, contain real addresses, and certain pointers describing data storage (memory links) use real addresses to point back to the PRT so that it is not relocatable. For this reason the core assigned to the PRT is marked as "nonoverlayable" or not available to dynamic memory management.

The initial program segment is then read into core, and the program is entered into the mix. At this point the base minimum for execution has been brought into core. Qualification for "active" may only be, therefore, not that the program be resident in core, but that its basic elements be present in core. Program segments are relocatable, as they are entirely address-independent, addressing being done through the PRT descriptors, which hold absolute addresses. Only the program segment actually in execution need be in core at any time.

The potential optimization of core usage aims at retaining in core only those portions of a program actually in use, freeing other core for occupancy by other programs and tending to increase true core utilization and the number of programs that can be "active." This is consistent with our earlier observations about system balance, since the B5000 (as the B6500, etc.) has "open channel" I/O subsystems, where any channel gets to any devices through an exchange external to the CPU.

The first aspect of a "liberal" core management policy is, therefore, to minimize the space required for a program to become active.

35.6. UNIT OF ALLOCATION

We notice that the unit of allocation on a B5000 is a segment and that segment sizes are variable. The system attempts to acquire space exactly equal to the amount of space required by the segment. There is no "overallocation" or internalization of dead space so that space is withdrawn from the system without real usage.

Such overallocation is an inherent feature of fixed page machines, which assign real space to fixed pages. The degree of overallocation is a function of the cleverness with which programmers and compilers "pack" program elements into a page. This packing is optimal when closely related elements (those that have a high probability of being referenced when others of the set are referenced) are clustered on a page and use all page space.

The packing of pages is an art very closely related to dynamic storage management. Algorithms for "predictive paging" in dynamic systems and for packing are similar in theoretical basis and relate very closely to algorithms that attempt to isolate parallel executable islands of code.

The packing of related clusters tends to minimize paging by increasing the probability that a reference to a page implies other references to that page. The "locality" of reference for a program over a given time frame is best represented in a minimum number of pages.

With variable segment sizes the overallocation problem is eliminated. As we recall from an earlier section, a segment is a logical structure loosely bound into a program. A page is an informal structure already integrated into a program; in a virtual memory system it is a slice of the linear memory space.

A segment may have attributes (protection, e.g.) and, particularly, it may have a size attribute. In page/segment systems, the size of a segment is given in the number of pages it contains; in a simple segment system it is represented in words (or bytes, or bits, or whatever the basic memory element is). In the B5500 memory allocation to segments is a software function. In the continuum between fixed-size page allocation machines and completely variable-size machines are machines that allow for a fixed number of allocation sizes (initially the GE 645 and the CDC 3300), usually in powers of two—512, 1024, 2048, 4096, for example.

A very conservative use of this feature is to allow as a system parameter (or a machine configuration parameter) an installation to select its own desired page size and to live with the page size it has selected. More flexible usage actually allows mixed page sizes in the system. The various sizes may then be thought of as quarter, half, three-quarter or full pages. The basic allocation unit is the smallest unit, and space is allocated in contiguous multiples for pages greater than the basic unit.

The possibility of varying page sizes, of course, reintroduces the real core fragmentation problem, and a more elaborate mechanism of control over allocation is required.

The second element of a liberal allocation policy is to allocate storage with the closest fit to the actual size of the element a process is requesting.

35.7. TOO LITTLE SPACE

The question of how much space, however, is not satisfactorily answered by "minimum" needed to operate. The constraints in the lower limit of allocated space derive from the "thrashing" problem. If a maximum number of programs are packed into core, each guaranteed only minimum space, the tendency for programs to require paging or roll-in operations becomes so severe as to limit the probability that a program can effectively use the CPU, because the tendency to require a roll-in to core becomes dominating. Each program makes an immediate request for memory load, and conceivably all programs enqueue on the roll-in mechanism and idle until roll-in completes. There is, evidently, a balance point that tends to constrain the number of active programs by insisting on some conceptual minimum amount of space for an active program in order to avoid an actual decrease in CPU utilization due to forced idles.

The balance point may be a function of the relative speed of the backing store response rate and the CPU/real core speeds. One study suggests that serious performance decays will occur if the loading level of the backing store channel is greater than 100 percent to any serious extent.

The greater the loading on the backing storage unit and the greater the average response time, the higher will be the probability of the CPU's being forced to idle because no programs have required pages in core.

Manipulation of the hardware resource is one solution to the low-CPU-utilization characteristic of overpaged systems. This relief is available in two forms: (1) increase of primary store to allow for more of a program to be in core (hardly a solution to the store utilization problem), and (2) dramatic reduction in the time it requires to deliver a page to core when it is requested.

Another simple solution lies in partitioning a system into foreground and background sections by software. In a time-sharing system the foreground sections are those programs experiencing some degree of interaction with an on-line terminal. Dynamic core allocation is required for this set of programs. In the background is a resident (or residents) that is always available in core and which uses the CPU when the dynamically managed terminal programs cannot.

The advantage to having this background population oriented toward CPU-boundedness is manifest, since its contribution must be to CPU usage. A difficulty, of course, develops in a "mixed" system, where there are other (nonpaging) channels whose utilization should be maintained. If core size allows, a CPU-bound and an I/O-bound program should be resident. If size does not allow and the choice must be made, the chance is purely an economic decision relative to the cost of CPU vs. channel idle in a system.

Many even medium-sized systems, such as IBM 360, model 50's running the disc operating system (DOS), allow for multiple background areas. A larger system like EXEC VIII supporting multiple services ("demand" and batch) also use independent core management strategies for terminal and nonterminal jobs. Nonterminal (batch jobs) are relatively stable in core, with only the pressing needs of a deadline job forcing roll in/roll out, whereas demand jobs (residents of the core swap queue) are characteristically brought in and out of core.

The expansion of hardware and the partitioning of the system, however, are not direct solutions to the problem of what constitutes reasonable residence in core. What must be determined is the behavior of programs under different conditions of storage availability, the patterns by which programs truly reference instructions and data.

A growing number of studies on the performance of programs in a paging environment are appearing in the literature of the field. In general, these studies have attempted to determine the effect on performance of such things as "programming style," page replacement strategies, and page sizes. Although there is some disagreement in specific interpretations of the results, there exists broad agreement in general tendency; certainly agreement is sufficient that we can undertake some generalizations in the area of how much space should be allocated to a program.

35.8. HOW MUCH?

Given that a program is of a certain size and has a given page organization, how much space should it have, and should it always have the same amount of space?

There is now general agreement that, in order to run well, a program requires more space than one had hoped. By "running well" one means that the program runs in some reasonable approximation of its time, given all the core it requires, and that the time that it can run without requiring a paging operation is long enough so as not to cause the program characteristically to be unable to use the CPU for whatever amount of time the system would like for it to do so.

Over a long enough period of time, any program given less than its full allotment will require paging. The match that must be made is to allow the program to operate for its "time-slice," or, in a mix sense, to allow sufficient space for programs so that for any scheduling cycle or period the system will be adequately used by programs in core.

The fundamental concepts are sufficiency and locality. The time-dependent nature of proper space is reflected in the locality concept. For any given amount of time in execution a program will reference a certain subset of its name space. The smaller the amount of time, the more "local" the references will tend to be. The contents of core during that execution interval should be the locality (region or neighborhood) that the program is going to reference. A sufficient amount of core is the amount of core required to represent that locality.

This "neighborhood" tends to be stable in size over time. Programs apparently tend to build up to a size and to stabilize, although neighborhoods may shift. The informal concept of the "parachor" is defined to be that number of pages that is the point at which a program will tend to overpage—the number of pages fewer than which performance will fall off radically.

One study places this number at around half of a program's total pages. As small storage space is expanded to larger storage space, the program is in an almost constant state of paging; the number of executions between page calls is small. During this period of time a program is accumulating its "parachor."

The time dependency of a parachor derives from its basis in the concept of locality. It follows from this that a system might make use in its scheduling and allocation policies of the simple relationship; programs that have more storage will tend (strongly) to operate without page faults for longer periods of time than programs that have less store.

35.9. HOW TO GET SPACE

If it is given that elements of a program will be moving in and out of core on the basis of reference patterns more or less discrete, more or less predicted, what policy should be applied to their placement? Paging schemes support a maximum flexibility of relocation. In the absence of a paging mechanism, OS/360 MVT requires that an element of a program once rolled out must be returned to the position from which it was removed. This is because of the address resolutions that take place, which fix references to absolute locations in the system. For a similar reason, PRT's are not moved in the Burroughs machines.

A requirement for dynamic relocation is the insensitivity of code to position in real core. In machines lacking paging hardware, a degree of address sensitivity is achieved by indirect addressing. In fact, paging is really a highly organized use of indirect addressing. The question of policy with regard to finding actual space concerns itself with page replacement and space control techniques and is the second question: How does a program get space?

The acquisition of space in real core is a process somewhat insensitive to when it is being done (the conservatism of allocation policy) and how much of it is to be gotten. Space is acquired for a program when a scheduling mechanism requests it, or when the program itself requests it either directly or indirectly.

A direct space request occurs when a task (or a scheduler on its behalf) requests the allocation of a given amount of core space; an indirect request for space occurs when a task (perhaps assisted by the operating system) makes a formal or informal reference to a function or to an address in virtual space that is not bound to a real core location. A direct request for space is exemplified by the execution of GETMAIN macro in OS/360.

Direct and indirect requests may or may not actually result in the acquisition of more physical core by a program. Initially in OS/360 MVT a program was limited by its region size; in the current system it is possible truly to grow by acquiring the region of a program designated (by itself) as "rolloutable." Similarly, an indirect request for a nonresident element may bring that element into truly newly allocated space or may cause it to overlay elements existing in already allocated space.

In any multiprogram environment where programs will asynchronously terminate, freeing space and opening the possibility of activating new programs, or where programs may acquire and release core, it is necessary to retain a mapping of free and used core space. This is required whether or not an address translation mechanism of any type exists in the system, whether or not space is allocated in fixed- or variable-sized blocks, or whether or not repacking of memory is undertaken by the system.

The memory map will, of course, vary in complexity in different environments. In fixed-size allocation schemes the map may be reduced to a bit representing each of the standard-sized blocks into which memory has been partitioned. In variable allocation schemes the map is considerably more complex, containing at least the origin and size of each area and possibly pointers to the next free area, previous free area, areas of same size, etc.

The MCP of the B5500 is an example of a variable-size allocation scheme of the type that would be useful in nonpaging systems, or in systems allowing variable page sizes.

Core is initially organized into nonoverlayable and overlayable storage with frequently used system elements and PRT's in nonoverlayable sections. The high probability of use (low interreference time) associated with system function and the cost of moving PRT's dictate residence and stability of position. The remainder of store is the overlayable area in which tasks in the mix compete for core.

This is a common approach of store management. The stability of high-usage areas may be assured in this way or in more dynamic ways, such as the association of a "sticking priority" with a function. System routines that service the user population may be accorded a "sticking priority" related to the frequency of usage. They are swapped out of core only on a core-required basis in order of sticking priority. All current batch systems define a basic system area where some routines are resident and others transient. In OS/360 MVT the "link pack" area may also be used to contain user specified functions. That the size and traffic through such areas is a critical element of system performance has been demonstrated by published observations of GECOS II (for the GE 635), which reports nontrivial speedup of the system by expansion of the size of this area.

In the overlayable area a distinction is made between program area and data area. This is because program segments are invariant, and data segments by their nature are highly changeable. An overlay of a program segment requires no write operation, since an exact copy exists in the system and return of the element can be achieved by reading that copy into core. Data overlays, however, must involve a write of the element, since that element is assumed to have changed during its residence in core. Many paging schemes contain a representation associated with each page, reflecting whether it has been merely referenced or referenced and changed during an interval; they prefer the overlay of nonchanged pages when a swap is necessary.

The B5500 memory map is actually two linked lists distributed through core. This distribution is not fundamental to the scheme. A linear table with an entry for each area properly maintained could serve equally well. The architecture of the machine, however, and the presence of link following instructions makes the linked list approach convenient for this machine.

The two lists are the available storage list and the in-use storage list.

The available storage list contains entries distributed through core at the head of each available store area. An entry contains an indication that this is an available area, the size of the area, the origin of the preceding area of storage, the origin of the preceding area of free storage, the address of the following area of storage, and the origin of the following area of free storage. Memory areas are totally described by the links in the head of the area. It is possible to follow a path from any free area to any other free area and to follow a path in any direction to all areas, used or free.

In Fig. 35.3 we show nine memory areas and the linkages to each. The area beginning at 1000 is the first overlayable area. It points to the area at 1200 as the next storage area, and to the area at 1500 as the next free storage area. Similarly, it points to 0 as the beginning of its preceding area and indicates that there is no preceding free area. The in-use storage list is similarly organized.

In addition to the area links and size indication, however, it also indicates what program is using the area and whether the area is a data or program segment; it must also point to those descriptors in the PRT that point to this area as the location in core where the segment can be found. If an in-use area is to be overlaid, then the system must mark these PRT descriptions as referring to nonavailable, not-in-core elements. The marking of these descriptors causes in the B5500 a not-in-core interrupt (identi-

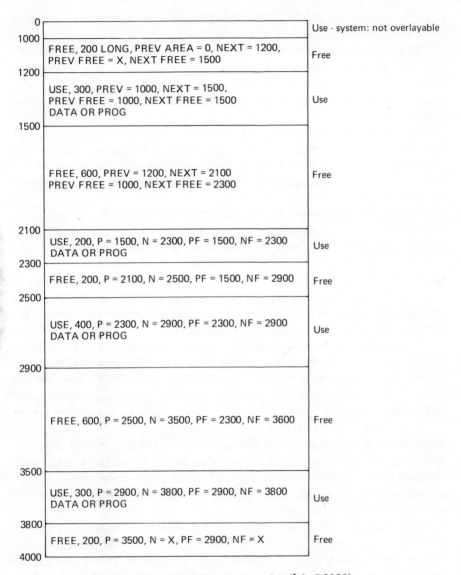

Fig. 35.3. Store map in address order (à la B5000).

cal in concept to a page fault in a paging machine), which begins the activity of bringing it in.

When a segment fault occurs, the descriptor whose reference caused it contains information relative to the allocation process. If the reference is to data accessed for the first time (data to be generated) for which no auxiliary storage image exists, this is indicated in the descriptor, as well as the size of the area required and whether the area is to be overlayable.

This information is handed to the allocation processor (GETSPACE) with an indication that the area is to be for data. When the area is allocated, the address of area is placed in the descriptor, the address of the descriptor is placed in the memory link, and the descriptor is marked present.

Allocation for a previously rolled-out data segment is identical, except that the program calling for the segment must be placed into a wait (blocked) state until the I/O operation associated with the roll-in of the segment is completed (equivalent to page wait status). Referencing a nonresident program segment is similar.

The memory allocator has information about size and type of area to be required. The first function of allocation is to attempt to find a sufficiently large free area in core. The list of available storage is searched until a section large enough to hold the segment is found. When a sufficiently large section is found, that part of it which is needed for the segment is removed from the list of available storage and added to the list of in use storage. Any space in the found section would be defined as a new available area and added to the free space linkages.

If no sufficient free space exists in core, then inspection is made for selecting an area to be swapped (overlaid). In selecting a segment for overlay, consideration is given to a number of possibilities: whether the segment is a program or data segment, whether the program interrupted might (in a program segment) have a resume pointer (interrupt return point) pointing to that segment, and whether (in a data segment) there might be absolute addresses in the stateword of the interrupted program. Swapping involves the finding of an area in use of sufficient size with preference for maintenance in core of those areas that have absolute addresses pointing within them or program segments that are next to be referenced on a program resume.

When an area is found, procedures are undertaken to disassociate the area from the program referencing it by resetting presence bits and converting entry points back to relative form. The area is then inspected for possible extension. The next area is inspected for availability, and, if available, is subsumed into the area about to be available by adjustment of memory links. The same is then done for the preceding area. If it is free, it subsumes the newly available area. If the newly available area is not subsumed by a preceding adjacent available area, it is marked as available.

With the provision of an available area, the process of roll-in continues as if an initial area had been found. The automatic compression of free areas in this way is not a fundamental. The compression may be delayed until a sufficiently large free area for a given request can be found. The compression occurs then only for the block that it is necessary to expand by forming a larger area from two contiguous areas. The rest of memory is left unaffected. The expansion of free memory areas is to be distinguished from packing, which attempts to force free areas together into a maintained contiguous block of core.

35.10. FINDING AND NOT FINDING

The techniques of looking for minimum required space for variable-size segments in the B5500 fashion introduces the problem of what to do when it cannot be found. Earlier in this chapter we described repacking as a solution, the intermittent rearrangement of used and unused areas to provide for a large enough space. Selected roll-out for making space (as in B5500) is another solution.

The two approaches can be combined into a single algorithm which (depending upon system goals) prefers one to the other, i.e., pack when possible to avoid roll-out, or roll-out when possible to avoid packing. Since packing involves all of the mechanisms of relocation (it is a comprehensive relocation of all segments), it is rather an expensive mechanism. Machine systems try to avoid packing. In the EXEC VIII packing is reduced by applying a best-fit-in-core criterion to job scheduling.

The problem may be handled by simply blocking a job until a block of sufficient size becomes available. In OS/360 MVT a STEP may be blocked because it cannot get a large enough contiguous region for its operation; the initiator associated with the STEP enqueues for FIFO-served core.

The concept of overlayable may be expanded to represent job class and/or priority. In looking for overlayable space, any area belonging to a program of higher priority than the program seeking allocation would be considered an unoverlayable area for that search.

In systems using a number of fixed-size segment allocations, allocation would be rounded up to the smallest available segment or contiguous collection of consecutive segments. In machines with boundary addressing constraints round-up to some legitimate address may also occur.

Among the considerations in assigning space from free storage is whether it is worthwhile to assign the first found block that is large enough, or to search for a block whose difference from actual request is minimum. In order for the latter to be easily done, the free storage list must be maintained in order of increasing size or searched if ordered by sequence, as in B5500.

There is some disagreement in the field as to whether keeping the list sorted by address or sorted by size is preferable. The penalty for address sorting is fragmentation due to the generation of more and more small blocks (remainders) in different parts of store; the penalty for size sorting is the inability to combine adjacent areas when they occur; the issue is where the more serious fragmentation occurs.

The ESP executor for the B8500 maintains free storage in a size order, and allocation is constrained by priority and class. Here class refers to whether the program is batch, conversational, or real-time. Like EXEC VIII, ESP will attempt to balance service between these types of programs. A listing of the considerations applied to the selection of a block to be overlaid is given in an early description of the system.

1. Does block belong to a running program?
2. Is I/O active?
3. Size.
4. Program or data.
5. Priority and class.
6. Number of users.

If space cannot be found, the requesting program is deferred and placed on an unallocated space queue that is periodically scanned. If the system recognizes overload, a job will be selected for termination, on the basis of priority class and percentage complete. In such a case its representation on the initial high-level scheduler job queue is modified to reflect the job as it was initially seen. Initial allocation of memory, as in B5500, is a minimum, consisting of PRT, a working storage area, and an initial program segment. Other elements are brought in on reference.

In another system, a dynamic calculation of a job's priority is undertaken (in the absence of a given priority) on a resource-required basis, with the ability for a user to modify system parameters to favor certain kinds of jobs. The priority of a job is the sum of its resource load for each resource modified by certain constants. The priority is a priority for the acquisition of resources. Peripheral resources are allocated first, after an initial check that core will be available on some basis.

The peripheral allocator feeds into a core queue; allocation proceeds until a maximum back-up on core queue is achieved. A program on the core queue with highest priority may cause swapping; lower priorities may cause packing.

Some aging process is used in the core queue to insure the eventual preference of older jobs with large core needs over newer jobs with smaller core needs. In the absence of age precedence, smaller jobs are preferred. Dispatching is oriented toward giving I/O-bound jobs preference. Time-sharing is run in a partition with its private scheduler and allocator.

35.11. PAGE REPLACEMENT

Although potential memory utilization is superior in terms of potential true memory utilization because of the avoidance of internalized dead space, the problem of selecting an area for roll-out (page-out, swap-out) is essentially the same in fixed- or variable-size allocation schemes.

Much attention has been paid to the development of page replacement algorithms for paging machines. The goal of such algorithms is to reduce the page traffic by reducing the probability that a page selected for overlay is a page that will be needed shortly, that is, to attempt to ensure that the set of pages in memory represents for a scheduling interval those pages that have the highest probability of being referenced.

Various algorithms have been developed, presented, and tested; they differ largely in the amount of page usage information they utilize to select a page and in their intimacy with the scheduler.

Any replacement scheme is basically rooted in the concept of locality, the assumption that a program does not distribute its memory references randomly throughout its name space, but tends to work in given areas for intervals of time. Without the concept of locality, a technique that randomly selected a page for replacement would be as efficient as any other, since nothing could be inferred about future usage of a page on the basis of some knowledge of its past usage.

Even the simplest nonrandom scheme attempts to make such a prediction. An optimum scheme would require that the times of next reference to a page be known and that pages be selected for replacement on the basis of these known next times of reference. A working system never has such information, although in exercising scheduling decisions the scheduler, by its choice of operation for the next scheduling interval, can define a set of pages used by programs not scheduled in the interval, which can be immediately released.

The policy of TSS 360 is to reduce core competition by limiting active programs, and an inactive program has no rights of residence in core. When a program releases the CPU, all of its pages are released, because the system knows that those pages have a zero probability of reference. To compensate for this, programs that have accumulated large populations of pages are accorded longer running times.

In resolving contention for space among active programs in the interval, however, only past performance, based upon locality, is a usable predictive measure for page replacement.

One simple replacement scheme is first in-first out (FIFO). This algorithm selects the oldest page in memory for replacement. The assumption of FIFO is that over time a program will change its locality and that pages that have been in memory longest will drop out of the locality of the program. If the program abandons pages before FIFO removes them, the incidence of bad choices can be low. If locality does not shift cyclically in this way, then the requirement to pull back replaced pages is very high.

FIFO is basically classified as a STATIC algorithm, because it applies a replacement rule uniformly applied and independent of the observation of actual page usage in the system. FIFO is the least effective replacement algorithm, and it would seem to suffer particularly in a multiprogramming environment.

With no adjustment in the scheme to reflect the pattern of CPU acquisition and release, the independent localities of programs in the mix cannot be reflected by the age of a given page in core. One method of using FIFO in multiprogramming is to fix the memory size of each program and constrain a program to paging only within its memory. This very conservative approach does not support the essential spirit of dynamic memory management, which is to allow programs to exchange space as required.

In free competition for space, larger programs will tend to lose more pages than smaller ones; having more pages, they will tend to be active when a need for a replacement occurs. An adjustment to the FIFO algorithm used in an experimental system is called BIFO (biased FIFO). This modification favors a program in the mix (actually a given virtual machine) for periods of time, protecting its pages from replacement. For some number of page calls a page of the preferred program cannot be replaced; other programs lose pages to it. The preferred program never replaces a page of its own; therefore, it cannot lose a page to which it might have a short next reference interval. During the period of time in which it is not preferred, it will lose pages to other programs. The desired fluidity of memory size is achieved. Some reduction (10 percent) has been reported in page replacements by the authors of this system (IBM M44/44X).

The variability of memory size for a program is significant not only for system throughput due to high utilization, but for the program itself. We have previously said that there is a "parachor" or "threshhold" of space that represents a minimum size beneath which a program will seriously overpage. This threshhold, however, is not constant for all execution intervals, and one wishes to allow variation around it.

The constant expansion of ability to run with more memory space is universally reported by all workers. In the FIFO algorithm a surprising increase in the number of page-ins in the presence of increased space has been observed on the IBM system. Although observed only with a single program, the surprising behavior has been shown to be possible over a large number of various reference patterns and results from the cyclic nature of the FIFO algorithms. FIFO paging and its effect are shown in Fig. 35.4.

Reference = 1, 2, 3, 4, 4, 3, 1, 2, 3, 1

	Three pages				Four pages			
Reference	Block 1	Block 2	Block 3	Reference	Block 1	Block 2	Block 3	Block 4
1	①	x	x	1	①	x	x	x
2	1	②	x	2	1	②	x	x
3	1	2	③	3	1	2	③	x
4	④	2	3	4	1	2	3	④
4	4	2	3	4	1	2	3	4
3	4	2	3	3	1	2	3	4
1	4	①	3	1	1	2	3	4
2	4	1	②	2	1	2	3	4
3	③	1	2	3	1	2	3	4
1	3	1	2	1	1	2	3	4
Page insertions = 7				Page insertions = 4				

Fig. 35.4. Page movements—FIFO.

35.12. DYNAMIC ALGORITHMS

A number of replacement schemes attempt to make use of some recent page history in selecting a page for replacement. The scheme described for residence in the associative memory of the IBM 360/67 is an approximation of such a rule—least recently used.

This scheme attempts to remove from store the page that has not been referenced for a period of time, or, in its strongest form, not referenced for the longest period of time. The assumption is that that page has dropped from the locality of a program or of the set of programs in the mix. Figure 35.5 shows the page sequence used for FIFO operating under a least recently used rule, which uses a constant reordering of reference.

The procedure we described above for the 360/67 is an approximation of the least recently used rule that eliminates the necessity for a constant reordering (or renumbering) of the in-store list of pages. The setting of the reference bit for a page marks it as used in the interval.

The initial interval associated with the 67 mechanism is the period of time during which a program has control (we recall that all bits are zeroed when a task switches). Until memory is saturated, all pages have the referenced bit set to ON, since they would not be in unless they had been referenced.

When available space is depleted, all reference bits are reset except the page that filled memory, establishing a point at which only the last referenced page is marked as recently used and all other space is available. As references are generated, each page referenced is marked as such. Each

3			Reference	4			
1	2	3		1	2	3	4
①	x	x	1	①	x	x	x
1	②	x	2	1	②	x	x
1	2	③	3	1	2	③	x
④	2	3	4	1	2	3	④
4	2	3	4	1	2	3	4
4	2	3	3	1	2	3	4
4	①	3	1	1	2	3	4
②	1	3	2	1	2	3	4
2	1	3	3	1	2	3	4
2	1	3	1	1	2	3	4

Page insertions = 6   Page insertions = 4

Fig. 35.5. Page movements—LRU.

marked page, therefore, represents a page most recently used, or more precisely used in the interval since resetting. When all space is referenced, all referenced bits are again reset.

During the interval between a reference to all pages, there will be some pages marked "referenced" and some marked "unreferenced." The candidates for replacement come from those which have not been referenced. The choice of which of the nonreferenced blocks is to be replaced may be random or may be sequenced such that the block "next" in memory to the last overlaid block is selected. This latter scheme is described for the overlay strategy of the B8500 ESP system. The search for a memory area to overlay is begun from the point at which the last overlay occurred.

The assumption of the approximation of LRU is that frequently used pages will have a higher probability of reference during an interval between resettings, and that at any time the favored pages are represented by those which have been referenced.

An extension to LRU that can be made also to FIFO (or any algorithm) is to introduce a change bit reflecting whether or not the page has been changed since it has been brought into memory. This is the same distinction as Burroughs makes with program and data segments, except that Burroughs assumes that a data segment brought into store has been changed and always rolls out such a segment.

The goal, of course, is to minimize page movement by avoiding writes to auxiliary storage when possible by observing whether in an interval a page has lost its correspondence to a disc image write/read. The model 67 associates a change bit with each page in this fashion. All pages, therefore, have one of four states—unreferenced in the interval and not changed since entry, unreferenced in the interval but changed since its entry to core during a previous interval (when the change indicator is set, it is not reset at the time the reference indicator is set), referenced in the interval but unchanged, or referenced and changed in the interval. In the specific LRU algorithm known as AR-1 (Belady), choice of selection is in the order listed here.

As an alternative to the use of LRU as a predictor for least probable referenced page, a count of references to a page could be used. This assumes that heaviness of usage is a useful measure of heaviness of future usage. Pages are selected for replacement on the basis of the count on the assumption that infrequently referenced pages have been discarded.

Local implementations of this scheme exist in various operating systems. It is useful particularly for elements of shared system code. In this use it often degrades to a count of users rather than a count of usages. The difficulty with the algorithm in pure form is that it contains no index of the currency of usage. Further, it requires more extensive support to count references than to reflect a reference in both hardware and software mechanisms. Various schemes have been discussed for introducing an aging factor into the frequency count. The simplest approach would be to modify the count of each reference by a time stamp, so that later references contribute more to the count value than earlier ones. Given the hardware to accomplish support of an algorithm of this type, it would appear to be very promising.

35.13. SENSITIVITIES

The performance of page replacement schemes is sensitive to the size of memory and to the size of pages. If memory is partitioned into larger, and therefore fewer, areas, the various probabilities of finding good candidates becomes less. The degree of multiprogramming also affects behavior by defining the load and consequently the paging rate of the system.

The LRU application to a multiprogramming mix will tend to force larger programs to lose more pages than smaller programs. This is because their reference pattern is more diffuse; the interreference time for each page is larger, and the probability of a reference during an interval is less.

There is considerable temptation to conclude that direct usage of the change bit in selecting a page for replacement improves system efficiency by reducing the page traffic and by reducing waiting time for a program to read rather than write/read time.

A number of problems arise, however. It is clearly preferable to record whether a page has changed in order to determine when a write can be avoided, but it is not clear that selecting a page for replacement on this basis is profitable.

Each time a changed page is preferred for maintenance in core, a procedure page must be flipped. Read only has the effect of assuming that any data page is more probably to be referenced than any procedure page. In effect, it defines a hierarchy of page types and prefers to maintain a given member. The space occupied by these pages is effectively withdrawn from use by other members, which must now compete for a more limited resource, and the size of this space grows.

As in any fixed-priority system, bad choice of priority will impair efficiency. To the extent that the priority given to data pages is not based upon a valid assumption, its use in selecting a replacement page can make little contribution to efficiency.

35.14. A MULTICS EXPERIMENT

An interesting page replacement mechanism lying "in between" FIFO and pure LRU is described for the GE 645 MULTICS system. MULTICS is a large time-sharing system similar in concept to the IBM 360/67. The GE 645 is a modification of the GE 635, providing for the support of virtual memory in a two-level segment/paging scheme similar in a general way to the 67. (The scheme provides initially for 1K and 64-word page sizes. A word is a 36-bit structure. In general, users of the IBM 7090/4 will find the 635/645 recognizable machines.)

The MULTICS machine is seen to be a multiprocessor with two CPU's and two generalized I/O controllers, having capability to reach any disc through mutual access to disc controllers having access to disc units. Other devices are specifically associated with a GIOC and are externally switch-able only.

The described MULTICS page replacement scheme involves a circular list of all pages in store marked as first used, not recently used, not modified, recently used not modified, not recently used modified, and recently used modified. With the exception of first usage status, we have seen these before as the AR-1 LRU statuses. A commutator revolves around these pages to select a page for replacement. On each page fault, the status bit of the page pointed to by the commutator is brought to a register of length k associated with each page. (This is similar to a hard-ware description of the implementation of working set by Denning and to another mechanism described by Randall).

The length of the register determines the attributes of the paging scheme. If the register is of length 0, the algorithm is FIFO. This is because there is no place in the system for the representation of the reference status history of the page. Since the commutator moves cyclically around the list of pages, the first pulled-in page, the oldest, will be pointed to by the commutator, and each page pointed to will be automatically selected for replacement, since there is no retaining mechanism. When a replacement is not necessary, of course, space is acquired from a free block list. A block taken from the free block list is placed upon the page pull list, scheduled (with its disc address) for roll-in from the drum. When a page is pulled in, it enters at the head of the replaceable page list with a first reference bit placed on.

It is a policy of MULTICS to maintain the free block list at a certain minimum size. Whenever, at the time of a page fault, the system determines it must add to the free block list, it starts at the first next page of the replaceable page list and brings up the status of the page. If the status is unreferenced, the page is selected for replacement. This must always happen in FIFO.

When selected, the block is placed at the end of the replaceable page list, exposing the next block, and placed on either the free storage list or the push list (scheduled for roll-out.) When roll-out is complete, the block is placed on the free list.

Notice in the status of the replaceable list as the commutator cycles through, that page 1 is selected for freeing first (it is the oldest in core); on freeing, page 1 is the latest in core and page 2 is selected. When 2 is freed, 3 is the oldest in core, 2 the newest, 1 the next to newest. When 3 is freed, 4 is oldest, 3 newest, 2 next, and 1 after that.

When k is greater than 0, then there is a possibility that a page will not be selected for replacement, its status indicating that it has been referenced since the last inspection by the commutator (since last inspection, because the status bit is set to zero at the time of inspection). The algorithm is no longer FIFO. It is an approximation to LRU not unlike AR-1, but varying from it in that page bits are not immediately reset when all pages are filled. If k is greater than 2, a right shift occurs, and the current status bit is placed in the high-order position. A necessary condition to free a page is a k register value of 0.

Notice that, as shifts are made, older states drop out of the history of the page and newer references are recorded. A certain amount of overhead

Replacable page list:

(1) ⟶ 1, 2, 3, 4, 5
(2) ⟶ 2, 3, 4, 5, 1
(3) ⟶ 3, 4, 5, 1, 2
(4) ⟶ 4, 5, 1, 2, 3
(5) ⟶ 5, 1, 2, 3, 4
(6) ⟶ 1, 2, 3, 4, 5

Fig. 35.6. MULTICS commutator—FIFO.

is paid for this selective retention, since the commutator will characteristically be forced to go much further (inspect more pages) to free a given number of block-free requirements.

As k becomes large, the algorithm approaches LRU, and the number of 0 high-order bits represents the number of cycles through the list since the page was last referenced.

An apparent anomaly occurs when k achieves a certain value. Since k is supposed to be a better and better approximation of LRU, a result showing larger page traffic for $k = 7$ than for $k = 4$ is interesting. One possible reason is that too many preferred pages have been defined reducing the size of the free block area. In working set, described below, this is equivalent to a working set time too large so that working set size is artificially large.

When $k = 1$ the algorithm is a FINUFO (first in not used first out). Corbato finds the dramatic difference in paging rate to lie between FIFO and FINUFO. The algorithm here has a one-at-a-time forgetting characteristic, whereas the other approximation of LRU that we described forgets all references when memory is saturated. (The last P-bit is set.)

35.15. IBM 360/85 HARDWARE LRU

A number of hardware implementations to LRU or working set algorithms have been presented. An interesting LRU technique which does not involve a shift register and which achieves perfect LRU is the replacement algorithm built into the hardware of IBM's 360/85.

The LRU algorithm is implemented by use of a matrix representing the elements of core. Whenever a block is referenced, the row representing the number of that block is set to 1's, except for the column corresponding to the block (see Fig. 35.7). Notice from the figure that a completely ordered list is maintained at all times and that the least recently used may always be selected. This hardware implementation is actually at the interface between a transparent fast core and a standard memory box for the system.

35.16. MULTIPROGRAM PAGE COMPETITION

Most testing and simulation of replacement algorithms have been done on a uniprogrammed basis testing the behavior of programs running in less than their name space. We have, in connection with FIFO and LRU, made passing comments about behavior in a multiprogram environment. There

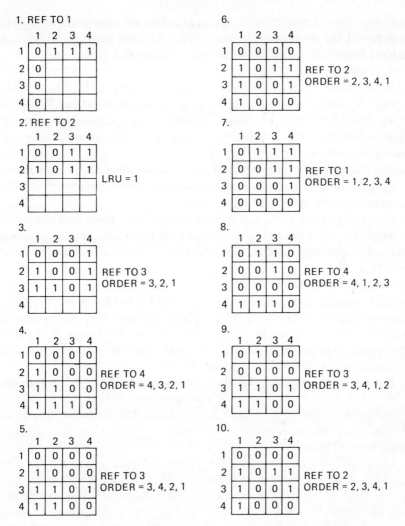

Fig. 35.7. IBM model 360/85 cache control (four sectors shown).

are a number of policy decisions that roughly accord to the levels of conservatism initially described for core management policy in connection with page algorithms. In the most conservative strategy a program is allowed a given amount of space, effectively a partition, and its parachor space, and is prohibited from causing replacement of any of its own or anyone else's pages. This effectively reduces management to avoiding fragmentation, since its space can be effectively distributed throughout memory. If a program exceeds its size, it is aborted, and all pages are released.

Beyond this a program is allowed to steal from itself operating in a fixed size. Page replacement is the application of some algorithm against the pages of the program. The final step is to allow programs to grow and contract on the basis of some criteria. Meaningful criteria are deadline, priority, and system utilization.

We have mentioned the effects of applying FIFO and LRU to pages in a system without distinguishing whose pages belong to whom. The lack of a distinction between ownerships undermines any scheduling policy that might be implemented in the system. The distinction of ownership seems essential for any management policy.

The priority of a program and the schedule over an interval can then be applied as initial criteria, with programs of lower priority more apt to lose pages than programs of higher priority.

One can predict low usage over an interval of a program's pages if one can predict a low probability of running that program. This is essentially what is accomplished in TSS. The effect of this is to reduce the number of pages that are unutilized in core at any time and to replace them with utilized pages. For the programs that are to lose pages in an interval (be inactive), a number of tactics may be applied: (1) One may immediately release all pages. (2) One may release pages, but take advantage of the fact that some pages may accidently remain in core so that at next start-up those pages that are in core may be used. (3) One may adopt a policy concerning which pages should be left in core.

Since we are assuming that the dynamic contraction and expansion of memory is a basic motive for core management and that the maintenance of a schedule is the primary goal to which the management of core contributes, we shall take interest here only in approaches that allow for dynamic expansion.

Certainly, in undifferentiated systems applying a single replacement algorithm to all pages, such expansion is possible and is a function of the time intervals given to programs. One wishes to control page replacement more distinctly. The BIFO algorithm is a basic way of allowing expansion based upon knowing what pages are allowed by which program. The program bias can be applied to any algorithm and need not be done on a cyclic basis.

Programs will steal from each other under heavy memory load when it is permitted. The Working Set approach, like the parachor approach, attempts to avoid this by refusing to run a program unless there is a minimum amount of space available to it. This amount of space may be a given parameter, or it may be dynamically observed. If dynamically observed, it can be observed once and then established as the necessary size of the program, or it can be observed repeatedly over intervals of time, therefore, the minimum size may vary over this history of the program.

The minimum size to run may be a system parameter; the size to run may be guaranteed and sustained for each entry to the running list, or the size may be allowed to vary based on performance observed at intervals while the program is on the running list as in working set. The close observation of what working size is at a given time allows the closest utilization of memory, since the load to be placed on memory by the program over the next interval develops from its usage during the previous interval. The avoidance of memory overload by restricting the number of active programs to those that will fit into memory is balanced by the dynamic reduction of requirements where possible. It is an approach in which one can visualize dynamic movement from a standby to an active queue, based upon memory loadings.

The basic working set scheme, as described by Denning, allows for the expansion and contraction of memory requests and attempts to limit the degree of interference that one program will place upon another by "stealing its pages." Its effectiveness is based upon the observation that system efficiency improves when the best set of pages for a program to have in core at any time can be determined by observation. This will tend not only to reduce page traffic, but to allow smaller program spaces and consequently to increase the number of active programs and memory utilization.

The specific scheme is an extension of LRU that involves a cyclic investigation of the referencing patterns over a selected interval. The "checker" (a scheduled active process) scans the page references of all programs active since the last check to determine what pages were referenced during that interval. All unreferenced pages are considered to have left the working set of the process, and a counter representing the size of the set is reduced. (During control of CPU a page reference to a missing page causes an increment of the counter.) The page unreferenced can be removed from storage. For any program, no page will be removed from memory so long as it has been referenced in the interval (is a member of the working set).

The interval of running checker is a function of the number of service intervals given to each active program and represents the time at which a working set is defined. As mentioned before, proper selection of time is critical. In pure working set no program can steal from another, since a guarantee of working set size is a requirement for activation. If the working set cannot be satisfied by existing space, a program can be deactivated or can "cannibalize" itself. At its next activation working set size will be larger. The approach tends to give priority of execution to small programs, whose working set sizes can more often be guaranteed.

Page stealing might be permitted in a basic working set environment if priority were introduced. The exceptional value of a given program might allow it to take space. The dynamic nature of release, however, might tend to make this an improbable event. Since processes are marking as available page space that has dropped from the working set, it is actually fluctuating constantly, and the need actually to displace a current member of an active program's working set would be infrequent. The constraint, therefore, is to keep a program from displacing a "favored page" of another program, and not from acquiring space as needed. There seems to be nothing implied in the theory that enforces some concept of preferability to use space released by pages falling out of the working set, nor to enforce some minimum working set size related to the initial value of the set.

The size of space required at a next placement on the active list is the last value of working space counter, and not an average value. Since the average size of the working set is not a known number, no attempt can be made to constrain loss of space to this number or to any other. If serious changes in the working set occur or if the set reduces to too few pages, some imbalance in the system is inferred, such as an inappropriate selection of the interval or a bad page size.

35.17. PREPAGING AND WORKING SET

The dynamic demand paging of TSS 360 is a pure form of demand paging. There is no predictive mechanism whatsoever for the precall of pages. In an experimental system which made heavy contributions to IBM's CP67, a virtual machine system which is in operation at IBM's Boston Programming Center, a predictive mechanism was investigated with results far less impressive than the effect of changes in programming style.

The concept of the working set is closely related to the concept of parachor. Parachor attempts to guarantee that a program does not run unless there is sufficient room for the number of pages that permits a program to run, spending less than half its time in page wait. Working set attempts to guarantee that a program will not run unless there is room in memory for its working set. The size of working set, unlike parachor, will vary over the life of a program in the system.

The working set is defined to be the set of pages referenced in an interval of time. That interval of time should be small enough so that the set does not grow too large and large enough so that the set is reasonably complete. That the validity of a given set is higher over smaller intervals than over larger ones we can see. The tendency of a process to complete its work in an area and go on to something else is greater over longer, as opposed to shorter, intervals of execution time. The paging rate of a program is related to the quality of the working set, the probability that there will be a missing page.

The maintenance of the working set in storage is properly a function of page removal and replacement techniques, and no prepaging of the working set is necessarily implied by the approach.

Peter J. Denning, author of the concept, describes the usefulness of prepaging as meaningful only in those instances where blocking techniques applied to the working set could reduce the access time to pages to some significant degree. The great exposure in prepaging is the relative high cost of a wrong guess, resulting in channel and memory cycle utilization that contributes nothing to the performance of the system.

In attempting to maximize memory utilization and reduce page traffic, the prepaging of pages can result in an undesirable residence in core of pages that are not truly active. If we remember the space/time memory utilization discussion of the previous chapter, we shall recall that minimization of the space-time product for a program's residence is a desired result of management policy.

A system implemented by RCA on the Spectra 70/46 reflects "working set" concepts. The number of pages estimated to be required by a task at each interval is obtained by using a count of the pages used in its previous active period. The authors of a description of the system mistakenly ascribe prepaging as an attribute of Denning's work. This seems to be a prevalent misconception, which this author encourages his readers to avoid. The underlying ideas of the "adaptive" technique used by RCA and the working set is the predictability of how much space is required by a program. There does appear to be, however, a limited prepaging mechanism in the 70/46 system.

The high cost of overpaging leads to the definition of minimum space considerably larger than the basic minimum logically possible. Some final comment should be made about the amount of store that is allocated to a mix of jobs.

Previously we discussed the beneficial effect of "overloading" CPU and memory and the possibility of dynamic balancing of a system by use of dynamic monitoring of the current load of programs, with contribution to target utilizations used as an activating criterion by the low-level scheduler. A working set is a statement about the demand that a program will place on the core while it is active or in page wait status.

In order to control thrashing, the limit of memory demand represented by those programs that are to be active over the next interval of time should not exceed 100 percent. Admission to active membership (from the ready but not active list) is on the resource-oriented basis of maintaining balance between CPU and memory utilization for the next interval. This is a subset of the loading vector that we discussed in connection with multiprogram scheduling.

35.18. PROGRAMMING STYLE

The question of "programming style" is an interesting one. Authors disagree as to the desirability of imposing on the programmer considerations about the nature of programming for a virtual memory machine. Critics of programmer awareness point out that the definition of overlay structures, the organization of formalized segments, and the planning necessary to run a program in a limited core space are things which virtual memory hoped to lift from the programmer's shoulders. Proper packing of pages and planning good virtual machine usage is a job every bit as complex.

Further, programmers writing in higher-level languages may not have the control required to accomplish good packing. Critics also point out that certain programming techniques that are known to be "good" techniques, such as the separation of procedure and code and a high degree of program modularity, are bad in a fixed-page-size environment in that they tend to increase the number of pages. By so distributing the program, they increase the probability that a page will make an off-page reference, causing a page pull into core (and a consequent possible page push—a roll-out of a page). Basically the critics feel that paging must be controlled by more sophistication in the hardware or by scheduling techniques invisible to the programmer.

There exist also a number of people who are not yet convinced about the general contribution of "demand paging" to system performance. This is a two-part argument. In one aspect it is a question put to the effectiveness of dynamic memory management as a throughput technique. Does the ability to roll in and out in different locations contribute to the utilization of a machine in a way significant enough to overcome its expense?

The other aspect concerns the level of "surprise" that a system can tolerate. There has been some comment that preferable to the rigors of page packing and planning is some capability for predictive page clustering. In such an approach the programmer might write code relatively independent of page pack considerations but be given some way of expressing "superpages." An awareness of page organization might be a requirement, of course.

Essentially, the programmer associates pages with each other, so that, whenever he refers to a page, the system, through his description of a "cluster," knows what pages he will be referring to within the next interval of time. An IBM experimental system provided for such predictive information in a conversational way. A user could inform the system of what pages he desired to use in the near future. Such a natural cluster of pages is, of course, the segment.

Defenders of programmer packing seem to feel that the more dynamic hardware-supported approaches to page management, particularly the replacement algorithms, have either marginal payoff or are impossible to implement because they require knowledge that the system cannot have, such as "time until next reference this page."

In a sense, the discussion centers about when it is proper and useful to define a "working set." If the cost of this definition is moved to the programmer, then paging algorithms can be simple; if it is left to the system, considerable expansion of the state of the art would seem to be necessary before true efficiency can be gained. Further, the complexity of virtual memory-oriented programming is held to be very low.

Whether or not the cost of programmer adjustments to the virtual memory system are supportable in terms of increased programmer cost, decreased modularity, etc. is an open issue. What is no longer an open issue is the effect of "programming style" on performance. This is a local optimization which is of uncontestable value.

In our previous reference to the desirability of local optimization, we raised the question on the optimization of resource allocation for a program and the general performance of a multiprogramming system. The otpimization here is not a question of the proper distribution of channels or buffers across a given load, but of the absolute reduction of the load by a reduction in number of pages and/or number of page movements.

It is always very difficult to assess the impact of any given performance element in a general way. The various statistical (analytical or empirical) studies that are in abundance in the field often vary an element or elements, holding other factors constant.

Since the other factors often represent components unique to a system with unique interactions, it is impossible to state the pure effect of any element across various systems. For example, in a published analysis of GECOS II, the authors encountered only a 5 percent burden for memory repacking to avoid fragmentation delays. But of course this is a function of allocation and scheduling strategy for this given system, and it cannot lead to a conclusion that memory repacking is, in general, in a large multiprogramming system that undertakes it, trivial.

Various published results in the effect of dynamic relocation on general throughput also lead to conflicting conclusions. We can discuss the impact of programming style tentatively, with the forewarning that differences of opinion about its desirability (and even its effectiveness) do exist in the field. The preponderance of evidence seems to be on the side of those who claim it is reasonable and effective.

The approach taken by experimentation has been to run programs not initially oriented for paging machines on paging machines, to observe the results, and then to investigate the nature of improvements. These improvements are of the following types:

1. Given a knowledge of page size, align program elements so as to place dependent routines in the same page.
2. Given a knowledge of paging behavior, order memory loads so as to minimize paging.
3. Reorganize the storage of data structures, using variable dimensioning and taking care about when to store in rows and when in columns.
4. Duplicate code rather than branch to subroutines.

In general, the decreases in the paging rate are dramatic for many classes of programs, including scientific, commercial, and systems programs, and these improvements tend to be independent of the scheduling and page replacement algorithms.

In some instances it was found that certain techniques were inherently bad for virtual memory machines and that other algorithms should be chosen. This is also true in developing algorithms for parallel processors. Still the impact of small modification to style in the same algorithm seems impressively large.

The point is taken that to develop programs that run well on a paging machine some attention must be paid to paging. An interesting side observation is that the programmer is under no absolute obligation to pay this attention; he can perform in a degraded way if he undertakes no optimization at all. The penalty he experiences for not doing so is overpaging (relative to other optimized program) and, in some systems, relatively bad treatment by the low-level scheduler and/or dispatcher (to the point even of being removed from active status).

35.19. FINAL REMARKS

Memory management is a complex area indeed. We have attempted in this chapter to introduce the problem and to discuss various policies and tactics associated with dynamic memory allocation. In the real world batch multiprogramming systems tend to be more conservative, allowing less flexibility than real-time or sharing systems, because the observed need for dynamism is less. The potential impact of supporting batch systems with more dynamic memory techniques has been observed in experiments running MFT-II (OS/360) on a model 67.

There is wide disagreement in the field as to the relative impact of good replacement strategies on program performance. TSS 360 developers point out that they have achieved enormous improvements in their system by honing and tuning without any change at all in the page replacement strategy, which is as simple as it can be. When a process is active, they bring in the pages it asks for; when it loses the CPU, it loses its core. Other workers report significant variations under various paging algorithms.

What does seem apparent is that different programs will behave differently under different algorithms; particularly, very large programs written for batch processing will behave differently than will small programs written for conversational systems. List processing programs, matrix inversion, and sorts of various types display different behavior, more or less satisfactory. The element of programming style seems to make at least as much difference to performance as replacement algorithms. The reason for this is easy to see—as programs are developed for virtual machines, the developers will choose algorithms and coding styles suitable for those machines; they will tend to be less diverse in form and reference pattern, and all algorithms will work better (except those completely insensitive to program structure and history). The concepts of good paging algorithm, of retaining local history, and of dynamic monitoring of that history are concepts that we have seen before: first in look-aside memories, then in dynamic running priorities, and finally in paging algorithms.

The principles of resource-oriented monitoring, loading, etc. appear again and again in all systems considerations at all levels in hardware and software design.

Paging schemes, like other optimization attempts, must be implementable within cost constraints; very excellent algorithms just cost too much to implement or imply too much overhead, and the set of implementable algorithms are approximations of ideal schemes. The need for any optimization is related to an observation as old as the computer industry and reflected in the discussion of interaction between resources in the previous chapter. There is a cost in time for the reduction of space requirements. Some time costs are so enormous relative to system speeds that they are a priori rejected (mounting and dismounting tapes so that a system with one tape unit can do a three-way merge); some are so low as to be below the threshold, except perhaps in the largest and most ambitious high-performance systems (rounding a last digit of a floating-point operation); some are right in the middle. The page replacement time or overlay time for a program operating in core smaller than its full size is just there, and that is why it receives so much attention.

The problem solution would lie in the acquisition of much larger

stores or large back-up core stores, but the current cost of system aug-
mentations of this type precludes them, and techniques that involve no
cost to the potential user of a system (as reflected in his monthly rental)
are searched for by vendors. As with many other areas in the computer
art, we are dealing with the economics of the use of a system whose cost
of operation relative to the dreams of its users is still comparatively high.

PART SEVEN:
SOURCES AND FURTHER READINGS

Arden, B.W., Galler, B.A., O'Brien, T.C., and Westervelt, F.H., "Program and Addressing Structure in a Time-Sharing Environment," *Journal of ACM* (January, 1966).

Batson, Ju and Wood, "Measurements of Segment Size," *ACM Second Symposium on
Operating System Principles* (October, 1969).

Belady, L.A., "A Study of Replacement Algorithms," *IBM Systems Journal,* Vol. 5,
No. 2 (1966).

Belady, L.A., Nelson, R.A., and Shedler, G.S., "An Anomaly in Space-Time Characteristics of Certain Programs Running in a Paging Machine," *Communications of ACM*
(June, 1969).

Bensoussan, Clengen, and Daley, "The Multics Virtual Memory," *ACM Second Symposium on Operating Systems Principles* (October, 1969).

Bobrow, D.G., and Murphy, D.L., "Note on the Efficiency of LISP Computation in a
Paged Machine," *Communications of ACM* (August, 1968).

Bovet, D.P., "Memory Allocation in Computer Systems," Clearinghouse For Federal
Scientific and Technical Information AD 670499 (June, 1968).

Brawn, B.S., and Gustavson, F.G., "Program Behavior in a Paging Environment,"
AFIPS 33, Fall Joint Computer Conference (1968).

Bryan, G.E., "Dynamic Characteristics of Computer Programs," Clearinghouse For
Federal Scientific and Technical Information AD 658819 (August, 1967).

Burroughs Corporation, "Master Control Program for the B5500."

Coffman, E.G., and Varian, L.C., "Further Experimental Data on Behavior of Programs in a Paging Environment," *Communications of ACM* (July, 1968).

Comeau, L.W., "Study of the Effect of User Program Optimization in a Paging System," *Proceedings of First ACM Symposium on Operating System Principles* (October, 1967).

Corbato, I.J., "A Paging Experiment with Multics Systems," Project MAC Docu-
ment-M-384, (1969).

Daley, R.C., and Dennis, J.B., "Virtual Memory, Processes and Sharing in Multics,"
Communications of ACM (May, 1968).

Denning, P.J., "Working Set Model for Program Behavior," *Communications of ACM*
(May, 1968).

Dennis, J.B., "Segmentation and the Design of Multiprogrammed Computer Systems," *Journal of ACM* (October, 1965).

Fine, G.H., Jackson, C. W., and McIssac, P.V., "Dynamic Program Behavior Under Paging," *Proceedings of ACM* (1966).

Fotheringham, J., "Dynamic Storage Allocation in the Atlas Computer," *Communications of ACM* (October, 1961).

Illiffe, J.K., and Jodeit, J.G., "A Dynamic Storage Allocation Scheme," *Computer Journal* (October, 1962).

International Business Machines Corporation, "IBM System 360 Model 67, Functional Characteristics," Form A27-2719.

International Business Machines Corporation, "IBM System 360 Time-Sharing System Concepts and Facilities," Form C28-2003.

Jodeit, J.B., "Storage Organization in Programming Systems," *Communications of ACM* (November, 1968).

McCullough, J.D., Spierman, K.H., and Zurcher, F.W., "Design for a Multiple User Multiprocessing System," *AFIPS Proceedings,* Fall Joint Computer Conference (1965).

McGee, W.G., "On Dynamic Program Relocation," *IBM Systems Journal,* Vol. 4, No. 3 (1965).

McKeller, A.G., Coffman, E.G., and Rosin, R.F., "Organization of Matrices and Matrix Operations for Paged Memory System," *Communications of ACM* (March, 1969).

O'Neil, R.W., "Experience Using Time-Sharing, Multi-Programming System with Dynamic Address Relocation Hardware," *AFIPS Proceedings,* Spring Joint Computer Conference (1967).

Oppenheimer, G., and Weizer, N., "Resource Management for Medium Scale Time-Sharing Operating System," *Communications of ACM* (May, 1968).

Randell, B., "A Note on Storage Fragmentation and Program Segmentation," *Communication of ACM* (July, 1969).

Randell, B., and Kuehner, C.J., "Demand Paging in Perspective," *AFIPS,* 33, Fall Joint Computer Conference (1968).

Randell, B., and Kuehner, C.J., "Dynamic Storage Allocation Systems," *Communication of ACM* (May, 1968).

Smith, J.L., "Multiprogramming under a Page on Demand Strategy," *Communications of ACM* (October, 1967).

Varian, L.C., and Coffman, E.G., "Empirical Study of Behavior of Programs in a Paging Environment," *First ACM Symposium on Operating System Principles* (October, 1967).

Wallace, V.L., and Mason, D.L., "Degree of Multiprogramming in a Page on Demand System," *Communications of ACM* (June, 1969).

Weill, J.W., "A Heuristic for Page Turning in a Multiprogrammed Computer," *Communications of ACM* (September, 1962).

EPILOGUE

The bulk of attention in the development of parallel algorithms has been in areas of mathematical processing beyond the scope of the assumed knowledge of this book. Many of these studies and results have been listed in the bibliography. It is important to realize that large scientific processes need not form the major thrust of interest in multiprocessing or other parallel techniques. Indeed, a process that seems most ideal for parallel processing is the payroll. This exhibits all of the properties of a parallel algorithm, in that each man's check is calculable independently of any other man's, and although there are several sums and balances to be made, the bulk of a payroll could be effectively processed by N independent processors with little interference, where N is the number of men on the payroll.

Numeric techniques have dominated interest because they are usually associated with those very large problems whose requirements are at the edge of technology—vast problems, such as weather prediction whose solution requires greater magnitudes of processing power than that available from the most advanced of current systems. Work is developing in connection with these problems, but little seems to be done to reinvestigate the nature of commercial work for parallel opportunity. The thrust here will develop in the next few years in response to the data-base communications-oriented systems, where sequential processing of data will tend to be more and more replaced by on-line data systems and the modules used in these systems will of necessity be forced to be highly parallel with regard to each other.

In general, in the next few years we may expect the following developments:

1. A general availability of multiprocessors in the UNIVAC 1108 or IBM MP 65 sense.
2. A proliferation of array and vector processor designs of the SOLOMON or ILLIAC IV type offered as special-purpose elements of more conventional processing systems.
3. The development of more basic parallelism in high-performance uniprocessors with attention particularly devoted to achieving a processing (instruction retirement) rate of one instruction every machine cycle.

Various elements of technology will tend to cause the redistribution of function and storage elements throughout the machine, with some differences of opinion as to whether system elements should be functionally differentiated or homogeneous. We may expect associative memories, microprogramming, and highly intelligent I/O subsystems to make contributions to system design.

In the area of software we may expect a proliferation of parallel explicit forms, as well as the refinement of compilers capable of automatic element and segment definition.

These projections are easily made. What no one can anticipate is the "surprise"—the insight that lurks somewhere in the genius of a man that will transform our machines into something unrecognizable to us even during our careers and make the journey from the ENIAC to the ILLIAC IV seem only the "twentieth part of the first syllable of the beginning."

INDEX